Leibniz

A Biography

Frontispiece G W Leibniz. An eighteenth-century copy of a painting by Andreas Scheits, 1711. Courtesy of the Historisches Museum, Hanover.

LEIBNIZ

A Biography

E J Aiton

Senior Lecturer in Mathematics
Didsbury College of Education
Manchester Polytechnic

Adam Hilger Ltd
Bristol and Boston

British Library Cataloguing in Publication Data

Aiton, E. J.
Leibniz: a biography.
1. Leibniz, Gottfried Wilhelm 2. Philosophers
—Germany—Biography
I. Title
193 B2597

ISBN 0-85274-470-6

Consultant Editor: **Professor A J Meadows**
University of Leicester

Published by Adam Hilger Ltd
Techno House, Redcliffe Way, Bristol BS1 6NX, England
PO Box 230, Accord, MA 02018, USA

Phototypeset by Quadraset Ltd, Midsomer Norton, Bath
Printed and bound in Great Britain at The Bath Press, Avon

Contents

Preface

Among the scholars of Europe Leibniz was famous in his own lifetime and after his death a number of obituaries appeared in the learned journals. Towards the end of the eighteenth century, the townspeople of Hanover, where he had lived for over forty years in the service of the ruling princes, erected to his memory a circular temple with a bust in white marble and the simple inscription 'Genio Leibnitii'. Leibniz exemplified the Renaissance ideal of the universal man and at the same time he ushered in the Age of Enlightenment. The existence in Hanover today of a Leibniz-Society, which organises International Leibniz-Congresses and Symposia besides publishing the journal *Studia Leibnitiana*, bears testimony to the continuing interest of scholars in the work of Leibniz, who is seen as having played a central role in the history of European thought. Even if he had only contributed to one field, such as law, history, politics, linguistics, theology, logic, technology, mathematics, science or philosophy, his achievement would have earned him a place in history. Yet he contributed to all these fields, not as a *dilettante* but as an innovator able to lead the specialists.

In the field of mathematics, Leibniz's outstanding achievement was the independent invention of the differential calculus, which he and his friends developed into a powerful tool for the solution of geometrical and physical problems. Through the edition of Erdmann, his project of a universal characteristic and the ensuing logical calculi, which had not been published in his lifetime, played a significant role in the history of modern symbolic logic. On the other hand, his pioneer work on determinants only came to light after their properties had been re-discovered.

Some of his notable contributions to natural philosophy were the concept of matter as active, his demonstration of the relativity of space and time, the establishment of the relation between *vis viva* and the height to which it can raise a body (in effect the convertibility of kinetic and potential energy) and the demonstration of the conservation of *vis viva* in collisions (despite the apparent loss in inelastic collisions). Another fruitful contribution was his deduction of the laws of optics, for in this he had demonstrated the utility of stationary principles in physics, which were later applied to good effect by Euler, Lagrange and Hamilton. Several of Leibniz's general principles, including the principle of continuity (which he had employed to refute the Cartesian rules of collision), the principle of sufficient reason, and above all, the distinction between necessary and contingent propositions (that is, truths of reason and truths of fact), had a bearing on scientific methodology. The last provided a sound basis for the combination of theory and experiment, while the principle of sufficient reason permitted scientific explanation both in terms of physical causation and in terms of teleological principles such as he had used in optics. Not least among his contributions to science and technology were the invention and development over a period of forty years of his calculating machine and the foundation (in effect) of the Berlin Academy of Sciences, of which he was the first President.

On completing his university studies, Leibniz declined the offer of a professorship, seeking instead to enter the service of a powerful prince. For it was only in this way that he could hope to achieve the utilitarian aim of his various endeavours by actually bringing about the application of new knowledge in science and technology for the benefit of a universal Christian society living in peace and harmony. His first appointment, in the service of the Elector of Mainz, led him to Paris, where he would have remained if the opportunity to exercise his influence as a paid member of the Royal Academy of Sciences had presented itself. However, the Academicians did not wish to appoint another foreigner (in addition to Huygens and Cassini) and Leibniz reluctantly accepted an invitation to enter the service of the ruler in Hanover, where he spent the rest of his life apart from short periods in Berlin and Vienna. At the Courts of Hanover, Celle, Wolfenbüttel, Berlin and Vienna, he met all the leading statesmen, diplomats and generals, besides forming close friendships with a number of princes and princesses, especially Anton Ulrich of Wolfenbüttel, Sophie, Duchess of Hanover, and her daughter, Sophie Charlotte, Queen in Prussia. In his last years, when all three of these friends were dead and he was increasingly out of favour with the last prince of Hanover under whom he served, George I of England, the priority dispute with Newton concerning the invention of the calculus was at its height. Despite his trials and tribulations, however, he never lost his optimism, so that his secretary Eckhart could write after his death that he always spoke well of everyone and made the best of everything.

Following the principle adopted in the definitive Akademie-edition of Leibniz's works, the biography is divided into a chronological sequence of chapters with a systematic treatment of each topic within the chapters. In this way it is possible to trace the evolution of Leibniz's thought, within the context of the social, political and intellectual background, while at the same time characterising the significant stages in the development of the separate themes. The early years present a particular difficulty in that there is a scarcity of reliable information. In an attempt to present a continuous account, use has been made of Leibniz's own reminiscences, written years later, to fill the many gaps between the known facts. These retrospective statements, however, are always identified as such, so that the reader can exercise the necessary caution. For while it is interesting to see the picture of his early life built up by Leibniz in later years, it is of course unlikely that this represents a completely accurate and objective account.

The primary sources on which this biography is based consist of the writings, works and correspondence of Leibniz, mainly printed editions but including also some manuscripts. An attempt has been made to present an account of the life of Leibniz and the development of his ideas, especially in mathematics, science and philosophy, which is in broad agreement with the results of modern Leibnizian scholarship, that is, the findings of fully documented research based on primary sources. While detailed points of current controversy concerning interpretations are avoided, for these will be of interest only to specialists, a number of myths which have grown up surrounding Leibniz have been eliminated by demonstrating the unreliability of the evidence for them.

On many occasions at Conferences and Symposia, I have had the opportunity to discuss the work of Leibniz with a number of friends. Three of these, Dr Heinz-Jürgen Hess of the Leibniz-Archiv, Niedersächsische Landesbibliothek, Hanover, Professor Dr Eberhard Knobloch of the Technische Universität, Berlin, and Professor Frederick C Kreiling of the Polytechnic Institute of New York, have read a first draft of the typescript and made a number of very helpful comments. It is with great pleasure that I acknowledge their invaluable assistance in enabling me to make some significant improvements. My grateful thanks are also due to the Administrators of the Museum für Geschichte der Stadt Leipzig, Historisches Museum, Hanover, Staatliche Schlösser und Gärten, Schloss Charlottenburg, Berlin, the Niedersächsische Landesbibliothek, Hanover, British Library, Bodleian Library, and also to Professor E Shimao, Doshisha University, Kyoto, Professor Dr Karl Arndt, Georg-August-Universität, Göttingen and Dr E A Fellmann of the Euler-Archiv, Basel, for so generously supplying illustrations.

E J Aiton
Manchester
Easter 1985

Plates

Abbreviations

The following abbreviations are used for frequently cited works.

A = Leibniz G W *Sämtliche Schriften und Briefe* (Berlin: Akademie Verlag, 1923–). The number of the Series is shown in roman numerals and the volume number in arabic numerals

AE = *Acta Eruditorum* (Leipzig)

D = Dutens L (ed) *G. G. Leibniz Opera omnia* 6 vols (Geneva 1768)

DS = Guhrauer G E (ed) *G. W. Leibniz, Deutsche Schriften* 2 vols (1838–40, reprinted 1966 Hildesheim: Olms)

FC = Foucher de Careil L A *Oeuvres de Leibniz* 7 vols (reprinted 1969 Hildesheim: Olms)

FCa = Foucher de Careil L A *Lettres et opuscules inédits de Leibniz* (1854, reprinted 1975 Hildesheim: Olms)

GBM = Gerhardt C I *Der Briefwechsel von G. W. Leibniz mit Mathematikern* (1899 Berlin)

GM = Gerhardt C I *G. W. Leibniz, Mathematische Schriften* 7 vols (1849–63, reprinted 1971 Hildesheim: Olms)

GP = Gerhardt C I *Die Philosophischen Schriften von G. W. Leibniz*, 7 vols (1875–90, reprinted 1978 Hildesheim: Olms)

HO = *Oeuvres complètes de Christiaan Huygens* 22 vols (1888–1950 The Hague: Société Hollandaise des Sciences)

K = Klopp O *Die Werke von Leibniz* 11 vols (1864–84, volumes 7–11 reprinted 1970–73 Hildesheim: Olms)

LH = Leibniz Handschriften in the Niedersächsische Landesbibliothek, Hanover

MK = Müller K and Krönert G *Leben und Werk von G. W. Leibniz. Eine Chronik* (1969 Frankfurt am Main: Klostermann)

NC = *The correspondence of Isaac Newton* 7 vols (1959–77 Cambridge: published for the Royal Society)

NP = Whiteside D T *The mathematical papers of Isaac Newton* 8 vols (1967–81 Cambridge)

P = Pertz G H *Leibnizens Gesammelte Werke* Reihe I, vol **4** (1847 Hanover)

RJ = Leibniz G W *Reise-Journal* (1966 Hildesheim: Olms)

SL = *Studia Leibnitiana* (Wiesbaden: Steiner)

W = Gerhardt C I *Briefwechsel zwischen Leibniz und Christian Wolff* (1860, reprinted 1971 Hildesheim: Olms)

ZHN = *Zeitschrift des historischen Vereins für Niedersachsen* (Hanover)

Introduction
The social, political and intellectual background

In 1648, two years after the birth of Leibniz, the Peace of Westphalia brought the Thirty Years' War to an end. Germany had been the main theatre of operations of this conflict, which started with a Protestant revolt in Bohemia and then spread to involve most of the countries of Europe as they entered ostensibly in support of the Protestant or Catholic cause but generally in pursuit of political advantages for themselves. Such was the havoc wrought upon the civilian population of Germany by the ill-disciplined and largely untrained armies of mercenaries, living off the land and devastating industry and agriculture, that the population decreased from twenty-one million to thirteen million during the course of the war.

France emerged as the strongest European state. While the Habsburgs had gained supremacy as rulers in their hereditary lands—Austria, Bohemia and Hungary—and were regarded collectively by France as a dangerous rival, the Holy Roman Empire, headed since 1437 by a Habsburg Emperor, became little more than a legal fiction without real power, though retaining the outward forms of authority, such as the Diet of Ratisbon and an Imperial Chamber for the administration of Justice. The politics of the Empire were complicated by the fact that some German princes ruled countries outside its boundaries (the Elector of Saxony, for example, was also King of Poland), while foreign rulers possessed territory within it (the King of Denmark, for example, was also Duke of Holstein). Since 1356 the Emperor had been chosen by seven German princes known as Electors, who claimed to be successors of the Roman Senate. These Electors consisted of the Prince–Archbishops of Mainz, Cologne and Trier, and four lay princes, the Catholic Duke of Bavaria and the Protestant princes of Saxony, Brandenburg and the Palatinate. The Upper Palatinate, together with the title, having been lost to Bavaria in the early stages of the war, a new

Electorate of the Lower Palatinate was created for the former Elector's son, as part of the settlement. Besides the Electorates, there were some 350 separate political entities, most of them very small, all nominally holding territory from the Emperor. At the end of the war, however, two major members, the United Provinces (Holland) and Switzerland formally seceded from the Empire.

The chief result of the Peace of Westphalia was the consolidation of Protestantism in northern Germany and Catholicism in the south. According to its terms, both Catholic and Protestant rulers were obliged to allow freedom of worship where it had existed in 1624, while Calvinism, at the insistence of the Elector of Brandenburg, was for the first time recognised as a lawful religion in the Empire on the same terms as Lutheranism had been since 1555.

Following the devastation of the Thirty Years' War, the need for reforms in government, diplomacy and national defence—above all the means of raising, provisioning and disciplining soldiers—was generally recognised. The required strong, centralised and effective government was provided by a new style of absolute monarchy, which in turn led to a stultifying conservatism. The theoretical foundation for this form of government had been laid by Jean Bodin writing in France after the massacre of the Huguenots on St Bartholemew's Day 1575. The model for the 'benevolent despot' was provided by Louis XIV, so that to ape the French came to be seen as aspiring to the height of modernity. In theory, the monarch was bound by the divine law. After 1648, most European states began to raise professional standing armies, which could effectively quell rebellions as well as defend the state. To finance their armies, the princes looked increasingly to industry, trade and commerce as sources of wealth from which to draw taxes. Anxious to avoid the horrors of another Thirty Years' War, many rulers adopted a foreign policy of preserving the stability that had been achieved by taking action against any individual prince who attempted to upset the balance.

France under Louis XIV adopted an aggressive foreign policy, using claims based on very dubious legal arguments as pretexts for military conquests. Following the death of Philip IV of Spain in 1665, Louis laid claim, on behalf of his wife, to the Spanish Netherlands. This brought him into conflict with Holland, and after a truce which gave him time for preparation, he occupied Lorraine in 1671, in order to clear his flank, and reached Utrecht after a short campaign in the following spring. Alarmed by these events, the Empire, Brandenburg, Spain and Lorraine formed an alliance, which was soon joined by Denmark, the Palatinate and other German states, for the purpose of preventing further French conquests. Nevertheless Louis devastated the Palatinate and captured numerous fortresses in the Spanish Netherlands before the conclusion of the Treaty of Nijmegen in 1678, which left Holland intact but Lorraine under French occupation.

The Treaty of Nijmegen had ceded towns and districts in frontier states 'with their dependencies', which were not defined. Having set up courts to identify these dependencies, Louis enforced their decisions by military action, occupying Strasbourg and Alsace besides other towns and bishoprics. The revocation of the Edict of Nantes in 1685, which deprived the Huguenots of all civil rights and caused large numbers of them to take refuge in Protestant countries, was followed in 1686 by the formation of a new alliance against France, known as the League of Augsburg. The members of the League, consisting originally of the Empire, Spain, Holland, Sweden and Saxony, and joined by Bavaria and Savoy in the next year, were committed to upholding the Peace of Westphalia. At the end of 1688, Louis XIV precipitated the War of the League of Augsburg by submitting the Palatinate to a second devastation. When William and Mary were crowned joint sovereigns of England in April 1689, England joined the League, which resulted in the defeat of the French fleet by the English navy in 1692. As the land warfare continued over most of Europe, bad harvests, famine and civil discontent made both sides anxious for peace, which was concluded by the Treaty of Ryswick in 1697. France kept Alsace and Strasbourg but restored Lorraine and gave up her conquests in the Palatinate.

Within four years, Louis XIV and Emperor Leopold I were again in conflict, this time concerning their rival claims to the Spanish succession. By September 1701, England and Holland had formed a secret alliance with Leopold against France, though their aims were somewhat different. While the Emperor wanted the undivided succession for the Habsburg claimant, England and Holland were content to leave the Duke of Anjou, the Dauphin's second son, as King of Spain, provided the French and Spanish crowns would never be united, but they promised the Spanish Netherlands and Spanish Italy to Austria. When the allies opened hostilities against France in April 1702, the Grand Alliance had been joined by Denmark, Prussia, Hanover and the Palatinate. Louis planned to attack Vienna through Bavaria (now his ally), but he was forced to retreat to the Rhine. Victories followed for the allies in the Netherlands, but when the Habsburg claimant to the Spanish crown became Emperor Karl VI in 1711, the Tory ministry in England ceased hostilities and opened negotiations with France, fearing that a Habsburg union of Spain and Austria would upset the balance of power just as much as a union of Spain and France. The Treaty of Utrecht, concluded in 1713, recognised the Duke of Anjou as King of Spain and the colonies, contained a declaration that the Spanish and French crowns should never be united, and assigned the Spanish Netherlands and Spanish Italy (apart from Sicily) to Austria. The Emperor rejected the settlement and continued to fight alone for a while, but the struggle finally came to an end with the Treaty of Rastatt in 1714. In the same year, the Elector of Hanover became King of England.

Another conflict of the early eighteenth century was the Great Northern

War. This started when Denmark, Poland and Russia believed they could take advantage of the inexperience of the young King Charles XII of Sweden to recover some of the territory they had lost in previous wars. Having entered into an alliance with King William III of England the previous year, Charles was able to call for naval assistance which enabled him first to invade Denmark and then, in November 1700, rout the Russians besieging Narva, before going on to invade Poland and Saxony, forcing the Elector in 1707 to give up his Polish crown and withdraw from the alliance with Russia. In 1708 he invaded Russia itself, but in June 1709, Peter the Great won an overwhelming victory. Charles XII escaped to Turkey and Russia became the dominant power in the Baltic. Twelve years earlier, Peter the Great had made a visit to western Europe, lasting eighteen months, in order to study resources and techniques and to recruit experts for service in Russia. Having learnt how trade, manufacture and knowledge could bring a nation power and prosperity, he was keen to reap these advantages in his own Empire, but he was ruthless in crushing any opposition.

Also in the early years of the eighteenth century the foundations of Prussian power had been laid. In 1701, the Elector of Brandenburg had crowned himself King in Prussia. Besides continuing the work of the Great Elector in developing the country by building roads and canals, as well as reclaiming marsh lands for agriculture, he exercised a fair degree of religious tolerance and sought to give Prussia a leading position in the promotion of German culture, notably by founding the University of Halle in Prussian Saxony for the teaching of the new studies of history and science.

Agriculture was still the dominant occupation, peasants making up three-quarters of the population of Europe. East of the Elbe, serfdom—that is, hereditary bondage to the land—was the rule. Although it still survived in many areas of western and southern Germany, it was the exception west of the Elbe. In Poland and Hungary, the serfs had rights to a plot of land and the means to cultivate it, but in Russia, their position was that of slaves who could be sold like cattle, lacking even the security of bondage to the land which at least kept families together. Many of the free peasants in the west leased their land rather than owned it. They were burdened with tithes (for the upkeep of the clergy) and taxes which took up a substantial proportion of their profits. All members of the family, even young children, had to work long hours in order to compete economically with the larger units of the nobility and the Church.

Towns were small, often surrounded by walls, so that overcrowding made them natural breeding grounds for diseases. Wages were higher than in the countryside. Although direct taxes were light, heavy excise duties on basic commodities were levied at the gates. The great majority of town-dwellers earned their living, the largest group being that of domestic servants. Capitals and regional centres had their social seasons, when prominent families increased their demands for servants and luxuries. Except in the most flourishing centres of trade and manufacture, the ruling classes provided the source of most employment.

There was a tendency for industry to move from the towns in order to be nearer the sources of power—wood and water—and to escape the restrictions imposed by the trade-guilds, though these were losing some of their power as the state became increasingly involved. A large part of manufacturing industry, including spinning and weaving for example, was organised according to the domestic system, whereby capitalist middle-men supplied the raw materials, and sometimes the machines, to craftsmen working at home. Besides printing works, saw-mills, breweries, tanneries and shipyards, there were also factories for the manufacture of candles, dyestuffs, sugar, chocolate, tobacco and cotton. Mining for iron, silver, tin and copper also required professional management and large-scale organisation.

Cheaper transport, mainly by water, facilitated the growth of commerce by the exchange of regional products and luxuries. This led to a wider distribution of wealth among the middle classes. While luxuries became more common for those who could afford them, the increased productivity was accomplished largely by processes and techniques already known. For almost a century, the size of population was static and prices were stable. There was little change in social conditions. Much of the trade and finance of major cities, such as Hamburg and Berlin, was in the hands of the Jews, who were debarred on account of their religion from official positions of power and responsibility. Almost everywhere the social elite despised trade as base and degrading. Besides the merchants and master craftsmen, the middle classes also contained professionals, such as doctors, teachers, civil servants and lawyers. The cost of training for these professions restricted their membership to the sons of middle class parents.

In a large town, up to a quarter of the population could be without regular employment or any fixed address. Although the Catholic Church had a fairly good organisation for the dispensation of charity and the poor were sometimes given relief in the form of employment on public works (in Denmark, for example), they did not fare so well in Protestant countries, where the poor relief was in the hands of parish officers who worked voluntarily.

The early seventeenth century was a time of revolutionary developments in science and philosophy. Besides discovering the laws of planetary motion, Johannes Kepler introduced a new approach to astronomy, abandoning the fictitious circles which earlier astronomers had used to represent the apparent motions across the celestial sphere in favour of an explanation of the real motions of the planets in terms of natural causes. Although his results came to be generally accepted as empirical laws, Kepler's physical theories had virtually no following, for the Aristotelian theory of motion on which they were based was soon superseded. Galileo laid the foundations for a new theory of terrestrial motion in his *Discourses on two new sciences*, published in Leiden in 1638. Francis Bacon saw himself as the Columbus of a new intellectual world in which science would be harnessed for the benefit of mankind. This, he thought, could not be brought about by the kind of reasoning employed by the Scholastics but only by the application of a new method of inductive generalisations from systematic observation and experiment, which he set out in his *Novum organum* (*New method*), published in 1620. His utopian *New Atlantis*, published posthumously in 1627, with its picture of an institution of organised scientific research (House of Solomon) was no doubt a source of inspiration for the establishment later in the century of the Royal Society.

The study of things themselves advocated by Bacon was greatly facilitated by the invention and development of scientific instruments, especially the telescope and the microscope. Galileo's telescopic discoveries of the satellites of Jupiter, phases of Venus, and the mountains and valleys on the surface of the moon served to disprove the Aristotelian division of the universe into two regions of entirely different nature. The heavens were evidently composed of the same elements as the earth and subject to the same natural laws. The range of biological knowledge was greatly extended through the use of the microscope, which revealed a previously unknown world of tiny creatures and enabled Antoni van Leeuwenhoek to observe the complete circulation of the blood which had been established theoretically by William Harvey.

Dissatisfied with Scholasticism, René Descartes, like Bacon, also sought a new method, which he then used to formulate an explanation of the natural world in terms of matter and motion alone. Although Huygens could describe Descartes as the author of 'un beau roman de physique'—for he erred in supposing that matter had no essential qualities apart from extension (the result of an over-optimistic belief that what is clear and distinct must be true)—his mechanical concept of nature achieved a great influence among philosophers and scientists for almost a century. In place of Aristotle's ordered cosmos he substituted a system of fluid vortices, each carrying a star or planet, extending indefinitely and homogeneously throughout space. Animals were conceived simply as machines or automata without conscious perception or thought. While he regarded the human body also as a machine, the mind (or soul) he believed to be a non-material substance. Thus he envisaged two distinct worlds, made up of spiritual and material substances respectively and somehow held together by the supernatural intervention of God, which his followers attempted to define more clearly in terms of the philosophical doctrine of occasionalism.

Cartesianism in its early stages was not without an influential rival. This was provided by Pierre Gassendi, who combined an atomistic natural philosophy derived from Epicurus with a Christian metaphysics. Although Descartes supposed matter to be divisible to infinity, he claimed to have demonstrated that the operation of the laws of nature had caused the matter created at the beginning to assume three forms or elements, in such a way that the parts (or corpuscles) of each element were approximately the same size and the speeds of the different elements inversely as their sizes. The first element, consisting of the smallest corpuscles and consequently in rapid motion, formed the sun and stars. The second element, consisting of spherical particles too small to be seen in the microscope, constituted the aether, while coarser and slower corpuscles made up the earth and planets. As the corpuscles of the first element were of indeterminate size and shape, they could fill up all the spaces between the corpuscles of the others so as to prevent the existence of a void. Leaving aside the metaphysical differences, the atomism of Gassendi and the corpuscular physics of Descartes could be regarded as essentially the same. This was the position taken, for example, by the chemist Robert Boyle.

There was a general recognition of the importance of mathematics for its possible applications in the development of the mechanical philosophy but the difficulty of designing experiments that would lead to results susceptible of mathematical analysis and the gap that existed between the complexity of physical problems and the limitations of the mathematical techniques available for their solution prevented the early development of a general mathematical physics. Progress was achieved in particular fields, notably geometrical optics and mechanics, though rarely by Cartesians, who were too rigid in their adherence to the ideas of Descartes. After the work of Galileo in laying the foundations of a science of motion, the next significant developments in the field of mechanics came from the great Dutch mathematician and natural philosopher, Christiaan Huygens, who did not accept the Cartesian philosophy, though he was influenced by it, and, in particular, approved of Descartes' banishment of the Scholastic substantial forms from the realm of physics.

In the first half of the seventeenth century, new techniques of mathematical analysis were developed by a number of outstanding mathematicians. While Kepler devised his own numerical procedures for the solution of the problems he encountered in astronomy, the more general innovations in pure mathematics originated outside Germany. In France, Descartes and Fermat produced analytical geometry besides methods for constructing tangents and the determination of maxima and minima. Frans van Schooten, who taught mathematics to Huygens in Holland, edited Latin versions of the geometry of Descartes and included results of Jan Hudde on maxima and minima and on the theory of equations in his own publications. In Italy, Bonaventura Cavalieri, who was a student of Galileo, introduced his method of indivisibles for finding areas and volumes, which was far simpler than the cumbersome procedure of Archimedes. Cavalieri's important work was not well written but fortunately, Evangelista Torricelli gave a clear exposition of his method. Isaac Barrow and John Wallis in England contributed respectively to the theories of tangents and quadratures. Among other mathematicians working in the field of analysis may be mentioned James Gregory in Scotland and Gregory of St Vincent in the Spanish Netherlands. Blaise Pascal, besides applying the method of indivisibles to the investigation of the cycloid, also designed the first successful calculating machine for the operations of addition and subtraction. The earlier invention of logarithms by John Napier had provided another useful aid to computation.

As the universities were still largely committed to the teaching of Aristotle, the practitioners of the new experimental science sought mutual support by meeting together informally. One group met at Gresham College in London under the leadership of John Wallis, who, on moving to Oxford, organised a similar group there, which held its meetings in the home of Robert Boyle. These informal assemblies led in due course to the foundation of the Royal Society of London in 1662. Neither funds nor buildings were provided by the King, so that the continued existence of the Society required the election of members of wealth and influence who would be willing to give their financial support. The nucleus of the new Society, however, consisted of the members of the London and Oxford groups, and the first meetings took place in Gresham College. Similar informal assemblies existed in Paris, the most prominent being that of Habert de Montmor, which maintained close links with the Royal Society. Both the Secretary of the Montmor Society, Samuel Sorbiere, and the wealthy amateur Melchisedech Thevenot, who was the leader of another group, appealed to the Chief Minister Colbert to secure royal patronage, without which they believed the organisation of experimental work was impossible. As a result, the Académie Royale des Sciences was established in Paris in 1666 with Christiaan Huygens in charge. Members received pensions and funds were made available for the building of an Observatory, which was completed in 1672. Few of those associated with the old Societies—Huygens was the most notable exception—became members of the new Academy, because the Cartesians and Jesuits were excluded on account of their rigid adherence to a particular philosophy and consequent lack of open-mindedness. Among the members, however, were several with Cartesian sympathies and the Baconian method was usually combined with the formulation of speculative explanatory hypotheses such as those proposed by the Cartesians.

The scene was set for the appearance of the young Leibniz on the international stage. But he was not alone, for at about the same time as he first met Huygens in Paris, the young Newton sent to Henry Oldenburg, the Secretary of the Royal Society, a new theory of colours.

1
Childhood and youth
(1646–1667)

Gottfried Wilhelm Leibniz was born at 6.45 PM on Sunday 1 July (NS) 1646 in the Protestant city of Leipzig, which had been a prominent seat of German learning and science since the Renaissance. The Court Chaplain Martin Geier, together with the lawyer Johann Frisch and Catharina Scherl stood as godparents at his christening, which took place in the St Nicolai Church at 2 PM on 3 July. His father relates that, as the deacon Daniel Moller held the baby on his arm, the infant raised his head and eyes as if to invite the sprinkling with water. This was just one of a number of signs which inspired in Leibniz's father great hopes for his son's future. Another incident, which gave rise to much talk in the town and invited the ridicule of his father's friends, took place when the young Leibniz was about two years old (*P*, p 165). In later life he remembered the incident as if it had occurred the day before yesterday. It happened on a Sunday morning after his mother had gone to the church service while his father lay ill in bed. He climbed on to a table and when his aunt tried to take hold of him, he stepped back and fell to the floor. When his father saw that his young son was uninjured, he recognised in this the hand of Providence and immediately sent a messenger to the church, in order that the custom of a thanksgiving prayer after the service would be observed.

Leibniz's father, Friedrich, had been born on 4 December 1597, the son of Ambrosius Leibniz and Anna Deuerlin, daughter of a nobleman of Leipzig. At the time his famous son was born, Friedrich was Vice Chairman of the faculty of philosophy and Professor of Moral Philosophy in the University of Leipzig, besides being in practice as a notary. He was evidently a competent though not an original scholar, who devoted his time to his offices and to his family as a pious, Christian father. Friedrich was married three times. From his first marriage in 1625 there was a son, Johann

Friedrich, and a daughter, Anna Rosina, who married a doctor of theology. Leibniz described his half-brother as a good-natured, pious man, who was satisfied with his lot as a schoolmaster. Friedrich's second wife died childless in 1643. In 1644, he married Catharina Schmuck, daughter of a celebrated Leipzig lawyer. This was Leibniz's mother. Born in Leipzig in 1621 and orphaned at the age of eleven, she was brought up in the home of Johann Hopner, Professor of Theology, and before her marriage lived in the home of her guardian Quirinus Schacher, Professor of Law. She was clever, pious and gentle. One of her sisters married a lawyer and another a doctor of theology. Leibniz's relations on both sides thus enjoyed good social standing and also scholarly reputations. Besides lawyers and theologians, his forebears had also included some musicians and mining engineers. The family had its roots in Saxony, as far as these can be traced, so that Leibniz's own belief in its Slavonic origin must be regarded as a myth (Kroker 1898).

Leibniz had a sister, Anna Catharina, born on 11 August 1648. She married the archdeacon of the Church of St Thomas in Leipzig and died on 3 March 1672, leaving a son Friedrich Simon Löffler, born on 19 August 1669, who became his sole heir.

Childhood and intellectual awakening

When the young Leibniz was only six years old, his father died. One of the things he remembered vividly was how his father had endeavoured with enthusiasm to inspire him with a love of secular and biblical history, and with such success that he anticipated distinction for him in the future. After the death of his father on 15 September 1652, Leibniz's mother devoted herself to the upbringing of her children. When Leibniz was only seventeen and his sister fifteen, however, his mother died of a respiratory illness. At her funeral it was said of her that she set an example not only to her own children but also to others as a model of piety. Striving to live with all in peace and harmony, she thought evil of no-one, and lightly forgiving those who had offended her, excelled all in patience. By this example, the young Leibniz had implanted early in his life the seeds of virtue and religion. The conspicuous traits in his own moral being almost agree with the character of his mother here portrayed.

As a child, Leibniz had little inclination to play, preferring instead to read history, poetry and literature. In July 1653 he entered the Nicolai School in Leipzig, where he remained until Easter 1661. The rector of this esteemed school, Johann Hornschuch, was also the Professor of Greek in the University. Leibniz recalls that, in the school, he began to learn Latin, and would no doubt have progressed as slowly as the other boys, if he had not chanced upon two books misplaced by a student living in the house. One was an illustrated Livy edition and the other the chronological *Thesaurus* of

E. E. Hochw. Raths
der Stadt Leipzig
Ordnung
Der Schulen zu St. Nicolai.

Publiciret den 1. Octobr. 1716.

Zum Druck gebracht
Von JOH. THEODORO BOETIO, und zu haben im Durchgange
des Rathhauses in der Boutique zum Contoir-Calender. 1717.

Plate 1 The Nicolai School in Leipzig. From the regulations for 1716. Courtesy of the Museum für Geschichte der Stadt Leipzig.

Sethus Calvisius. As he had a German book concerning general history, which frequently said the same thing as the Calvisius, he was able to understand this fairly easily. In the Livy, he read here and there the words underneath the illustrations. After repeated attempts to penetrate deeper into the sense, though without the use of a dictionary, most of the text, he later recalled, became clear to him. Thus he began to teach himself to read Latin, building on the instruction he received in school.

When his teacher found out what the young Leibniz was trying to read, he demanded of those responsible for his upbringing (that is, his mother and aunts) that he should be prevented from having access to books so unsuitable for his age. The conversations concerning this problem were witnessed by a learned nobleman, a friend of the family who happened to be in the neighbourhood. After questioning the young Leibniz to satisfy himself of his ability, this nobleman obtained a promise from Leibniz's relations that he would be allowed into his father's library, which had long been under lock and key. Leibniz thus gained access to his father's library in 1654 when he was eight years old. He read the Latin classics and Church Fathers as the impulse led him. By his twelfth birthday, according to his own recollection, he understood Latin fluently and began to stammer Greek, in which he had received instruction in school for perhaps two years. At Whitsuntide 1659, when he was thirteen years old, Leibniz composed and delivered at a school celebration a verse in Latin hexameters, the boy originally entrusted with the task having fallen ill three days before.

It was the instruction in the traditional syllogistic logic of Aristotle, given in the upper classes of the school, that first awakened the inventive genius of the young Leibniz. As he later recalled (*P*, p 167), not only could he easily apply the rules to concrete cases, which to the astonishment of his teacher he alone among his classmates could accomplish, but even at that time he also recognised some of the limitations of Aristotelian logic and occupied himself with new ideas, which he wrote down, in order not to forget them. Later he had taken much pleasure in reading again the things he had written down as a fourteen-year-old. The 'categories' held a special interest for him, so he examined many books on logic to see where the best and most exhaustive lists could be found (*GP* 7, pp 516–17). Of the Aristotelian categories—substance (with division into species and genera), quantity, quality, relation, place, time, position, state, action and affection—he found that the last two (or perhaps four) were either included in the others or were without application, while on the other hand many things were entirely excluded. Leibniz recalls how he used the categories as a means of guessing or recalling to mind something forgotten by narrowing down all irrelevant matters until the missing thing could be discovered. Nebuchadnezzar, he thought, could perhaps have reconstructed his forgotten dream in this way.

Some of his new ideas or questions Leibniz took to his teachers; among others, whether, since simple terms or concepts can be ordered through the known categories, it was possible to set up a new species of categories for the ordering of complex expressions or truths and of propositions themselves. At that time, Leibniz recalled later, he did not know that mathematical demonstration was what he was seeking. As he penetrated into this study, Leibniz tells us, he hit upon the wonderful idea that a certain alphabet of human thought could be found, and that from the combinations of the letters of this alphabet and analysis of the words built from them, everything could be discovered and also demonstrated. The elaboration of this wonderful idea later became one of his principal lines of enquiry.

Alongside the logical exercises performed in school, Leibniz pursued at home, in his father's library, the study of metaphysics, both scholastic and more recent, as well as theology, concentrating especially on the works of the famous Catholic and Protestant controversialists (*P*, p 168). His theological studies confirmed his acceptance of the moderate beliefs of the Augsburg Confession, while he found as much pleasure in the study of Zabarella, Rubius, Fonseca and other Scholastics as he had found previously with the historians. In particular, he recalled reading Francisco Suarez[1] with as much ease as a novel.

University studies

At Easter 1661 Leibniz entered the University of Leipzig, where he attended lectures on philosophy, especially on the philosophy of Aristotle, and introductory lectures on Euclid given by Johann Kühn. The mathematics lectures were so obscure that scarcely anyone but Leibniz understood them. He alone among the students engaged in discussion with the teacher and asked questions in order to make the theorems clearer to his classmates. Later he remarked on the low level of mathematics teaching at Leipzig and added that, if like Pascal he had spent his youth in Paris, he would perhaps have been in a position to enrich the sciences sooner. Leibniz was more fortunate in his philosophy teachers, especially Jacob Thomasius, whom throughout his life he continued to hold in high esteem.

Thomasius, who founded the scientific study of the history of philosophy in Germany, supervised Leibniz's dissertation for his bachelor's degree in philosophy. This essay, with title *Disputatio metaphysica de principio individui* (*Metaphysical disputation on the principle of individuation*) (*A* VI **1**, p 3; Quillet 1979), was defended and published in 1663, Leibniz's seventeenth year; it concerns a theme that was basic to the development of his mature metaphysics.

In his introduction to Leibniz's essay, Thomasius notes that the origin of the controversy concerning the principle of individuation was the problem

[1]On the influence of Suarez, see Robinet (1981) and Mccullough (1978).

of the distinction of minds proceeding from their common source. As interpreted by Thomasius, Aristotle distinguishes two kinds of individuals, the 'monadic', where each individual constitutes a species, as for example the immaterial planetary movers, and the 'sporadic', where innumerable individuals are embraced under the same species. The second kind applies to the sublunary world of the terrestrial elements, where Aristotle posed matter as the principle or cause of individuation. Aristotle's doctrines were accepted by St Thomas Aquinas, so that he supposed matter to be the principle of individuation within terrestrial species, whereas angels, being pure forms, he supposed to be by definition separate species. It is interesting to note that the term 'monad', later adopted by Leibniz in the definitive form of his metaphysics, was introduced to him at this early stage by Thomasius. Moreover, Leibniz later remarked in a letter to the Landgrave Ernst von Hessen-Rheinfels that he accepted the teaching of Aquinas concerning the angels more generally, provided the term species was taken not physically but metaphysically (*GP* **2**, p 131). In other words, on the metaphysical plane, each individual is monadic.

Boethius's question concerning the relation of universals to the individual things that exemplify them had given rise to two schools of thought among the Scholastics, the followers of Aristotle in the Middle Ages. According to the realists, the universals existed in their own right, while the nominalists regarded them simply as names. The interpretations of Aristotle taught in the Lutheran universities in the seventeenth century had been greatly influenced by the writings of Spanish Jesuits, especially Suarez, whose teachings Thomasius followed, and whose writings, it will be recalled, Leibniz said he could read as easily as a novel. Of the four general opinions concerning the principle of individuation, Leibniz adopted the nominalist standpoint of Suarez; namely, that in reality, the complete entity is the principle of individuation; in other words, the complete entity itself is what makes it an individual. Under the form of a syllogism, Leibniz demonstrates that unity or oneness adds nothing to being.

Major premise: Everything by which something is, it is by this that it is one numerically.
Minor premise: Each thing is by its entity.
Conclusion: Therefore each thing is one by its own entity.

The idea of the principle of the identity of indiscernibles, later enunciated by Leibniz, could be regarded as a simple corollary. According to this principle, there cannot exist two identical individuals.

Although the view of the principle of individuation adopted by Leibniz in his essay is not original, it is the one he continued to hold throughout his life. In the essay he skilfully defends this view with syllogistic demonstrations and disproves the opposing views, though his knowledge of these opinions seems to be based on the accounts in more recent writings rather than the original sources. In particular, he rejected what Duns Scotus called 'haecceity'—a quality or mode of being in virtue of which a thing becomes a definite individual—because, from the nominalist standpoint, an individual could not be conceived as composed of real universal forms. All individuals, he remarked, were individual in their totality.

There remained a problem concerning the relation, for example, of 'humanity' (a universal form and hence just a name) to the real property of 'humanity' possessed by individual human beings. Suarez also provided a basis for a solution of this problem in his ontology of relations, which seems to have influenced Leibniz in his definitive metaphysics of monads (Mccullough 1978, pp 254–70).

When the writings of the moderns fell into his hands, Leibniz had to choose between Scholastic philosophy and the new physics. As he later recalled, in a letter to Nicolas-François Remond, he had pondered this question while walking in a grove on the outskirts of Leipzig called Rosenthal (*GP* **3**, pp 605–8). His memory was probably at fault in fixing this incident at the age of fifteen. It seems more likely that it was about the time he took his bachelor's degree, or soon after, that he abandoned the substantial forms in favour of the mechanical philosophy. This decision, he remarked to Remond, led him to apply himself to mathematics. The rejection of the substantial forms or verbal categories of Scholastic philosophy as entities having explanatory power in natural science was in line with his nominalist standpoint and opened the way for acceptance of the validity of mathematical explanations.

Leibniz spent the summer semester of 1663 at the University of Jena, where he came under the influence of the Professor of Mathematics, Erhard Weigel (Moll 1978, pp 42–59). Earlier in the year, a student dissertation on the acceleration of falling bodies had earned a doctorate and it therefore seems likely that such topics in the field of natural philosophy were discussed with his students. Weigel was not simply a mathematician (nor indeed a practitioner of the first rank in this discipline) but also a philosopher, especially a moralist, who contributed original ideas in the field of natural law. The aim of his *Analysis Aristotelica ex Euclide restituta*, published in Jena in 1658, was no less than a reform of the whole of philosophy and science through a reconciliation of Aristotle and the moderns, such as Bacon, Hobbes and Gassendi,[2] based on the mathematical method; that is, the method of demonstration following the pattern of Euclid. This work exerted a profound and decisive influence on the philosophical orientation of the young Leibniz. Up to this time, he had followed the method of

[2]Weigel makes only critical comments concerning Descartes. At this time, however, the opposition to Cartesianism in the German universities, both Catholic and Protestant, especially from the theological faculties, made it impossible for any teacher to defend Cartesian doctrines publicly without losing his academic position (Hestermeyer 1969, p 51).

disputation of Scholastic philosophy, but later recalled how Weigel, by forcing his Scholastic opponents to repeat their empty terminology and definitions in plain German, made them appear ridiculous. Weigel's method of mathematical demonstration freed philosophy from the Scholastic battles of words, and by embracing the whole of philosophy in its systematic coherence as a *scientia generalis*, guaranteed the unity of the sciences. In Weigel's book, Leibniz remarks, he found many good ideas for the perfection of logic and for giving demonstrations in philosophy (Couturat 1903, p 179).

During his stay in Jena, Leibniz became a member of an academic society called *Societas quaerentium*. Under the presidency of Weigel, the members, consisting of professors and students, met weekly for exchange of views and discussion of both old and new books. Similar meetings of students which took place in Leipzig were also attended by Leibniz.

In October 1663 Leibniz returned to Leipzig for the beginning of the winter semester, when he began his specialist studies in law under the professors Quirinus Schacher and Bartholomaeus Schwendendörffer. As a result of the background knowledge acquired through his study of history and philosophy, Leibniz found the new discipline easy to understand, so that the theory presented no difficulty and he was able to devote his attention to practice. An assistant judge at the High Court in Leipzig, with whom the young Leibniz was friendly, often invited him to his home, where he would teach by examples how the verdicts had to be drawn up. Leibniz was attracted to the function of a judge but repelled by the intrigues of lawyers (*P*, p 168). This was the reason, he later remarked, why he had never wished to lead trials, although it was generally acknowledged that he wrote very well in his German mother tongue.

Early in February 1664 Leibniz graduated as Master of Philosophy with a dissertation *Specimen quaestionum philosophicarum ex jure collectarum*, published in December of the same year (*A* VI **1**, pp 69–96). In this work, Leibniz acknowledges his debt to his teacher Weigel. A study of the relations of philosophy and law, he asserts, will help to remove the contempt of students of law for philosophy. Moreover, without philosophy most of the questions of law would be a labyrinth without exit. Among the questions he considers are whether a person who is asleep is 'present' and whether bees, pigeons and peacocks are wild animals.

Just nine days after presenting his master's dissertation, Leibniz suffered the loss of his mother, who died of 'a catarrh which obstructed the respiratory tract'. Besides his sister and himself, his aunt, who was married to the famous legal scholar Johann Strauch, shared his mother's inheritance. Leibniz visited his uncle in Brunswick to sort out the rather complicated details. During this visit, Strauch recognised the outstanding abilities of his nephew and soon after sent him a learned letter on the subject of legal science, which Leibniz used in the preparation of a dissertation supervised by

Schwendendörffer, *De conditionibus*, which formed part of his studies for the bachelor's degree in law (*A* VI **1**, pp 97–150). Again there is a strong philosophical emphasis, for Leibniz develops a theory of hypothetical or conditional judgments applied to law. The hypothesis (antecedent) is called *conditio* and the thesis (consequent) is called *conditionatum*. Among the theorems are the following:

1 If the hypothesis is posed, the thesis follows.
2 If the thesis is suppressed, the hypothesis is suppressed.

Leibniz notes that a hypothetical judgment affirms nothing categorically, neither the hypothesis nor the thesis. In application to law, he considers the case of a law subject to a certain condition. If this condition is impossible, the law is null. If the condition is necessary (and therefore certainly satisfied), the law is absolute. If the condition is contingent or uncertain, the law is conditional. These results are set out in the following table, which is remarkable for the numerical values of 0, 1 and ½ given to laws which are null, absolute and uncertain respectively. The symbol ½, he notes, stands for some fraction between 0 and 1.

Conditio:	*impossibilis*	*contingens*	*necessaria*
	0	½	1
Jus:	*nullum*	*conditionale*	*purum.*

There is just a suggestion here of a calculus of probabilities. However, neither this novel idea, nor that of conditional judgments depending on other judgments (that is, the secondary judgments introduced again by George Boole in the nineteenth century) appear again in Leibniz's logical writings (Couturat 1901, pp 552–4).

After receiving his bachelor's degree in law, Leibniz worked on his *Habilitationsschrift* in the faculty of philosophy, with the title *Disputatio arithmetica de complexionibus*. This was the beginning of the *Dissertatio de arte combinatoria* (*Dissertation on the combinatorial art*), published in 1666 without further reference to the University. Among Leibniz's early works, the *Dissertatio de arte combinatoria* (*A* VI **1**, p 163) was outstanding for its originality. For the new logical and mathematical results of this work provided a common ground for his various philosophical interests and pointed the way to some of his greatest discoveries and projects. Written before he had penetrated into any real science (especially mathematics), Leibniz often regretted having published this youthful work; yet through it he announced his discoveries to the world and greatly enhanced his reputation among the scholars of the time.[3]

[3]When a second edition was published in 1690 without his knowledge, however, he felt it necessary to write an anonymous review, pointing out that the essay was a work of his youth (*GP* **4**, pp 103–4). To the Jesuit mathematician Kochanski he explained that if he had been consulted he would have made many improvements. See Knobloch (1973), p 55.

The combinatorial art

In *De arte combinatoria* Leibniz developed the wonderful idea that had occurred to him in his schooldays of an alphabet of human thought. For all concepts, he believed, were only combinations of a relatively small number of simple or fundamental concepts, as words and phrases were only the indefinitely varied combinations of the letters of the alphabet. By combining simple concepts, all the truths which their relations express could be discovered. Thus Leibniz considered the principal application of the art of combinations to be a logic of invention, which he distinguished from the traditional logic of demonstration embodied in the syllogistics of Aristotle. Leibniz adopted the view, which became a fundamental principle for the construction of his metaphysics, that all propositions consist of a combination of subject and predicate (or at least that all propositions could be reduced to this form). Hence the logic of discovery or invention concerns the finding of all the true propositions in which a given concept occurs either as subject or predicate; in other words, (i) given a subject, finding all the possible predicates, (ii) given a predicate, finding all the possible subjects.

Leibniz was inspired by the thirteenth-century Catalan encyclopedist Ramon Lull, whose 'Great Art' was a general method permitting the formation of all propositions that could be conceived. Lull formed a kind of table of categories, in six series; he distinguished nine absolute attributes, nine relations, nine questions, nine subjects, nine virtues and nine vices. Not knowing the arithmetical laws for combinations, he used a mechanical means of realising some of these combinations. Thus he constructed six concentric circles of increasing radius, turning independently on their centres, and inscribed the nine terms of each series on a different circle. Then, by turning the circles and in each position taking the six terms situated on the same radius, various combinations could be obtained. Leibniz mentions this mechanical method of Lull and refers to several other inventors who employed similar devices. He criticises the 'Great Art' for the arbitrary choice of concepts, the artificiality of fixing the number of each category at nine, and the fact that the virtues and vices are not primitive or universal concepts. The invention of Lull, he concludes, was a device useful to rhetoric rather than a table of categories appropriate to the needs of philosophy.[4]

Leibniz learnt about the known results on combinations chiefly through the work of Daniel Schwenter and Philipp Harsdörffer, though he also knew at first hand the commentary on the elementary textbook on astronomy known as the *Sphere of Sacrobosco* where the Jesuit mathematician and

[4]Lull's primary aim was to provide a method of rational argument for use by Christians in disputations with Muslims, for whom the authority of Scripture carried no weight.

astronomer Christoph Clavius deals with questions concerning combinations (Knobloch 1973, p 1). For permutations Leibniz used the term *variationes ordinis*, while he distinguished combinations of different classes, reserving the term *combinationes* itself for selections consisting of two objects. Selections of three objects he called *con3nationes* (*conternationes*) and so on, while combinations in general were described as *complexiones*; the numbers of the classes were called *exponentes*. The table of combinations (table 1.1) built up by Leibniz is essentially the triangle of Pascal,[5] although the form of the table differs from all earlier arrangements and Leibniz believed the method of generalising from $\binom{n}{2}$ by means of the relation $\binom{n}{r} = \binom{n-1}{r} + \binom{n-1}{r-1}$ to be original. In Leibniz's terminology, it may be noted, n is the number and r the exponent.

Table 1.1

Exponentes														Complexiones
0	1	1	1	1	1	1	1	1	1	1	1	1	1	
1	0	1	2	3	4	5	6	7n	8u	9m	10e	11r	12i	
2	0	0	1	3	6	10	15	21	28	36	45	55	66	
3	0	0	0	1	4	10	20	35	56	84	120	165	220	
4	0	0	0	0	1	5	15	35	70	126	210	330	495	
5	0	0	0	0	0	1	6	21	56	126	252	462	792	
6	0	0	0	0	0	0	1	7	28	84	210	462	924	
7	0	0	0	0	0	0	0	1	8	36	120	330	792	
8	0	0	0	0	0	0	0	0	1	9	45	165	495	
9	0	0	0	0	0	0	0	0	0	1	10	55	220	
10	0	0	0	0	0	0	0	0	0	0	1	11	66	
11	0	0	0	0	0	0	0	0	0	0	0	1	12	
12	0	0	0	0	0	0	0	0	0	0	0	0	1	
*	0	1	3	7	15	31	63	127	255	511	1023	2047	4095	
†	1	2	4	8	16	32	64	128	256	512	1024	2048	4096	

The idea of including combinations of the first class $\binom{n}{1}$ was indeed original with Leibniz. That the entries in the row for $r = 1$ (called *uniones*) were intended to be regarded as combinations (as well as the values of the *numeri*) is demonstrated by the inclusion of these values in the total number of combinations, $2^n - 1$, given in the row marked *. Another interesting

[5]Pascal's *Traité du triangle arithmétique* was published in 1665, though the 'triangle' had appeared in China in 1303 (Needham 1959, p 135).

feature of the table is the implication that $\binom{n}{0} = 1$, indicated by the row of ones corresponding to $r = 0$. As this row is separated from the array of *complexiones* and the ones in it are not included in the totals given in the row marked *, it is clear that Leibniz did not regard the null class as defining a class of *complexiones*. The row marked † simply indicates, of course, that when 1 is added to the total number of combinations of all classes, the result is 2^n.

Applications of combinations described by Leibniz include examples taken from law, the register of an organ and the Aristotelian theory of the formation of the elements from the four primary qualities.[6] Since the elements are formed by combinations of two qualities, the number of possible elements would seem to be $\binom{4}{2} = 6$. But contrary qualities are prohibited, so that two of the possible combinations are useless, leaving four, corresponding to earth, water, air and fire. Leibniz introduced a new idea when he considered permutations of relative place in a circle. Among other examples which he mentioned were the application of permutations with repetitions to the combination of letters into words and musical notes into melodies.

For Leibniz the philosophical applications were more important than the mathematical results. After mentioning the ingenious thought of Hobbes, that all reasoning is a calculation, Leibniz made his own first attempt to formulate the beginnings of a logical calculus. A certain number of simple concepts, he supposed, constituted the alphabet of human thought to which reference has already been made. He illustrates his ideas with an example in which the numbers 3, 6, 7 and 9 represent four simple concepts. These constitute the first class. A second class of new concepts is built by forming combinations of pairs of simple concepts in order; these are represented by $3 \cdot 6$, $3 \cdot 7$, $3 \cdot 9$, $6 \cdot 7$, $6 \cdot 9$ and $7 \cdot 9$. Similarly, a third class is formed by taking combinations of triples. One of these is $3 \cdot 6 \cdot 9$, which is also $1/2 \cdot 9$, $3/2 \cdot 6$ or $5/2 \cdot 3$ where p/q means the pth term of the qth class. Hence there are several expressions for the same thing, whose equivalence is verified by decomposition into the terms representing the simple concepts. Thus Leibniz supposes the complex concepts to be built up from the simple by a method of combination analogous to arithmetical multiplication.

As an example of the application of his logic of discovery, Leibniz offered an attempt to define the elementary concepts of geometry by combinations of primitive terms. For his set of primitive terms (labelled Class I) Leibniz takes twenty-seven items which include the following:

1 *Punctum*, 2 *Spatium*, 3 *Intersitum*, . . . , 9 *Pars*, 10 *Totum*, . . . , 14 *Numerus*, 15 *Plura*, . . . , 20 *Fit*, ...

[6]The four primary qualities were cold, hot, wet and dry.

From these he builds twenty-four classes of definitions, of which the following are samples:

Class II. 1. *Quantitas est* 14 τῶν 9 (15)

meaning 'Quantity is the number of parts (several)'.

Class III. 1. *Intervallum est* 2·3·10

meaning 'Interval is the total space included'.

Class IV. 3. *Linea*, ⅓τῶν 1 (2)

meaning 'A line is the interval of two points'.

It should be noted that cardinal numbers are placed in parentheses to distinguish them from the symbolic numbers, and that the number of the class of a definition is the sum of the denominators of the symbolic numbers, counting wholes as fractions with denominator one. Also it should be noted that Greek supplies the articles which Latin lacks.

Leibniz compared this to a universal writing or language, similar to Egyptian and Chinese, in which ideas are represented by a combination of signs corresponding to their component parts. Such a system was a first step towards the universal characteristic he was seeking, that would not only provide a direct representation of ideas but would also permit reasoning and demonstration by a calculus analogous to those of arithmetic and algebra. To achieve this end it would be necessary to replace the verbs, prepositions, articles and cases by signs. For Leibniz this should be possible in principle, since the items concerned represent relations, while relations in his view are reducible to predicates of subjects. The further analysis needed would have to extend even to Class I, since this includes terms representing relations as well as categories.

As an appendix to his essay, following the synopsis, Leibniz included a proof of the existence of God, setting out the cosmological argument in the pattern of a Euclidean demonstration. The inspiration for this probably came from Weigel, though strangely perhaps, his Jena teacher is not mentioned anywhere in *De arte combinatoria*.

Graduation as Doctor of Law

Despite his erudition and scholarly reputation in his own city and beyond, Leibniz was refused the doctor's degree by the University of Leipzig. His own account throws some light on the obscure circumstances of this strange decision. Besides the professors, the faculty of law in Leipzig included twelve assessors, appointed in order of seniority from the doctors of law of the University. As early admission to membership of the faculty would evidently help their career, some candidates for the doctorate intrigued to eliminate rivals by securing the deferment of graduation of the younger candidates. The majority of the faculty, Leibniz tells us, assented to his own deferment; consequently, he decided to change his plans and leave Leipzig (*P*, p 169). A variation of this story was given by his secretary, Eckhart, who claimed that Leibniz had told him many times that it was the Dean's wife who persuaded her husband to refuse the doctorate out of malice towards him (Ross 1974, p 222).

At the beginning of October 1666, Leibniz matriculated in the faculty of law of the University of Altdorf, situated in the small republic of Nuremberg. Without delay, he presented his dissertation, *De casibus perplexis in jure* (*On difficult cases in law*), which he had already prepared in Leipzig. This was published in November 1666 (*A* VI **1**, p 231). On 22 February 1667, he formally received the doctor's degree, as he noted, in his twenty-first year and with the approbation of all. Describing the circumstances of his oral disputation, he remarked that the audience admired the clarity and penetration of his exposition and even his opponents declared themselves to be extraordinarily satisfied. From Leipzig, he received the congratulations of his teachers Thomasius and Schwendendörffer.

Commenting in his dissertation on the idea of 'case', Leibniz draws a parallel (quoting Weigel) between the geometers, who first used the term, and the jurists. In opposition to those lawyers who believed no solution to be possible in the kind of difficult case he had in mind, or who advocated a decision by drawing lots, or accepting the personal opinion of an arbitrator, Leibniz held that the law always had an answer. For in cases that were uncertain, the natural reason should be brought in to help and the decision based on the principles of natural justice and international law, which limit and determine the civil law. Examples of such difficult cases are resolved by Leibniz with much technical skill.

The authorities in Nuremberg wished to acquire the services of the brilliant young scholar for the University of Altdorf. Leibniz relates how the minister responsible for education, Johann Michael Dilherr, indicated to him that, if he had the inclination, an early appointment to a professorship could be guaranteed. But Leibniz declined this offer, remarking that his spirit moved him in a wholly different direction (*P*, p 170). He had, it seems, concluded that a reform and improvement of the sciences after the pattern he had in mind could not be achieved within the confines of a university.

2
First steps into the world of politics and learning
(1667–1672)

Having completed his studies at the University, Leibniz tells us, he set out on his travels, with the intention of going to Holland and beyond, but as he was passing through Mainz, he made the acquaintance of the Elector, Johann Philipp von Schönborn, who appointed him a judge in the High Court of Appeal, the highest tribunal of the Electorate and Archdiocese, when he was scarcely twenty-four years old (*MK*, p 11).

Before he took up residence in Mainz, Leibniz had already found a patron and friend in the distinguished statesman Baron Johann Christian von Boineburg, who for a long time had been Schönborn's minister until a French intrigue led to his dismissal in 1664. Early in 1668, the time of Leibniz's arrival in Mainz, Boineburg was reconciled with the Elector through the marriage of his eldest daughter with the Elector's nephew, Baron von Schönborn. Leibniz had evidently met Boineburg before the reconciliation took place, for in a letter to Hermann Conring of 26 April 1668, Boineburg remarks that he knew Leibniz well and that the brilliant young scholar lived in Mainz on his recommendation. Yet the circumstances of the first meeting of Leibniz and Boineburg are somewhat obscure. According to one account, they met by chance at an inn in Nuremberg; according to another, Leibniz was introduced to Boineburg by an alchemist friend. Like many scholars of the time, Boineburg occupied himself a little with alchemy. It is certain that Leibniz himself was associated for a short time with an alchemical society in Nuremberg, though few reliable details of this episode in his life have survived.

In a letter written many years later, Leibniz related that he was introduced to the study of chemistry at Nuremberg and did not regret having learned as a youth things that should be regarded with caution (Ross 1974, p 242). For this knowledge had proved useful to him later on, when patrons of alchemy among the princes he met asked him to undertake chemical investigations. Leibniz's secretary and first biographer, Johann Georg Eckhart, relates how Leibniz often told him that he had gained entrance to the alchemical society by means of a trick. According to this story, Leibniz composed a letter using obscure alchemical terminology which he did not understand himself and sent it to the priest who was President of the society, requesting admission. Concluding from this letter that Leibniz was a true adept, the President not only gave him access to the laboratory but offered him a salaried position as Assistant and Secretary, which Leibniz accepted. Another version suggests that Leibniz was introduced to the society by Daniel Wülfers, whom Leibniz often visited in Nuremberg, where he also made the acquaintance of other learned men. It seems at least plausible that Leibniz in his later life may have invented the story of the trick in order to play down the seriousness of his early interest in alchemy. There is also no reason to believe that the society was Rosicrucian, as was first suggested in the nineteenth century, though it may have come into existence partly out of the splitting up of the Rosicrucian movement in the early part of the seventeenth century.

During the few months that he was associated with the alchemical society, Leibniz may have gone to live in Nuremberg which was a few hours' journey from Altdorf. By 25 November 1667, however, he had taken up residence in Frankfurt, perhaps at the instigation of Boineburg, who was living there at the time. On the journey, Leibniz composed his *Nova methodus discendae docendaeque jurisprudentiae* (*A new method for learning and teaching jurisprudence*), which he dedicated to the Elector of Mainz in the hope of obtaining a position at the Court. This work, as he remarks, contained good ideas, although it was written in inns, without the aid of books; sketched rather than written, as he expressed it. The work was published anonymously in Frankfurt at the end of 1667.

Among those who received the new ideas contained in Leibniz's work with approbation was the famous scholar Hermann Conring, to whom Boineburg sent a copy asking for his opinion of it. The work is in two parts; the first relates to learning and teaching in general and the second to jurisprudence in particular, including a philosophical analysis of the principles of law together with a proposed new course of study that would help students to acquire insight into the theoretical foundations as well as practical competence. The art of judging in general, Leibniz believed, was reducible to two rules (*A* VI **1**, p 279):

1 No term is to be accepted without definition.
2 No proposition is to be accepted without proof.

These, he remarked, were more absolute than the four Cartesian rules in the *Discours de la méthode* (having no books, he mistakenly placed them in the *Meditationes de prima philosophia*). Descartes' first rule—that whatever is perceived clearly and distinctly is true—he rejected as deceptive in endless ways. Applied to the interpretation of law, the two rules of Leibniz require first the determination of the sense of each word, and, second, the determination of the intention of the legislation.

In the second part, Leibniz adopted for his treatment of jurisprudence the divisions of theology, calling on the analogy of the two fields. Thus he considered in turn the didactic, historical, exegetical and polemical aspects of jurisprudence. Besides exegesis, in which he distinguished between philosophical analysis of principles and interpretation of the law based on such analysis, his proposed course of study also included polemics in the form of arguments on cases conducted in the manner of trials, an excellent preparation for practice (Kalinowski 1977).

Leibniz presented the work in person to the Elector of Mainz, who then invited him, for a weekly remuneration, to assist the Court Assessor and Privy Counsellor Hermann Andreas Lasser with the improvement of the Roman civil code for the needs of the state. Together they produced in 1668 a programme for the great undertaking and as the work proceeded, Leibniz evidently took up residence with Lasser, for Boineburg remarked to Conring, in a letter of 22 April 1670, that Leibniz lived in Mainz with Lasser. The payments were made for a while but then stopped, so that early in 1669 Leibniz had to remind the Elector of his promise and request the payment of arrears he had not received (*A* I **1**, p 20).

The Polish plan

Besides assisting Lasser, Leibniz was also occupied by turns as Boineburg's secretary, assistant, librarian, lawyer and adviser, while being at the same time a personal friend of the Baron and his family. Soon after his reconciliation with the Elector of Mainz and his return to the Court, Boineburg undertook an extraordinary diplomatic mission, for which his independence of political factions was an added qualification. King Johann Casimir of Poland having abdicated, the election of a new king was necessary. The Palsgrave von Neuburg, who had the support of the Elector of Mainz, asked Boineburg to undertake a mission to Poland in support of his candidature. Leibniz spent the winter of 1668 cataloguing the Baron's library and preparing a document for his Polish mission. This took the form of a writing entitled *Specimen demonstrationum politicarum pro rege Polonorum eligendo* (*Model of political indications for electing a king of Poland*) (*A* IV **1**, pp 3–98) supposedly published in Vilna, Lithuania, by an unknown Polish nobleman, Georgius Ulicovius Lithuanus, setting out the claims of the Palsgrave. Both Boineburg and Schönborn must have given their approval to this deception. Having referred in the preface to the ideal of mathematical demonstration introduced by Galileo, Descartes, Hobbes and Bacon in science, the fictitious author applies this method to the political question of the election, concluding in favour of the Palsgrave von Neuburg (Voisé 1967). In the course of the argument, ideas of ethics and politics are considered as elements of a calculation of probabilities and in later years Leibniz valued the writing for this alone.

The work was published in Königsberg, not Vilna as alleged on the title page (*A* IV **1**, p xviii). Moreover, as a result of unforeseen delays, it appeared only after the decision had been made on 19 June 1669, but Boineburg's speech in support of the Palsgrave von Neuburg was based on Leibniz's arguments (Guhrauer 1846 **1**, appendix p 13). The misprinted date 1659 on the title page has led to the suggestion that Leibniz intended the reader to believe that the work was written ten years earlier with prophetic insight. But this idea is certainly mistaken, for at every step there are references to things that occurred after 1659 and would be known to every reader.

Theological and philosophical studies

As Boineburg was preparing for his mission to Poland, he asked Leibniz to prepare a reply to a letter he had received in 1665 from the Polish Socinian Andreas Wissowatius.[1] In the response that he composed for Boineburg, entitled *Defensio Trinitatis per nova reperta logica* (*Defence of the Trinity restored by new arguments*) (*A* VI **1**, pp 518–30; cf Korcik 1967), Leibniz did not in fact produce new speculative grounds for the mystery of the Trinity but uncovered errors in the dialectic of Wissowatius. Boineburg was himself a pious convert to Catholicism who had for years worked without success for the reunion of the Lutheran and Catholic Churches in Germany. Leibniz remained a Lutheran but also desired the reunion of the Churches. Both were firm in their resolve to work against the common enemies of Christianity and religion. Soon after their first meeting, Leibniz had sent to Boineburg an essay without title or signature, which he had written down in an inn without preparation, defending the existence of God and the immortality of the soul against atheists and materialists. Two years earlier, it will be recalled, he had given proof of the existence of God a prominent place in his *De arte combinatoria*. From Boineburg the essay passed through many hands until it finally reached the Augsburg theologian Gottlieb Spitzel, who published it in 1669, without knowing the identity of the author, in his *Epistola ad Reiserum de eradicando atheismo*, under the title *Confessio*

[1]Andrzej Wiszowaty was born in 1608 in Lithuania. On his mother's side he was a grandson of Faustus Socinus, the Italian theologian who, with his uncle Laelius, founded the doctrine known as Socinianism, which resembles modern Unitarianism. Educated in Leiden, he became a minister in Volhynia, Poland but later became a refugee in Hungary, the Palatinate and finally Holland, where he died in 1678.

naturae contra atheistas (*Confession of nature against atheists*) (*A* VI **1**, pp 489–93).

In the first part of the essay, concerning proof of the existence of God, Leibniz seeks to establish that bodies cannot exist of themselves without an incorporeal principle. At this time he had accepted the mechanical philosophy in the general sense envisaged by Robert Boyle, who regarded the doctrines of Gassendi and Descartes as constituting a single corpuscular philosophy, to which Galileo, Bacon and Hobbes also adhered. Defining body as something that exists in space, he proceeds to show that the qualities of size, shape and motion cannot be deduced from the nature of bodies and concludes that these qualities cannot therefore exist in bodies left to themselves. While it is true that the size and shape arise from the limits of the space in which a body exists, there is nothing in the definition of body—something that exists in space—from which the determinate size and shape can be deduced. Likewise from the nature of a body is born mobility but not its determinate motion, for which there must be a reason. To ascribe the qualities to the effects of other bodies does not help, for this simply leads to an infinite regress. Again the cause of cohesion of the ultimate particles is not found in the definition of body. It would be absurd to attribute cohesion to a void between the particles, for in this case, bodies once brought into contact would be inseparable. If hooks and rings are postulated, the problem is simply transferred to the explanation of the cohesion of the particles constituting these devices. So an incorporeal principle is needed to explain the quantitative properties of bodies, for if left to themselves, they would have neither determinate size nor shape, nor any absolute motion. The unity of this incorporeal principle, which Leibniz identifies with God, he supposes is the cause of the harmony of all things among themselves. It is worthy of note that the ideas of the important Leibnizian principles of sufficient reason and pre-established harmony are already in evidence.

In the second part of the essay Leibniz sets out his demonstration of the immortality of the soul in the form of a sorites chain. The argument may be summarised as follows. The activity of the soul is thought and thought (the immediate experience of consciousness) is without parts. If an activity is without parts, then it is not motion; for by Aristotle's demonstration and common agreement, all motion has parts. Since the activity of body is motion, something whose activity is not motion is not body. This establishes that the soul is incorporeal, from which Leibniz concludes that it is not in space. This means that (in addition to its activity not being motion) the soul itself is not movable; for motion is change of space. Hence it is indissoluble, for dissolution is motion of parts. Also it is incorruptible, since corruption is dissolution in time, and is therefore immortal. The argument is similar to that of Plato, based on the unity of the soul (*Phaedo* 100C–105E, *Phaedrus* 245C–246A).

Leibniz mentioned his essay in a letter of 30 April 1669 to his teacher

Jacob Thomasius. Having expressed his concern that so many brilliant men of his acquaintance were atheists, Leibniz remarked that, while he did not disapprove of the publication of his essay by Spitzel, he regretted the confusion into which his demonstration of the immortality of the soul had been thrown by changes introduced into what he had written. Leibniz asked his teacher especially for an opinion on this demonstration, adding that, since writing the essay, he had penetrated more deeply into the problems treated in it. Occasioned by the publication of the second edition of Thomasius's work on the history of ancient philosophy, Leibniz's letter is especially valuable in revealing the development of his thought, for it contains what might be described as a first sketch of his own philosophy, though some of the ideas were introduced in an earlier letter to Thomasius and in the *Confession of nature against atheists*.

The letters to Thomasius

In his letters to Thomasius (*A* II **1**, pp 10–11, 14–24) Leibniz aimed to show that Aristotle was remarkably in accord with the modern philosophers, with whose ideas he was generally in agreement, and that the obscurities sometimes attributed to Aristotle were the product of the hazy thought of the Scholastics. He rejected only Aristotle's claim of the impossibility of the vacuum, on which the moderns were divided. In his view, neither a vacuum nor a plenum was necessary. There was, of course, evidence for the existence of a vacuum in the well known experiment of Guericke, which had been performed in 1654. While he agreed with the modern philosophers that only magnitude, figure and motion should be used in explaining corporeal properties, Leibniz distinguished Descartes from other moderns such as Bacon, Hobbes and Gassendi[2], and declaring that he was not a Cartesian, rejected the Cartesian notion of matter simply as extension. Primary matter, Leibniz held, was mass itself, in which there was nothing but extension and impenetrability. Although he used the term mass (*massa*), there was no suggestion at this stage that matter possessed the property of inertia. Accepting Aristotle's definition of body as matter and form, Leibniz took form to be simply figure. Moreover, he claimed to be following Aristotle in supposing space, of which figure is the determination, to be substance. Thus Leibniz at this stage took space to be a real entity. Indeed, he regarded space as almost more real than body; for space can exist without body but not body without space. For Leibniz then, the substantial form of a body—that which

[2]Moll (1982) has argued that Leibniz received his first orientation on the tradition of ancient atomism from Gassendi and paved the way to his Aristotelian atomism by working on problems that Gassendi had himself either neglected or to which he had given only eclectic answers.

distinguishes it from other bodies—was its figure. From this notion of body and substantial form, he believed, could be derived a mathematical demonstration (that is, a proof after the maner of Euclid) of the necessity of an incorporeal mover. For since a body was nothing other than matter and figure, and the explanation of motion could not be derived from these concepts, it was necessary that the cause of motion was outside the body, and was therefore mind. Impenetrability, it should be noted, implies only mobility and not determinate motion itself. But Leibniz adds that, if incorporeal and quasi-spiritual substantial forms (such as those envisaged by the Scholastics) are admitted, by means of which bodies can move themselves, then the path to the demonstration of the existence of the first mover, God, is closed.

For Leibniz then, mind was the primary principle of motion. Figure, magnitude and motion were properties of bodies brought about by mind. The question for the mechanical philosophy was whether all changes in bodies can be explained in terms of these properties alone; that is, figure, magnitude and motion. Leibniz and the modern philosophers answered in the affirmative. First, Leibniz explained that, if primary matter or mass had been created discontinuous or separated into parts by an intervening void, there would at once be certain concrete forms of matter; that is, bodies. If, however, matter had been continuous in the beginning, then the division into bodies could only have been produced by motion. For the relative motion of parts would mark off the boundaries of bodies and thus give them determinate figure and magnitude. Motion could also explain changes in the motions of bodies. For, on account of the impenetrability of matter, a body must either give way when another body strikes it, or bring this other body to rest. When a body is moved as a consequence of a collision with another, Leibniz supposed the other body to be the cause of the impressed motion but the figure of the body itself to be the cause of the received motion. For example, a sphere gives way more easily than an equal body of another figure when it is struck by a colliding body. As Leibniz remarked, he admits that form is the principle of motion within its own body and that body itself is the principle of motion in another body. But the first principle of motion, he added, is mind, which is the efficient cause.

The modern philosophers, Leibniz believed, had demonstrated sufficiently that all changes in bodies could be explained in terms of local motions, so that heat and colour, for example, were merely the effects of subtle motions. As already remarked, however, motion itself cannot be derived from the nature of body. Hence, Leibniz concluded, there is no motion, strictly speaking, as a real entity in bodies. Motion was defined by Leibniz as change of space and it seemed to him that this could only be brought about by the re-creation of the body at every instant in assignable motion. The idea of continuous creation, he believed, was both novel and clearly necessary; moreover, it would serve to silence the atheists.

Mind, Leibniz remarked, supplies motion to matter in order to achieve a good and pleasing figure and state of things. This statement follows remarks concerning a beautiful harmony among the sciences and points the way to the definitive philosophy of the *Theodicy*. As Leibniz explained, theology or metaphysics deals with the efficient cause (mind), moral philosophy—that is, ethics or law, which as he had learnt from Thomasius were one and the same science—with the final cause (the good), mathematics with the form or figure, while physics deals with the matter of things and its unique modality, motion, which results from the combination of matter with the other causes.

Commissions for Boineburg

On his return from the unsuccessful mission to Poland, Boineburg set Leibniz to work on a new edition of the book *Anti-Barbarus seu de veris principiis et vera ratione philosophandi contra Pseudophilosophos* (*On the true principles and reasoning of philosophy against Pseudo-philosophers*) by the Italian Marius Nizolius, which had been published first in 1553. The new edition, with a preface by Leibniz, which is especially interesting for his discussion of philosophical language, appeared at the Frankfurt Book Fair in April of the following year, 1670.

In August 1669, Leibniz accompanied Boineburg on a visit to Bad Schwalbach, where his patron had gone for treatment. Here he met the jurist Erich Mauritius, who drew his attention to the publications of Christopher Wren and Christiaan Huygens in the *Philosophical Transactions* concerning the collision of bodies, thereby occasioning the composition of the *Hypothesis physica nova* (*New physical hypothesis*). Leibniz began to write a first draft while still in Bad Schwalbach. This was revised and expanded, especially as a result of his intensive study of Hobbes in 1670, and the work was published anonymously (under the initials *G.G.L.L.*) in 1671. In its final form, it consisted of two complementary essays, entitled *Theoria motus concreti* and *Theoria motus abstracti*, dedicated respectively to the Royal Society of London and the Royal Academy of Sciences in Paris.

At the end of 1669, on the recommendation of Christian Habbeus von Lichtenstern, the Swedish Ambassador in Frankfurt, Duke Johann Friedrich invited Leibniz to Hanover. But he declined this invitation, evidently preferring to live and work with Lasser on the reform of the civil code, assist his friend and patron Boineburg in a multitude of tasks and meanwhile hope for a position at the Court in Mainz, which was realised soon after by his appointment to the High Court of Appeal. Again in August 1670 he accompanied Boineburg to Bad Schwalbach, where their discussions concerning the preservation of peace in Europe gave rise to the first formulation of the idea of the 'Consilium Aegyptiacum', a secret diplomatic project that was eventually to lead him to Paris.

The Nizolius edition

Having at Boineburg's instigation prepared the new edition of Nizolius, a work of only historical interest (in which he perceived many serious errors), Leibniz took the opportunity to use the publication as a vehicle for the clarification and propagation of some of his own ideas. Thus he declared in the preface that it was his intention, even through the work of another author, to contribute something to the establishment of the sounder philosophy that the concerted efforts of the greatest geniuses were so excellently advancing (*GP* **4**, pp 129–74). Leibniz had found in Nizolius two particular merits which could be used to justify a new edition of his work. First, his literary style had the qualities needed for clear philosophical discourse, and, second, like the modern reformers of philosophy, Nizolius was a nominalist. Nominalists, Leibniz reminded his readers, held that all things except individual substances were merely names. One of the most serious errors of Nizolius perceived by Leibniz, however, concerned the nature of universals. If, as Nizolius held, universals were just collections of individuals, then knowledge could not be attained by demonstration but only by induction. Nizolius actually drew this conclusion, but in Leibniz's view, knowledge would in this case be impossible. According to Leibniz, the aid of some universal propositions, which do not depend on induction but on a universal idea or definition of terms is needed. The following, he claims, will suffice:

1 If the cause is the same or similar in all cases, the effect will be the same or similar in all.
2 The existence of a thing which is not sensed is not assumed.
3 Whatever is not assumed is to be disregarded in practice until it is proved.

The first proposition corresponds to what later came to be known as the principle of the uniformity of nature. Even with the help of these propositions, Leibniz adds, perfect certainty can never be hoped for from induction.

The greatest blunder of Nizolius identified by Leibniz was his attribution of the errors of the Scholastics to Aristotle himself. To counter this misinterpretation of Aristotle by Nizolius, Leibniz printed at the end of his preface the letter he had written to Thomasius, thereby giving to the public this first sketch of his own philosophy.

Nizolius earned praise from Leibniz for urging that anything incapable of description in simple terms in the vernacular should be regarded as non-existent, fictitious and useless. Perhaps with the example of Weigel in mind from his student days in Jena, Leibniz related how certain philosophers invited their opponents to explain their meaning in a living language and that when they attempted this, it was remarkable how they became confused. The obscurities of the Scholastic style of philosophising evidently could not survive translation into the vernacular, and this was the reason, Leibniz suggested, why this style had become obsolete in England and France, where the vernacular was used by scholars. German, he believed, though compatible with true philosophy, was not well adapted to express the Scholastic fictions, and as these had only recently been rejected, had been slow to come into philosophical use. The greatest clarity, Leibniz maintained, whether in German or in Latin, was to be found in commonplace terms with their popular usage retained. Although he admitted that there was no science in which technical terms were not needed—for example, when the matters dealt with were not directly obvious to the understanding or in frequent use—such terms were to be avoided where possible, since they generally contained a certain obscurity. For the attainment of a clear philosophical style, Leibniz recommended, jargon should be avoided and concrete verbs chosen in preference to abstract nouns. For the most part, he claimed, Aristotle had followed these principles.

Essays on motion

In the *Hypothesis physica nova* (*GP* **4**, pp 177–240) Leibniz developed further the outlines of his philosophy contained in the letter to Thomasius. The idea of considering motion in two ways, abstract and concrete, was essentially Cartesian. While the complete problem was to explain concrete motion or motion as it happened in the world—for example, in the experiments of Wren and Huygens on elastic collision—both Descartes and Leibniz set out as a first step to formulate the basic principles of motion in itself, that is, abstract motion. Neither mass nor elasticity enters into the account of motion in itself. Primary matter, as in the letter to Thomasius, is taken to have extension and impenetrability, but in Leibniz's theory of abstract motion, a body at rest avoids penetration simply by moving; there is no suggestion of an inherent or inertial resistance to motion.

Leibniz begins his search for the fundamental principles of abstract motion by considering the problem of the continuum.[3] Since the continuum is divisible to infinity, he concludes that there is an actual infinity of parts; the indefinite of Descartes, he remarks, is simply an idea without corresponding reality. However, there is no minimum in space or body; that is, no part of which the magnitude would be zero. For the existence of such a minimum would imply that there are as many minima in the part as in the whole; in Leibniz's view a contradiction. Nevertheless there are indivisibles or unextended beings, for otherwise body and motion would have neither beginning nor end. Leibniz had in mind the indivisibles of Cavalieri, those beginnings of lines and figures smaller than any given magnitude

[3]The problem has a long history going back to the paradoxes of Zeno.

whatsoever. The basic problem of the continuum that Leibniz was struggling to resolve may be stated as follows. Geometrical points, considered as parts of space with zero magnitude, do not exist. In order, therefore, to define the beginning and end of a given space or body, these unextended geometrical points must be invested with some kind of reality. It was this problem that eventually led Leibniz to his metaphysical theory in which the real continuum was constructed from the unextended monads. At this stage, Leibniz found inspiration for the construction of the continuum of motion in an idea of Thomas Hobbes, whose work he studied intensively and to whom he addressed a letter on the subject in July 1670, which, however, remained unanswered. The term and mechanical idea he took from Hobbes was *conatus*, translated by Hobbes as 'endeavour'. Leibniz defined *conatus* as the beginning of motion and hence the beginning of existence in the place into which the body is striving.

Leibniz held motion to be continuous—that is, not interrupted by intervals of rest as Gassendi believed—for if a body is once at rest, it will remain at rest unless a new cause of motion occurs. Conversely, a body once moved will always move with the same velocity and in the same direction if left to itself. These principles were in fact generally accepted and had been formulated in the works of Descartes, Gassendi and Hobbes.

Whatever moves, Leibniz supposes, will propagate its *conatus* in full against all obstructions to infinity. When a body has been stopped, it endeavours to move and begins to move the obstructing bodies, however large. The effect of *conatus* is evidently virtual velocity. Quite clearly, one *conatus* can be greater than another, and since the instants or indivisibles of time are taken to be equal, Leibniz concludes that one 'point' (the indivisible of space 'traversed' in an instant by one body) can be greater than another ('traversed' by a slower body). Such paradoxes were inherent in the theory of indivisibles and could only be resolved later with the aid of the concepts of the infinitesimal calculus.

A body could possess several *conatuses*[4] simultaneously. Whenever possible, these *conatuses* would combine to produce a composite motion. In direct collisions, the stronger of the opposing *conatuses* prevailed, so that the effect was measured by the difference of the two.

The concept of *conatus* enabled Leibniz to propose what he believed to be a new explanation of the true distinction between body and mind. Body, he believed, was a momentary mind; that is, a mind without memory. The reflective perception of its own actions and passions was the basis of memory and thought in the mind. For without action and reaction (that is, opposition) and then harmony, there was no sensation. The actions and passions of a body were its own *conatuses* and the contrary *conatuses* arising from encounters with other bodies. But body could not retain its own

[4]This anglicised form of the plural is used for clarity.

conatus and the contrary *conatus* together for more than a moment, so that it could not perceive the resolutions of its own actions and passions into harmony, and without this perception, it could have no memory. This explained why a body moving in a curve had a *conatus* to move along the tangent; it had no memory of the motion in the curve up to its present position. In a letter to Henry Oldenburg, the Secretary of the Royal Society, Leibniz claimed that he was the first to demonstrate this important principle (*A* II **1**, p 167).

Having established the principles of abstract motion, Leibniz applied them to specific cases of collision. As already remarked, elasticity and mass (inertia) played no part in the determination of motion in itself, so that Leibniz took account only of speed. Bodies offered resistance only in so far as they possessed a contrary *conatus*, so that a body at rest offered no resistance. When a moving body therefore strikes a body at rest, the two move together after the collision with the original speed of the first body. When, however, two moving bodies collide directly, both move together after the collision with velocity equal to the difference of the original velocities. The predictions of the theory of abstract motion, as Leibniz knew well, were rarely in agreement with observation. Two problems in particular needed to be resolved: first, why bodies separate after collision, as the experiments of Wren and Huygens had shown; second, why the motion in the universe does not continually decrease and the diverse directions reduce to one. For in most collisions, there is a loss of velocity and after the first moment, two *conatuses* combine their directions into that of a single straight line. Answers to these problems are provided by the theory of concrete motion.

This theory introduced hypotheses that were not inconsistent with the principles established *a priori* in the abstract theory but implied those modifications that were needed to explain the phenomena. The theory developed by Leibniz is similar to that set out by Descartes in the third part of his *Principia philosophiae*. A universal aether, whose existence was attested by the transmission of light, served to explain all phenomena mechanically, though the actual explanations were often vague. To explain the observed dependence of the effects of collision on the masses of the bodies, Leibniz supposed that, instead of being continuous, bodies were composed of corpuscles separated by aether. Each corpuscle, he supposed, had its own *conatus*, so that the resistance was proportional to the number of corpuscles in the body; that is, to its size or mass. Again, the separation of bodies after collision was the effect of elasticity arising from the action of the aether. It should be noted that the resistance of bodies, as conceived by Leibniz, was still not the effect of a natural inertia but only of its *conatus* (which opposed the *conatus* of the striking body), so that a body at rest would offer no resistance whatever to the impact of another body. However, he supposed that a body completely without motion could not exist, for it would be indistinguishable from pure space (*GP* **7**, p 259). Both weight and elasticity were effects of the aether, the former arising from the circulation of the aether about the earth and the latter from the dispersive tendency of the aether in the spaces between the parts of bodies.

Immediately following the publication of the *Hypothesis physica nova* about the middle of 1671, the Royal Society brought out a new edition and the Secretary Oldenburg wrote to Huygens revealing the identity of the author and mentioning that he had included comments on the laws of collision discovered by Wren and Huygens himself (*HO* **7**, p 56). About the same time, Leibniz also made contact with Pierre de Carcavy, the Royal Librarian in Paris, and to him he mentioned his calculating machine for the first time. The theologian Johann Leyser reported to Leibniz that Carcavy and the mathematician Jean Gallois wished to obtain for him membership of the Paris Academy and that he should introduce himself in Paris as soon as possible (*A* I **1**, pp 158–9). He was invited by Carcavy himself to send the calculating machine to Paris so that it could be shown to Louis' minister Colbert (*A* II **1**, p 125).

Correspondence with Duke Johann Friedrich

In May 1671 Leibniz sent to Duke Johann Friedrich[5] two writings on religious themes, entitled *De usu et necessitate demonstrationum immortalitatis animae* (*On the utility and necessity of demonstrating the immortality of the soul*) and *De resurrectione corporum* (*On the resurrection of bodies*), together with brief explanations of some of the ideas contained in the *Hypothesis physica nova* (*A* II **1**, pp 105–17). Of particular interest is Leibniz's reference to a vital core of its substance in every body, whether of men, animals, plants or minerals, so subtle that it remains in the ash of burnt things and at the same time can contract itself into an invisible centre. Besides the obvious examples of confirmation, such as the regeneration of plants, Leibniz cites the experiences of people following the amputation of limbs, where the sensation often remains. This idea of unextended vital

[5]On the death of his eldest brother Christian Ludwig in 1665, Johann Friedrich took advantage of the absence of his elder brother Georg Wilhelm to seize possession of Lüneburg (Celle), under the pretext that his father's will entitled him to choose one of the two principalities Lüneburg (Celle) or Calenberg (Hanover). Through the mediation of the Kings of France and Sweden and the Electors of Cologne and Brandenburg, an agreement was reached that he should have Calenberg (Hanover) and give up Lüneburg (Celle) to his brother (Spangenberg 1826, pp 48–9). Assuming the role of an absolute ruler in his principality, from that time known as Hanover, he pursued a foreign policy of support for France in return for financial aid. Although he became a Catholic, the religious freedom of his predominantly Protestant subjects was in no way restricted (Hohnstein 1908, p 370).

centres which persist through change, such as generation and corruption, clearly foreshadows the concept of the monad.

In another letter to the Duke, written in October 1671—it seems probable that Leibniz met the Duke in person in Frankfurt earlier in the month—he gives a detailed autobiographical account of his researches in various fields, referring in particular to his project of an alphabet of human thought (which he called a *Scientia Generalis*) and to his ideas on motion (*A* II **1**, pp 159–65). In his explanation to the Duke that the cause of motion is mind, Leibniz clarified his ideas concerning the role of God. From the scientific point of view, it was necessary to explain how God regenerates motion at each instant. Leibniz explained that God is the cause of the universal harmony of things, which governs the laws of motion. Instead of being the direct cause of motion, God thus becomes the formal cause. A few months earlier, in a letter to Magnus Wedderkopf (*A* II **1**, pp 117–18), Leibniz had introduced some of the most important ideas concerning God and harmony that were to figure in his later metaphysics; for example the distinction between possibility and existence, and the principle of the best possible. Thus, Leibniz explains, God wills the things he understands to be best and most harmonious and he selects them, as it were, from an infinity of possibilities.

Leibniz confided to the Duke his desire to visit Paris, where he would find the best opportunity to apply the talents God had given him to the perfection of the sciences. Remarking that Colbert had already shown interest in his design for a calculating machine, he asked the Duke for recommendations that would help him to make the journey.

Visit to Strasbourg

Following a short stay in Bad Schwalbach in August 1671, Leibniz travelled to Strasbourg, probably on behalf of Boineburg, whose son Philipp Wilhelm matriculated at the university there in the winter semester. In Strasbourg he discussed Cartesian philosophy with the historian Johann Heinrich Boeckler, who had resided with Descartes at the Court of Queen Christina of Sweden. He returned by boat down the Rhine, where on both sides golden vines greeted the autumn. Two years later, in a piece entitled *Dialogus de religione rustici* (*Dialogue on peasant religion*) (*A* VI **3**, pp 152–4), he described his impressions of the journey through the beautiful landscape, conversing with the ship's company on religion. In the deep calm one could believe, he wrote, that the hills shouted for joy and the nymphs danced in the Black Forest. Yet he sensed that what he was experiencing symbolised the peace before the impending political storm.

On his return to Mainz, he set to work on the 'Consilium Aegyptiacum', the secret project for the preservation of peace in Europe that he had first discussed with Boineburg in Bad Schwalbach during the summer of 1670. At the same time he wrote his autobiographical letter to Duke Johann Friedrich, and following a discussion with Boineburg on the doctrine of the Eucharist, he addressed a letter to the Jansenist Antoine Arnauld in Paris. This letter, which remained unanswered, contains a description of his philosophical studies supplementing to some extent that given in the letter to the Duke (*A* II **1**, pp 169–81). Two points in particular merit attention. From the principle that no body is at rest, there follows, according to Leibniz, a demonstration of the Copernican hypothesis. Again, he deduced that circular motion around different centres cannot exist unless there is a vacuum. Thus, at a time when he was in correspondence with Otto von Guericke, Leibniz seems to have moved from the view that a vacuum was possible to a belief in its actual existence.

The Egyptian plan

In its original form, Leibniz's idea of turning Louis XIV's aggressive intentions away from Europe and towards Egypt was introduced as a pure political speculation at the end of a memorandum drawn up after his discussions with Boineburg in Bad Schwalbach in the summer of 1670 (*A* IV **1**, pp 167, 181). At this stage the idea took the form of a suggestion for a general crusade against barbarians and infidels. The idea was not new but represents a customary trend in European politics to transfer internal conflicts to the rest of the world; in fact, Leibniz used a work of the early fourteenth century in which the Venetian Marino Canuto had put a similar suggestion to the Pope. The discussions of Boineburg and Leibniz on the problem of the political stability in Europe took place following a meeting in Bad Schwalbach a month earlier between the Electors of Mainz and Trier to decide their response to the pressing demand of the Duke of Lorraine, with whom they were in alliance, for a guarantee of assistance in the case of a threat from Louis XIV. In agreement with Leibniz, Boineburg rejected the alliance as a 'fragile reed' and recommended the best understanding with France. A few weeks later, the Duke of Lorraine was driven out of his lands by the French. However, Leibniz and Boineburg kept the idea of the Egyptian project to themselves.

Apart from the political considerations, Boineburg had personal reasons for wishing to be in favour with the French king. For a long time he had striven unsuccessfully for the recovery of significant arrears of rent from property and also a pension that were owed to him in France. Now he had been given grounds to believe that both could be regained if he made a personal appearance at the French Court. From the beginning of 1671, the Elector of Mainz, advised by Boineburg, was in negotiations with France in an attempt to maintain good relations. Boineburg therefore had reason to hope that the Elector would send him on a mission to France. But various obstacles to such a mission arose. First, the French foreign minister died, so that any visit by Boineburg would have to be delayed until a successor was appointed. The new minister, Simon Arnauld de Pomponne, took up his office in January 1672. By this time, however, a diplomatic visit by Boineburg had become superfluous, for in December 1671 Louis XIV had himself sent an ambassador to Mainz to communicate his intention of attacking Holland and request from the Elector the free passage of ships on the Rhine and the promise to use his influence with the Emperor to persuade the German states not to interfere.

At the time of the French mission to Mainz, Leibniz had been working with Boineburg's encouragement on the details of the Egyptian project. Boineburg decided that Leibniz should take the plan in secret to the French Court and at the same time endeavour to secure payment of the rents and pension that were owed to him. To this end Leibniz drafted a brief note setting out the advantage which the King could draw from a 'certain undertaking' that the author of the project would be pleased to discuss in person with a representative nominated by the King. The primary aim of the project remained the prevention of an attack on Holland. Boineburg revised the note and translated it into French. He sent it to the King on 20 January 1672. The document (*A* I **1**, pp 250–8) is so vague that one could not detect in it any reference to Egypt or the East (*A* IV **1**, p xxv). A favourable reply, perhaps prompted only by curiosity, came from the foreign minister on 12 February 1672, who asked for further information either by a personal appearance of Boineburg at Court or in whatever way the Baron thought best (*K* **2**, p 115). In a letter of 4 March 1672, Boineburg advised the minister that he was sending Leibniz.

Preparations were made as quickly as possible, and, accompanied by a servant, Leibniz set out for Paris on 19 March 1672, arriving twelve days later. Boineburg provided him with full power of attorney in respect of the rents and pension, an advance of travelling expenses which he was later to repay (*A* I **1**, p 381) and a letter of introduction to the minister (*K* **2**, p 124). Boineburg impressed upon the minister the need for secrecy and asked that Leibniz should be maintained in quiet and comfort and his travelling expenses refunded. Boineburg added that Leibniz had 'nothing from his chief that he can contribute, if not his study, his fidelity and his application that he will employ perfectly to the execution of the orders of His Majesty'.

Although, as we have seen, Leibniz had been very keen to spend some time in Paris, it must have seemed something of a mystery to his friends and relations why he suddenly left Mainz for that city and did not return. For the Egyptian project was a secret between himself and Boineburg. At the time, it was generally supposed that he had accompanied the Baron's son to Paris as his tutor.

Just before his journey to Paris, Leibniz lost his sister Anna Catharina Löffler. On 22 January, only a month before her death, she had written to her brother warning him that there were rumours circulating in Leipzig that he had become a Calvinist and that, on account of the favour shown him by the Elector of Mainz, he had enemies at that Court who wished him harm (*A* I **1**, pp 231–2). No doubt her fears were exaggerated.

3
Paris
(1672–1676)

When Leibniz arrived in Paris at the end of March 1672, England[1] had already opened the war against Holland and France followed within a week, so that the primary aim of the secret diplomacy of Boineburg and Leibniz—namely, the prevention of such a war—was no longer attainable. Yet the project could still have a role in saving Germany after the conclusion of peace with Holland, which could not be long delayed. A revision would be needed to take account of the changed political circumstances but at this time Leibniz devoted no further effort to the development of the project. Evidently, he was given no opportunity to discuss the project with the foreign minister, Simon Arnauld de Pomponne, for, six months later, he was still seeking an appointment through the good offices of the minister's uncle, the Jansenist Antoine Arnauld,[2] with whom he had already discussed philosophical problems on several occasions (*K* **2**, p 139).

Leibniz remained in Paris—though in May he had considered returning (*A* I **1**, p 271)—in order to pursue the financial claims of Boineburg besides

[1]Charles II had agreed to join France in the war against Holland in return for an annual pension of £100 000.

[2]Cornelius Jansen, Bishop of Ypres, a friend of Saint-Cyran, one time spiritual director of the religious community of Port Royal, with which the Arnauld family and also Pascal were associated, completed the book on Augustine, which was the basis of his movement for moral reform, shortly before his death in 1638. By attacking the Jesuit theory of grace on the title page, Jansen's book, when published in 1640, provoked the hostility of the Jesuit Order, which eventually led to the destruction of Port Royal early in the eighteenth century. At this time, however, the Port Royal nuns exercised much influence on several noble ladies and others at the Court, while Arnauld, who had been expelled from the Sorbonne in 1656 for his Jansenist views, was received at Versailles.

his own studies, and to await the expected arrival of Boineburg and his son. While Leibniz was corresponding with the revenue officer Morel over Boineburg's rents, Boineburg had evidently decided to enlist the Elector's support for the Egyptian project. For when the Marquis de Feuquières arrived in Mainz in the first week of June as an emissary from the French Court, the Elector put the project to him in outline. This brought a reply from the foreign minister, in the name of the King, rejecting the idea. But Leibniz did not easily give up, for in the autumn he composed a more detailed document, with title 'Consilium Aegyptiacum', for Boineburg to discuss with the Elector (*A* IV **1**, p 383).

Early in November, Boineburg advised Leibniz of the impending arrival of his son Philipp Wilhelm and asked him to supervise and support his studies in Paris (*A* I **1**, p 282). The young Boineburg arrived on 16 November in the company of Melchior Friedrich von Schönborn, nephew of the Elector and Boineburg's son-in-law, who had come to Paris as representative of the Elector and Diet of the Empire to the official peace negotiations. The purpose of the mission was to obtain the agreement of Louis XIV to Cologne as the venue for a general peace congress. Leibniz drafted the letter seeking an audience with the French King. When they arrived in Versailles, Melchior Friedrich saw the King but Leibniz himself was denied any part in the proceedings, so that he had no opportunity to mention his Egyptian project.

Before the end of the year Leibniz was to suffer another personal misfortune. For on 15 December Baron Johann Christian von Boineburg died of a stroke. Later Leibniz described his patron, to whose influence and support he owed his position at the Court in Mainz, as 'one of the greatest men of the century, who honoured me with a very special friendship' (*A* I **1**, p 476).

First meetings with scholars in Paris

Besides the meetings with Antoine Arnauld, to which we have already referred, Leibniz also met Pierre de Carcavy, the Royal Librarian, during the summer of his first year in Paris. For Carcavy he wrote a report on Otto von Guericke's book concerning experiments with the vacuum (*A* II **1**, pp 221–2). No doubt encouraged by the interest Carcavy had already shown in his design of a calculating machine, Leibniz actively pursued this project, so that a working model was produced by the beginning of 1673 (*HO* **7**, p 244).

In the autumn of 1672 Leibniz visited Christiaan Huygens, the celebrated Dutch scientist whom Colbert had entrusted with the planning and organisation of the Académie Royale des Sciences founded in 1666. At this meeting, which probably took place in Huygens' apartments at the Royal Library, Leibniz mentioned that he had a method for summing infinite series. Besides recommending him to read the *Arithmetica infinitorum* of John Wallis and the *Opus geometricum* of Gregory of St Vincent, Huygens set a problem to test the young scholar, on whom he had received such a favourable report from Oldenburg (*HO* **7**, p 56). The problem was one that Huygens had himself solved in 1665; namely, to find the sum of the infinite series of reciprocal triangular numbers.

Leibniz's first major mathematical discovery in Paris arose out of his continuing interest in the method of demonstration as a means for the formation of a sound logical basis for philosophy. In principle Leibniz wished to admit only two types of unproved truths in demonstrations, namely definitions and identities. His immediate problem concerned Euclid's axiom that the whole is greater than the part, to which his attention had been directed by his reading of Hobbes. Since the axiom was neither a definition nor an identity, Leibniz regarded it as a provable theorem and sought a demonstration. This he gave in a manuscript 'Demonstratio propositionum primarum', where he described it as an excellent example of such demonstrations (*A* VI **2**, pp 479–86). It takes the form of a syllogism in which the major premise is a definition, the minor premise an identical proposition and the conclusion the given theorem.

Theorem: The whole *cde* is greater than the part *de*.

c d e

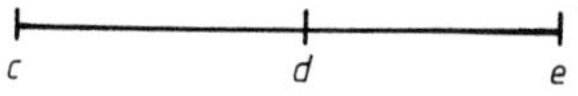

Figure 3.1

Major (definition):
Of two bodies, that is greater whose part is equal to the whole of the other.
Minor (identical proposition):
The part *de* of the whole *cde* is equal to the whole *de* (that is, to itself).
Conclusion: Therefore the whole *cde* is greater than the part *de*.

From the axiom of identity Leibniz derived the general method of summing series which he communicated to Huygens at their first meeting (*GM* **5**, p 396). Taking $0<a_1<a_2< \cdots <a_n$ and $b_0=a_1-a_0$, $b_1=a_2-a_1$, $\ldots$, $b_{n-1}=a_n-a_{n-1}$, it follows from the identity $a_0-a_0+a_1-a_1+a_2-a_2+ \cdots +a_n-a_n=0$ that $b_0+b_1+b_2+ \cdots +b_{n-1}=a_n-a_0$. Thus the sum of a series of differences is equal to the difference of the extreme terms of the original series. An example often cited by Leibniz—in a letter to Princess Sophie for instance (*K* **8**, p175)—is the sum of a series of consecutive odd numbers expressed as a difference of two squares. By means of his general method Leibniz believed that it should be possible to sum any series of terms formed according to some rule; even an infinite series provided that it converged.

Immediately following his meeting with Huygens, Leibniz borrowed the book of Gregory of St Vincent from the Royal Library and succeeded in adapting the procedure explained there for summing geometric series to the application of his difference principle. Instead of placing the line-segments representing the terms of the series end to end, as Gregory had done, Leibniz placed them so that they all started from the same point (figure 3.2). The differences of successive terms were then proportional to the terms of the original series. The recognition of this property was in itself a considerable insight but generalisation of the procedure led to new and interesting results; not only the sum of the reciprocal triangular numbers, which Huygens had accomplished, but also to the systematic derivation of the sums of other reciprocal polygonal numbers.

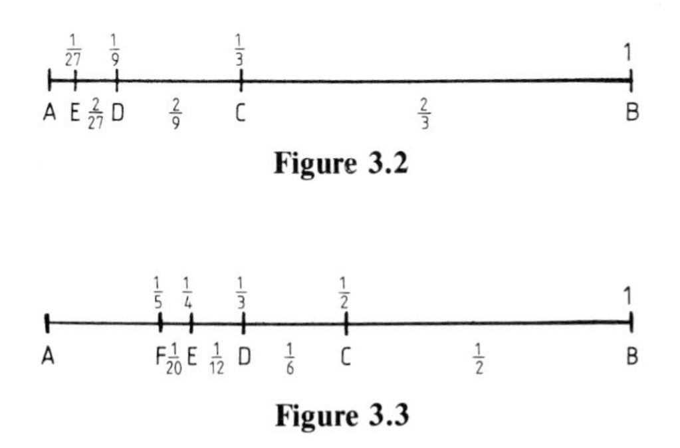

Figure 3.2

Figure 3.3

Starting with the series (figure 3.3) $AB=1$, $AC=1/2$, $AD=1/3$, $AE=1/4$, the differences are $BC=1/2$, $CD=1/6$, $DE=1/12$, $EF=1/20$ and $BC+CD+DE+EF+\ldots=AB$, so that $1/2+1/6+1/12+1/20+\ldots=1$. Multiplying by 2 gives the sum of the reciprocal triangular numbers $1+1/3+1/6+1/10+\ldots=2$. Similarly, starting with the series $AB=1$, $AC=1/3$, $AD=1/6$, $AE=1/10, \ldots$, the differences are $BC=2/3$, $CD=2/12$, $DE=2/30$, $EF=2/60, \ldots$, so that $2/3+2/12+2/30+2/60+\ldots=1$, or $1+1/4+1/10+1/20+\ldots=3/2$, which is the sum of the reciprocal pyramidal numbers. Clearly this process may be extended indefinitely.

Leibniz wrote up his results, though without the demonstrations which he described as difficult and requiring a number of lemmas, intending to send them to Jean Gallois for publication in the *Journal des Sçavans* (*A* III **1**, pp 1–20). However, this journal ceased publication at the end of 1672 and when publication was resumed in 1674, Leibniz had come to recognise that the only originality he could claim in his 'Accessio ad arithmeticam infinitorum' was his method of summation and not the results themselves, so he withheld publication.

The visit to London

As the political mission to France had failed, it was decided that Melchior von Schönborn, accompanied by Leibniz, should go to London in order to put the same proposal for a peace congress to the English Government, and then return to Mainz through Holland, taking the opportunity to seek support for the plan in that country. After a delay of several days in Calais owing to bad weather, the legation arrived in Dover on 21 January 1673 and reached London three days later. At the earliest opportunity Leibniz visited Oldenburg, who arranged for him to demonstrate the wooden model of his calculating machine that he had brought with him at a meeting of the Royal Society on 1 February. On this occasion the machine was inspected carefully by Robert Hooke (Birch 1756–7 **3**, pp 72–3). In a letter to Oldenburg, Huygens had described the machine as a promising project, though at this stage it was far from perfect. A week later Leibniz attended another meeting of the Royal Society, when Oldenburg read the famous letter from Sluse on tangents.[3] After this meeting he met Robert Moray, who informed him of Samuel Morland's calculating machine. A day or two later, a meeting took place between Leibniz and Morland in the presence of Oldenburg, when both calculating machines were demonstrated. The two were not comparable, for while Leibniz's machine was designed to perform the four basic arithmetical operations, Morland's machine relied on the application of Napier's bones for multiplication and division.

On 12 February, Leibniz visited Robert Boyle, who introduced him to the mathematician John Pell. When Leibniz explained that he had a general method for representing and interpolating series by constructing a series of differences, Pell remarked that similar results had been obtained by François Regnauld and given by Gabriel Mouton in his book *Observationes diametrorum solis et lunae apparentium*. Morose and ill, Pell was clearly not sympathetic to the young German and Leibniz could detect a veiled accusation of plagiarism. The next day he looked at Mouton's book in the library of the Royal Society and found Pell's assertion to be correct. On the advice of Oldenburg, he immediately wrote an explanation of the affair to deposit with the Royal Society (*A* III **1**, pp 22–9). As an example of his general method, Leibniz described the representation of the series of cubes. First, the differences were written down. Then adjoining the arithmetical triangle, adapting it to determine binomial coefficients using the additive process described in *De arte combinatoria* (which he wrongly asserted was

[3]René François de Sluse gave a rule for finding the tangent to any curve whose equation could be written in the form of a polynomial $f(x,y)=0$. In modern notation, the rule is equivalent to $\frac{\mathrm{d}y}{\mathrm{d}x} = -\frac{\partial f}{\partial x} \Big/ \frac{\partial f}{\partial y}$. Sluse discovered the rule between 1655 and 1660.

not to be found in Mouton or Pascal), he arrived at the representation written in modern notation as $0\binom{n}{0}+1\binom{n}{1}+6\binom{n}{2}+6\binom{n}{3}=n^3$. Finally, he described his summations of the infinite series formed from the reciprocal figurate numbers.

Leibniz was not present at the next meeting of the Royal Society on 15 February to hear Hooke's derogatory remarks about his calculating machine. When Oldenburg informed Leibniz of this criticism, he indicated that he should not be too disturbed, pointing out that Hooke was generally known to be cantankerous, but he advised his young compatriot to speed up the technical improvements of his machine.

Leibniz and his companion had been in London for less than a month when news came of the death of the Elector of Mainz, Johann Philipp von Schönborn, on 12 February, an event that was to have decisive significance for Leibniz's future career; for the present it enabled him to return to Paris instead of travelling through Holland back to Mainz. Arrangements for the departure were made so quickly that Leibniz had been unable personally to take his leave of Oldenburg, who sent him a brief note enclosing a letter for Huygens and a copy of the *Philosophical Transactions* containing Sluse's tangent method for Leibniz himself. On the day of his departure, 20 February, Leibniz wrote to Oldenburg requesting admission as a member of the Royal Society (*A* III **1**, pp 33–4).

The return to Paris

On his return to Paris, Leibniz resided in the Rue Garantière, Quartier de Luxembourg in the Faubourg St Germain (Guhrauer 1846 **1**, app, p 20). To Melchior Friedrich von Schönborn he expressed the hope that the new Elector, Lothar Friedrich von Metternich, would leave him in his office and pay the salary still owing to him for the last two years (*A* I **1**, pp 312–13). Moreover, he requested a current salary that would enable him to return to Mainz at least once a year, while serving the Elector in Paris, using his connections there to report on political, scientific and cultural affairs (*A* I **1**, p 316). On the recommendation of his friend Melchior Friedrich von Schönborn, Leibniz received in May the Elector's permission to remain 'for a while' in Paris without danger to his post (*A* I **1**, p 349). But his salary remained in abeyance and he did not receive the further political or scientific commission he had sought (*A* I **1**, p xxxi).

Whatever the uncertainties of his position, Leibniz had clearly formed the intention to remain in Paris as long as possible. If he had wished for greater financial security, the opportunities presented themselves. First, his friend Christian Habbeus von Lichtenstern wrote to him on 25 March 1673 with the offer of an appointment as secretary to the chief minister of the Danish king at a handsome salary with free board and lodging. His remarks concerning the conditions under which he might find such a post congenial—conditions which quite clearly would not be conceded—may be seen as a polite refusal (*A* I **1**, pp 415–18). Besides, his chief desire was not money but the freedom needed to do something useful for the general good. A second invitation, equally attractive in financial terms, came from Duke Johann Friedrich of Hanover, who wrote on 25 April 1673 offering him a post as Counsellor. About a month earlier, he had written to the Duke describing his visit to London and his activities in Paris (*A* I **1**, pp 487–90). This was a natural thing to do, for it will be recalled that he had earlier described his researches and projects to the Duke, who was especially interested in religion. Both in Paris and in London, he tells the Duke, he had the opportunity to meet people interested in religion and politics as well as science. He mentions especially Antoine Arnauld, who was similar to the lamented Boineburg in having a special interest in religion. The highlight of his visit to London had been the demonstration before the Royal Society of his calculating machine. Although only an imperfect model had been seen, the machine was acknowledged in London, as also in Paris, to be one of the most considerable inventions of the time. When completed, it would be useful in the performance of many tasks, both in the fields of government and of science. Already it had cost him not a little time and money, having been modified over a hundred times, but he assured the Duke, too confidently, indeed, that he hoped to perfect it within a few weeks.

Leibniz was still in the service of the Boineburg family and in April 1673, Anna Christine von Boineburg gave him authority to arrange and supervise the further education of her son Philipp Wilhelm (*A* I **1**, p 336). The plan proposed by Leibniz set out a programme of studies that would occupy the young Boineburg from 6 AM to 10 PM with very little free time (*A* I **1**, pp 332–3). Although the Boineburg family gave their approval and the seventeen-year-old baron took up residence with Leibniz, he had little inclination for study, preferring to enjoy himself with his young friends (*A* I **1**, p 338). It was inevitable that friction should arise, so that within a few months Leibniz had to complain to his relations concerning his lack of diligence (*A* I **1**, pp 369–73). In his report Leibniz explained that he had reasoned with the young baron, who had ability, but lacked the will, seeking a thousand pretexts for his negligence. Towards the end of the year, Leibniz asked for arrears of expenses and also requested the payment of an annual allowance for himself so long as he was responsible for the young baron's study in Paris (*A* I **1**, pp 379–81). At the same time he complained that he had not received the slightest recompense for all his service to the elder Boineburg in Mainz. Friction and misunderstandings between Leibniz and the Boineburg family continued until he was coolly dismissed from the Boineburg service by the young baron's mother on 13 September 1674 (*A* I **1**, p 396). It seems that Philipp Wilhelm himself left Paris in 1676. Later his achievements in high political offices were more worthy of his father, who had been Leibniz's first patron and respected friend.

Family affairs

Leibniz had vanished from Mainz for over a year when he received a letter from his brother-in-law, Simon Löffler, inviting him to his wedding to Regina Koch (*A* I **1**, p 412). His first wife, Leibniz's sister Anna Catharina, had died just before Leibniz departed from Mainz. It seems that Leibniz's letters to his relations during the intervening year must have gone astray, for the only news they had received, as Löffler remarks, came from an acquaintance in Mainz who told them that Leibniz had been sent by the Elector on a mission to the Courts of France and England and would soon return (*A* I **1**, p 411). Löffler received a reply to this letter, in which Leibniz asked his brother-in-law to take up with the Exchequer of Sachsen-Altenburg a claim that his parents had bequeathed to their children. Löffler thought there was little chance of success in the circumstances of Leibniz's absence in Paris (*A* I **1**, pp 419–20).

It was understandable that suspicion concerning his religion and patriotism should have arisen while he was in the service of a Catholic prince in Mainz, and that the mistrust of his friends and relations in Leipzig should have grown as a result of his prolonged stay in Paris for reasons which were not clearly known to them. When they failed to receive letters from him, this must have appeared to them as a confirmation of their fears that he had deserted both family and fatherland. After he had been in Paris for two years, his brother Johann Friedrich reproached him for his silence, reminding him of his moral obligations to his family and fatherland (*A* I **1**, pp 420–1). In his reply, which unfortunately also went astray, Leibniz assured his brother that he had written more than once to his relations, enquiring with careful interest after every member of the family, including the youngest children and especially the old aunt who had helped to bring him up and had witnessed his miraculous escape from injury when, as a young child, he had fallen from the table (Guhrauer 1846 **1**, p 160).

After more than another year of silence, Leibniz's brother wrote again, on 6 May 1675, with further reproaches, remarking that half a year had passed since he had informed him of the death of his brother-in-law Simon Löffler (*A* I **1**, pp 423–4). Leibniz replied on 11 October 1675, telling his brother that before receiving this letter, he had not known of Löffler's death, news of which had caused him not a little grief (*A* I **1**, pp 431–3). In this letter to his brother and in a letter to another relation, Christian Freiesleben (*A* I **1**, pp 427–31), Leibniz explained that those who questioned his religion or patriotism did him an injustice. For he would not change his religion for gain, while he had taken every opportunity to speak well of his fatherland.

Also, he defended his activities in Paris, which had been both profitable and honourable, and declared his intention to make a journey to Italy in the following spring, taking the opportunity to visit Leipzig, where he hoped to see his relations and friends, before returning to Paris through Holland and England.

Intensive study of mathematics

Although Leibniz was unanimously elected a Fellow of the Royal Society on 19 April 1673 (Birch 1756–7 **3**, pp 82–3), his meetings with the English scientists had not been an unqualified success. First, he had not been able to demonstrate that his calculating machine could actually perform automatic multiplication and division, as it was designed to do. Second, his claims concerning infinite series had revealed his ignorance of current work in mathematics and, even worse, exposed him to the suspicion of plagiarism.

On returning to Paris he lost no time in trying to repair the damage to his reputation, seeking Pell's opinion concerning his declaration to the Royal Society on series, especially in relation to Mengoli (*A* III **1**, p 43). A detailed reply was prepared for Oldenburg by his adviser on mathematics, John Collins, who had not met Leibniz during his visit to London (*A* III **1**, pp 50–63). First, he described what had been achieved by English mathematicians in regard to Regnauld's theory of interpolation. Then he explained that the summations of the reciprocal figurate numbers were given in Mengoli. Collins also claimed that he could sum the harmonic series and the reciprocal squares and cubes, though Mengoli could not, and he gave an example of a series of a hundred terms related to repayment of loans with interest that could be summed by his method. From this example Leibniz gained the impression that Collins was claiming for Mengoli only the summation of finite numbers of terms of the reciprocal figurate numbers. Again, he reasoned that Huygens had set him the problem and would not have done so if it had already been solved. So he was fairly confident in claiming the summation of the infinite series of reciprocal figurate numbers as his own (*A* III **1**, pp 83–9). Alas, Oldenburg had to tell him that Mengoli had also found the sums of the infinite series (*A* III **1**, pp 96–9). Of course, Leibniz's originality was in the method, but as far as the English mathematicians were concerned, it was the results that counted. There was just one claim left that Leibniz could make; he had shown that the harmonic series diverged (*A* III **1**, pp 92–5). Collins informed him that Mengoli had shown this too. Moreover, he added, the English could find partial sums of the harmonic series, a reference to Newton's logarithmic approximation. Leibniz could not guess that what he had been told concerned only an approximation to the sum and his efforts to find an exact solution were, of course, unsuccessful.

Leibniz had become painfully aware of his lack of knowledge of higher mathematics. In order to fill the gaps, he devoted a whole year to intensive study, meanwhile breaking off his correspondence with Oldenburg, who had repeatedly reminded him of his promise to complete the calculating machine as soon as possible.

On his return to Paris, Leibniz had met Jacques Ozanam, through whom he became acquainted with the problems of indeterminate analysis and number theory (*A* III **1**, pp 34–8). He succeeded in solving one of Ozanam's problems, namely to find three numbers whose sum is a square and the sum of whose squares is a fourth power (Hofmann 1969, p 107). A more difficult problem proposed by Ozanam—to find three numbers x, y, z, such that $x-y$, $x-z$, $y-z$, x^2-y^2, x^2-z^2, y^2-z^2 are all squares—defeated him, however. Leibniz was later impressed with what he learnt of James Gregory's solution of this problem (Hofmann 1974, pp 89–93).

The letter Oldenburg had given him for delivery to Huygens gave Leibniz an excellent opportunity to make another call on the Dutch scientist. On the occasion of this visit, Huygens presented to him a copy of his newly published *Horologium oscillatorium* concerning the study of pendular motion, and explained to him how everything went back to Archimedes' methods for centres of gravity (*GM* **5**, p 398). As a reading list to help him understand these problems, Huygens recommended the works of Pascal, Fabri, James Gregory, Gregory of St Vincent, Descartes and Sluse. Leibniz immediately set to work on the study of these sources, borrowing the books from the Royal Library.

The first success achieved by Leibniz was a generalisation of a result of Pascal. To calculate the moment of a quadrant of a circle about the x axis (that is, in modern notation, $\int_0^{\pi a/2} y\,\mathrm{d}s$), Pascal made use of the similarity of the characteristic triangle $\mathrm{d}x$, $\mathrm{d}y$, $\mathrm{d}s$ and the triangle y, $a-x$, a (figure 3.4). From this similarity it follows that $y\,\mathrm{d}s=a\,\mathrm{d}x$, so that $\int_0^{\pi a/2} y\,\mathrm{d}s=\int_0^a a\,\mathrm{d}x=a^2$. Leibniz recognised that the method could be generalised to any curve, replacing the radius of the circle by the normal to the curve. Huygens remarked to him that he had used the same method in other particular cases; for example, in his determination of the surface of the paraboloid of revolution (*HO* **14**, p 234).

It was also the work of Pascal that inspired the discovery of Leibniz's first significant new theorem in the geometry of infinitesimals. Using this theorem, which will be referred to as Leibniz's transmutation theorem, he found the area of a segment of a cycloid. This result he communicated to Huygens (*A* III **1**, p 115) and Oldenburg (*A* III **1**, p 120) in the summer of 1674, but a detailed proof of the theorem itself was first given a year later in letters to La Roque and Gallois (*A* III **1**, pp 342–4, 347–50, 360–1). The transmutation theorem, he explained, was one of the most general and fertile in the whole of the geometry of infinitesimals, providing a means of demonstrating all the known quadratures and a foundation for the

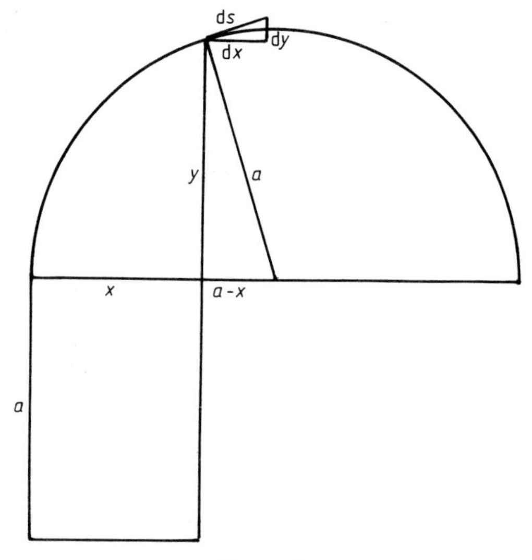

Figure 3.4

arithmetic of the infinite that Wallis had only been able to establish at the beginning by induction (*A* III **1**, p361). While Cavalieri and others had resolved the figures into rectangles of the same infinitesimal width, Leibniz resolved them also into triangles concurrent in a point. From the similarity of the triangles OWT and PNQ (figure 3.5), it follows that $z\,dx = h\,ds$. Then, since the area of triangle OPQ $= \frac{1}{2}h\,ds$ and the area of rectangle RSVU $= z\,dx$, it follows also that area OPQ $= \frac{1}{2} \times$ area RSVU. Consequently, Σ OPQ $= \frac{1}{2}\,\Sigma$ RSVU, so that the area of the segment OAPO (figure 3.6) equals half the area under the dotted curve, whose ordinate is z. Since z can be computed using Sluse's tangent rule, Leibniz's theorem establishes a relationship between the theory of tangents and the theory of

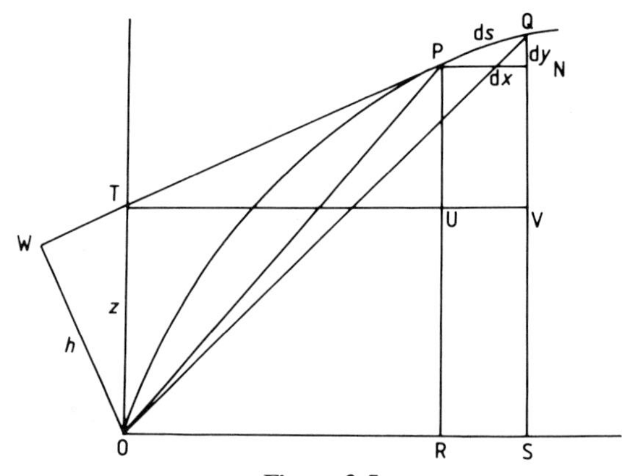

Figure 3.5

quadratures. At first Leibniz proved the theorem in the case where PR and QS are taken perpendicular to the axis but he later extended the theorem to the general case where PR and QS are drawn parallel but inclined at any given angle to the axis. Evidently the theorem involved certain affine transformations in which area is conserved.

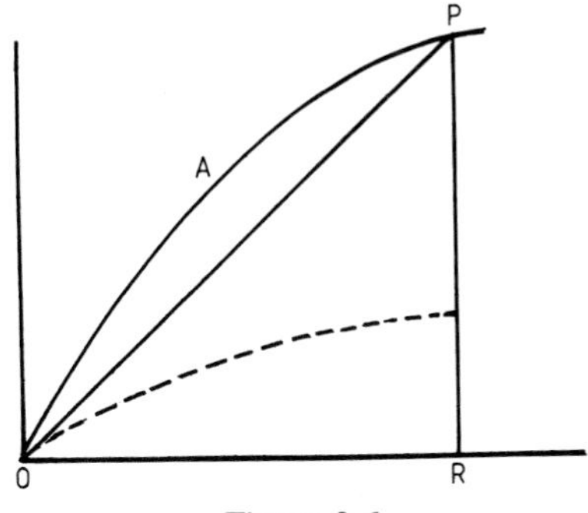

Figure 3.6

The basic idea of Leibniz's theorem is quite simple and was immediately obvious to Huygens when Leibniz showed it to him. The method in essence transforms an integral by substitution into an integral of a rational function, which may then be evaluated by expressing the integrand as a series and integrating term by term. One of the finest results Leibniz derived using the transmutation theorem was his arithmetical quadrature of the circle.

Application of Sluse's tangent rule to the equation of the circle (figure 3.7) $y^2 = 2ax - x^2$ gives

$$z/a = x/y = \sqrt{\frac{x}{2a - x}}$$

so that $x = 2az^2/(a^2 + z^2)$. This curve is shown as a dotted line. From $y^2 = 2ax - x^2$, we have $y/(2a - x) = x/y$, and since $z/a = x/y$, it follows that $ay = z(2a - x)$. Hence, the area of triangle OPC $= \tfrac{1}{2}ay = \tfrac{1}{2}z(2a - x)$. Also, it is clear from the diagram that $\int_0^x z\,dx = xz - \int_0^z x\,dz$. By the transmutation theorem, the area of the segment OUPO $= \tfrac{1}{2}\int_0^x z\,dx$. Hence the area of the sector COP = area of triangle OPC + area of the segment OUPO

$$= \tfrac{1}{2}ay + \tfrac{1}{2}\int_0^x z\,dx$$

$$= \tfrac{1}{2}z(2a - x) + \tfrac{1}{2}\left(xz - \int_0^z x\,dz\right)$$

$$= az - \int_0^z \frac{az^2}{a^2 + z^2}\,dz.$$

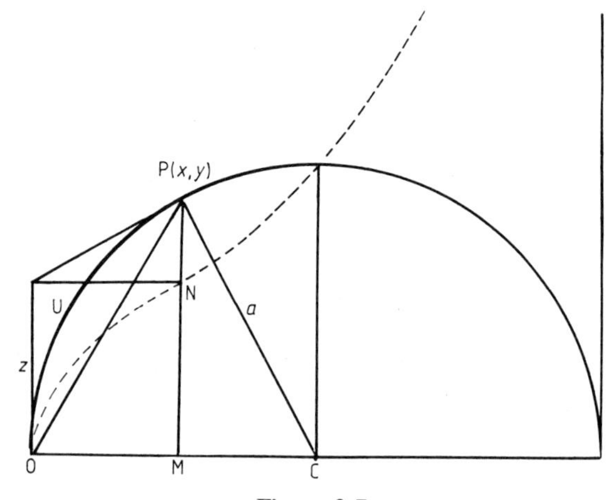

Figure 3.7

Now, following the method used by Mercator and Wallis in the quadrature of the hyperbola, Leibniz evaluates the integral by long division and integration term by term to obtain the result that the area of the sector COP of the circle $= az - z^3/3a + z^5/5a^3 - z^7/7a^5 + \ldots$. Putting $z = a$, this reduces to $\pi/4 = 1 - 1/3 + 1/5 - 1/7 + \ldots$ or, equivalently, $\pi/8 = 1/(1\cdot3) + 1/(5\cdot7) + 1/(9\cdot11) + \ldots$. Leibniz also considered the series $1/(2\cdot4) + 1/(6\cdot8) + 1/(10\cdot12) + \ldots = \frac{1}{4}\log 2$, depending on the quadrature of the hyperbola achieved by Brouncker, thus establishing a relationship between the hyperbola and the circle.

In response to Huygens' request for an explanation of the arithmetical quadrature of the circle, Leibniz prepared drafts of a proof in Latin and French, and also a fair copy, intended for the *Journal des Sçavans* (*A* III **1**, pp 141–69). In his letter to Oldenburg (*A* III **1**, p 120) he simply referred to his general theorem and the quadrature of the circle in terms of an infinite series, without giving details. When Oldenburg's reply failed to reach him, Leibniz wrote again on 16 October 1674 (*A* III **1**, pp 126–31). This time he received a reply, written on 18 December 1674 (*A* III **1**, pp 171–5). Oldenburg writes that both Newton and Gregory had progressed to general methods of quadratures applicable to all geometrical and mechanical curves and, in particular, the circle. If the circle could really be squared by arithmetic, then Leibniz, he added, was to be congratulated. But since Gregory was occupied with a proof of the impossibility of the exact quadrature of the circle, it might be necessary to consider the matter more carefully. Oldenburg's failure to perceive that Gregory and Leibniz were making claims concerning the circle that were not necessarily incompatible may be attributed to his limited knowledge of mathematics, but there is no doubt that the tone of the letter is rather cool. This is not difficult to understand. Leibniz had been silent for over a year and he had failed to fulfil the promise he had made at a meeting of the Royal Society, of which he had been elected a Fellow, to send a finished model of his calculating machine. Also, Oldenburg would not have forgotten the embarrassing affair with Pell. He could not have known that, since his visit to London, Leibniz had developed into a knowledgeable and creative mathematician.

Uncertainties and seeds of discord

At the beginning of 1675 Leibniz had, after many trials, developed his calculating machine to a stage that permitted a successful demonstration in the Paris Academy of Sciences (*A* I **1**, p 495). This brought orders from Colbert for three production models, one for the King, one for the Royal Observatory and one for Colbert's finance office. Describing the machine in a letter to Duke Johann Friedrich of Hanover (*A* I **1**, pp 491–3), Leibniz explains that multiplications and divisions of large numbers require only a few turns of a wheel, while additions or subtractions of whole pages of numbers can be performed in less time than it would take to write them down, all without any work or thinking. If the Duke would like to have one for himself, Leibniz added, this could be produced at the specified cost.

The main purpose of Leibniz's letter to the Duke was to accept, rather tardily and evidently without much enthusiasm, the Duke's invitation to enter his service in Hanover. This decision was no doubt seen by Leibniz as the only one possible in view of his failure to enter the service of the Emperor as he had hoped and the apparent lack of opportunities for paid employment in Paris (*A* I **1**, p 393). Remarking that he hoped to finish his work in Paris during the winter, Leibniz sought to leave the Duke in no doubt concerning what he expected from his patron in Hanover. While he had been in Paris, Leibniz tells the Duke, he had exchanged the pursuit of law, *belles lettres* and controversies, which had mainly occupied him in Germany, for the new study of mathematics. Freedom to pursue his own studies in arts and sciences for the benefit of mankind, he added, was all he desired.

Besides the calculating machine, Leibniz also introduced another invention to the Paris Academy of Sciences early in 1675; namely a chronometer which he had designed four years earlier (*A* III **1**, p 245). The sequence of events leading to the appearance of this invention started in January when Huygens discussed with Leibniz his own invention of the balance spring, which uses the isochronous property of elastic oscillations to regulate the movement of a clock. Colbert granted a patent to Huygens, who also registered his invention with Oldenburg by means of an anagram (*HO* **7**, p 400). When the clockmaker and others claimed that they had contributed to the invention, Huygens revealed the secret of the anagram, sending to Oldenburg a copy of the description he intended to publish in the *Journal des*

Sçavans. When Huygens' letter and enclosure were read to the Royal Society on 18 February 1675—the first meeting attended by Newton—Hooke immediately claimed that he had made the invention a long time before. There followed a bitter dispute in which Hooke first accused Oldenburg of betraying his fundamental idea to Huygens, and then, more intemperately, accused Oldenburg of being a French spy and informer (*HO* **7**, p 513). Eventually the Royal Society supported Oldenburg in this affair (Birch 1756–7 **3**, p 324).

Leibniz was dragged into the dispute when he sent to Oldenburg a copy of the paper he had prepared for the *Journal des Sçavans* describing his own invention (*A* III **1**, pp 208–13). This device for the regulation of a clock, which he points out is wholly different from the acclaimed invention of Huygens, employs two springs, arranged so that each releases the other on return to its original position, where it is held in its state of tension until released again. Oldenburg published an English translation of Leibniz's paper in the *Philosophical Transactions* (*A* III **1**, pp 192–201) but had to inform him of doubts raised in the Royal Society concerning his invention. This was no doubt a polite reference to Hooke's criticism that the invention was of no practical use. On this occasion, at least, Hooke's judgment was not far from the truth.

In reply to Oldenburg's comments concerning his arithmetical quadrature of the circle, Leibniz explained that this was quite different from the problem of analytical quadrature that had occupied Gregory (*A* III **1**, pp 208–13). In the first draft of the letter (*A* III **1**, pp 201–5), though not in the version sent to Oldenburg, Leibniz remarked that analytical quadrature had to do with rational, irrational or even transcendental numbers, a term introduced here for the first time. Arithmetical quadrature, however, involved an infinite series of rational numbers. Brouncker's quadrature of the hyperbola had been the first arithmetical quadrature while the arithmetical quadrature of the circle Leibniz claimed for himself. To Oldenburg he explained that his quadrature was achieved by a far-reaching method which opened up the way to many new propositions. He offered to barter his quadrature of the circle for Collins' method of summing finite numbers of terms of the harmonic series and series of inverse squares and cubes. In the account he intended to publish, he would refer to the results of Newton and Gregory in the preface. Leibniz then boldly requested from Oldenburg information concerning the methods used by Newton and Gregory. On 22 April 1675 Oldenburg replied, giving Leibniz a Latin version of an English draft prepared by Collins describing in detail the current researches of Gregory and Newton (*A* III **1**, pp 217–43). From this communication, Leibniz became aware that Collins had come to take him seriously as a mathematician and was prepared to disclose information concerning unpublished work of Gregory and Newton. But he was told only their results and not their methods. Even the results were not communicated accurately. For example, Gregory's series for the circle was incorrectly transcribed by Oldenburg and Gregory's mistake in the tangent series remained uncorrected by Collins and Oldenburg. Even if there had been no mistakes, it is difficult to see how Leibniz could have inferred anything about methods from the results alone. Yet this letter was later used in evidence against Leibniz in the course of the priority dispute with Newton concerning the invention of the calculus. In particular, it was alleged that Leibniz produced his own discovery of the inverse-tangent series only after receiving this letter from Oldenburg. But he had in fact referred to his own discovery in letters to Oldenburg written the previous year, while his method of derivation, communicated to Huygens also in the previous year, owes nothing to Gregory (*A* III **1**, pp 141–69).

Towards the end of 1675, Leibniz had hopes of being able to remain in Paris after all. For the death of Gilles Personne de Roberval in October of that year had left a vacancy in the Paris Academy of Sciences, for which he was recommended by Jean Gallois and Colbert's son-in-law the Duke of Chevreuse, to whom he had demonstrated his calculating machine (*A* II **1**, p 556). For some trifling reason, however, the touchy Gallois dropped his support for Leibniz, who then had little prospect of securing the appointment. Leibniz later recalled how the Duke of Chevreuse had told him of the view held by those with influence that, with the Dutch Huygens and the Italian Cassini, the Academy already had enough foreigners in salaried posts.

Outside the Academy, Leibniz became acquainted with the Oratorian Nicolas Malebranche, who besides being a mathematician, was at that time the leading Cartesian philosopher. His major work, the *Recherche de la vérité*, first appeared in two volumes in 1674–75 and five further editions, with corrections and additions, were published during the author's lifetime. Malebranche's chief contributions to Cartesian philosophy were the development of the doctrine of Occasionalism, which had been introduced by Cordemoy and De la Forge in an attempt to solve the problem of Cartesian dualism, and his conception of the subtle matter of Descartes as small elastic vortices, which enabled him to give plausible explanations of the phenomena of heat and light (Aiton 1972b, pp 69–71). The subject of Leibniz's first conversations with Malebranche was the Cartesian doctrine that extension is the essence of matter (*GP* **1**, pp 321–7).

Friendship with Tschirnhaus

At the end of August 1675 there arrived in Paris a young nobleman from Saxony, Ehrenfried Walther von Tschirnhaus, with letters of recommendation from Oldenburg to Huygens and Leibniz (*A* III **1**, p 275). He soon established a friendly relationship with Leibniz, who reported to

Oldenburg that he found much pleasure in his company and recognised in him a young man of keen intellect and great promise (*A* III **1**, pp 326–34). This young compatriot of Leibniz had studied in Leiden, where he became attached to Cartesianism and developed considerable skill in formal algebraic calculations. During a stay in Amsterdam in 1673, friends of Spinoza introduced him to the doctrines of the hermit in The Hague, whom he may have met personally in 1674. Then in May 1675 Tschirnhaus visited England, where he had a friendly reception. With a letter of introduction from Oldenburg he met Wallis, to whom he reported on his algebraic results. On 9 August 1675, he met Collins to discuss mathematical topics. Although Collins recognised Tschirnhaus's ability in algebra, he believed that he went too far in claiming that the new contributions of Sluse and Barrow to the solution of equations and the whole theory of quadratures, rectifications and determination of centres of gravity were nothing but deductions from Descartes. While Collins formed the impression that Tschirnhaus possessed comprehensive methods for the solution of higher equations—in fact the equations he selected for study were contrived by working backwards from solutions—Gregory was far less impressed by a person who so grossly overvalued Descartes.

In the months following their first meeting towards the end of September 1675, Tschirnhaus and Leibniz exchanged results and ideas in mathematics besides making a number of joint studies. Their collaborative work included examination of the manuscripts left by Pascal. Although Leibniz strongly recommended publication—the treatise on conics, he remarked, was already in a state suitable for printing—this was never realised (*A* III **1**, pp 587–91). The manuscripts themselves are now lost, so that all that remains is Leibniz's summary of their contents. Another subject of collaboration was an unsuccessful attempt to find partial sums of the harmonic series. While Leibniz recognised and pointed out to his imaginative but uncritical friend the limitations of his work in algebra, Tschirnhaus on the other hand failed to appreciate the meaning and significance of Leibniz's infinitesimal methods. At the time of their first discussions of mathematical topics in November 1675, Leibniz was already in possession of the principles and notation of his infinitesimal calculus. For example, a manuscript note of this time records the rule for differentiating a product as a remarkable theorem valid for all curves (Hofmann 1974, p 175). How little Tschirnhaus had understood is illustrated by the fact that Leibniz had later to remind him that, at the time of their discussions in Paris, Tschirnhaus had dismissed the new notation as useless signs serving only to obscure and had not been prepared to listen when Leibniz offered to show him an example (*GBM* **1**, p 375). This has a bearing on the later priority dispute between Newton and Leibniz. For it seems unlikely that, in view of his evident lack of understanding, Tschirnhaus could have effectively communicated to Leibniz anything he may have been told in London concerning the work of the

English in the field of analysis. There is certainly no documentary evidence that Leibniz learned anything significant from him about the work of English mathematicians. In the summer of 1676, Tschirnhaus received from Collins, through Oldenburg, information about English infinitesimal methods, including Newton's tangent rule. By this time, of course, the information would have been too late to have been of any use to Leibniz in the invention of his calculus. Yet this correspondence was used in evidence against Leibniz by the simple expedient of antedating it by one year (Hofmann 1974, p 171).

The invention of the calculus

Leibniz made the decisive steps leading to the invention of his new calculus during a few days in October 1675. Since 1673 he had been preoccupied with infinitesimal problems, including the determination of centres of gravity, quadratures and rectifications of curves and also the inverse problem of tangents. The last type of problem, requiring the determination of the equation of a curve, given the law defining its tangent, he recognised to be reducible to quadratures. It was in the context of an inverse tangent problem that, in October 1675, he first replaced the abbreviation omn (for *omnes*) by $\int$ (for *summa*) (Child 1920, p 60). At this stage, all integrals were regarded as definite, though there was no special notation for the limits, so that the notation $\int y\,\mathrm{d}x$ (originally written $\int y$) represented what would now be written as $\int_0^x y\,\mathrm{d}x$. Recognising that the operation $\int$ raises the degree, Leibniz initially represented the converse operation by a d in the denominator, emphasising that the operation lowered the degree. Thus $\int y = z$ was equivalent to $y = z/d$. A little later he changed the notation x/d to $\mathrm{d}x$, which was then retained in all future work.

From the examples used by Leibniz to illustrate the new notation, it is evident that some of the rules governing the new calculus were already clear to him. For instance, constant factors were taken in front of the integral sign and the sum of integrals was taken to be equivalent to the integral of the sum.

Early in November, Leibniz was concerned with five problems involving the derivation of curves with given subnormals. First he considered the curve whose subnormal $p = y\,\mathrm{d}y/\mathrm{d}x$ is a^2/y. Evidently, $\int a^2\,\mathrm{d}x = \int y^2\,\mathrm{d}y$, from which he inferred the equation of the curve to be $a^2x = \frac{1}{3}y^3$, following his usual practice of taking the curve to start at the origin. Having derived the solution, Leibniz made a check using Sluse's rule. The next problem concerned the curve with subnormal a^2/x, which was easily found to be $\frac{1}{2}y^2 = a^2 \int \mathrm{d}x/x$, a transcendental curve, Leibniz remarked, which could be constructed using the logarithmic curve. Two of the remaining problems are especially interesting in relation to the development of Leibniz's ideas. In one of these, concerning the curve whose subnormal is $(a^2/y) - x$, he made a mistake by supposing $\int dx/y$ to be a logarithm, thus failing to distinguish between the integrand and base variable. Also, it was in the course of solving this problem that he changed from the notation x/d to dx. Finally, when Leibniz found the curve with subnormal $\sqrt{(x^2+y^2)}$ to be $\frac{1}{2}y^2 = \int_0^x \sqrt{(x^2+y^2)}\, dx$, he attempted to find an approximation to the integral in the form of a series.

The new algebraic symbolism provided for Leibniz a general viewpoint for the consideration of infinitesimal problems and greatly facilitated their solution. For example, when the mathematician Claude François Milliet Deschales set him the problem of determining the part of a circular cone cut off between its base and a plane parallel to the axis, he was able to find the solution the same evening, showing that it depended on the quadrature of the circle and hyperbola (Hofmann 1974, p 195). Again, he was able to determine the quadrature of a sector and the tangent of Bertet's curve (which Ozanam had drawn to his attention), generated from a circle by adding to the radius the arc length from a given point on the circumference, and to generalise the results, replacing the circle by an arbitrary curve. Yet again he reduced the quadrature of the equilateral hyperbola to an integral that could be evaluated in terms of an infinite series.

Leibniz's own perception of his recent achievements in the solution of infinitesimal problems may be gleaned from remarks he made in letters to Gallois and Oldenburg at this time. On 2 November he expressed to Gallois, whom he had been unable to visit the previous week owing to illness, his intention to publish what he had so far achieved in letters to well known persons, including Gallois himself (*A* III **1**, pp 304–6). The promised letter to Gallois was written towards the end of the year (*A* III **1**, pp 356–60). In this letter Leibniz described his arithmetical quadrature of the circle, making the interesting remark that the quadrature was not geometrical, for he did not claim to describe a square equal in area to a circle, but it was more than mechanical, for, beyond approximation, it gave the true and exact ratio of the circle to the circumscribed square, in so far as this could be expressed in rational numbers. This concept of a quantity defined by a converging infinite series, Leibniz claimed, represented a considerable advance in the analytic mode of investigation of the properties of curvilinear figures. The reason for his success, Leibniz told Gallois, was his employment of infinitesimal triangles besides the traditional rectangles in the resolution of figures. For by this means he had been led to his transmutation theorem, a result of great generality and unifying power, permitting, for example, the proof by a single demonstration of all the quadratures that Wallis had only been able to find by induction, besides a number of others he had found himself.

Writing to Oldenburg, also at the end of 1675, Leibniz promised his solution by a new method of approach of a geometrical problem that had hitherto proved intractable (*A* III **1**, pp 326–34). From this solution, which he would communicate when he had time to complete it, Oldenburg would recognise, so he hoped, that he had not only solved problems but also invented new methods, by which he meant the infinitesimal calculus. It was methods rather than results, he emphasised to Oldenburg, that he prized the most. The actual problem Leibniz had in mind for the purpose of impressing Oldenburg was probably one of several inverse-tangent problems that had been proposed by Florimond Debeaune.

The last months in Paris

At the beginning of 1676 Leibniz was in the middle of writing a letter of greetings for the New Year to Duke Johann Friedrich when he received from the Court official Johann Karl Kahm the formal offer of a position as Counsellor, so he ended the letter by confirming his acceptance of the post (*A* I **1**, pp 504–5). Yet he had not given up hope of being able to remain in Paris, for on the same day he addressed a letter to Colbert himself, whose nephew had favoured his aspiration to the vacancy in the Academy, asking support for his scientific work and reminding Colbert that this had not been entirely without success (*A* I **1**, p 457). Then the new Elector of Mainz, Damian Hartard von der Leyen, confirmed the appointment as Counsellor he had held up to that time. It seems that Leibniz saw in this circumstance at least a possibility of prolonging his stay in Paris as a political emissary of both the Elector of Mainz and the Duke of Hanover. For, in his reply, Leibniz mentioned his new appointment in Hanover while asking the Elector to consider his willingness to render further services from time to time (*A* I **1**, pp 398–9). On 11 February, however, his friend Melchior Friedrich von Schönborn, who had accompanied him on his first visit to London, informed him in confidence that, despite his own support, which should have carried considerable weight on account of the high offices he had held, he could hold out no hope. Assuring Leibniz of his own continued regard for his qualities, Schönborn commented that, in those times, the liberality of princes did not exceed the ruin of their states (*A* I **1**, pp 400–1).

Leibniz was formally appointed Counsellor to Duke Johann Friedrich on 27 January 1676 (*A* I **1**, p 508). In the letter of appointment, Kahm listed the principal members of the Court who would be Leibniz's senior colleagues and informed him that no promise could then be given concerning the prospect of appointment as a Privy Counsellor. A letter to Christian Habbeus von Lichtenstern, who had first recommended him to the Duke, shows clearly that Leibniz had no intention of settling permanently in Hanover (*A* I **1**, pp 444–6). To his friend Leibniz remarked that he believed he would be like an amphibian, sometimes in Germany, sometimes in France, having by God's grace the means to stay for some time here and there until he had the opportunity to settle. Concerning the appointment, he simply mentioned that the Duke had granted him a considerable favour, the nature of which he would describe on another occasion.

When Kahm informed Leibniz on 28 February (*A* I **1**, pp 510–11) that the Duke wished him to come to Hanover as soon as possible, pointing out that his salary would be paid from 1 January, Leibniz pleaded for two or three weeks to conclude his affairs in Paris. A month later he was still there and the Duke reluctantly gave him permission to stay until Whitsuntide (*A* I **1**, p 515). Even then he took no steps to leave Paris. Indeed, he made a last desperate attempt to find employment in the Academy by writing to Huygens in the middle of June seeking his support (*HO* **22**, p 696). Since the end of 1675 Huygens had been ill and received Leibniz's letter just as he was about to return home to The Hague in order to convalesce. Having no time before his departure to speak to Leibniz personally, he left him a letter in which he assured him that he had used his influence with Colbert and Gallois (who, as previously noted, had turned against Leibniz) and was still hoping for success.

The dilemma facing Leibniz was a consequence of his dedication to the pursuit of science in the service of mankind. According to his own perception, the best results would be achieved by remaining in Paris, in close contact with the Academy and the many mathematicians and philosophers who lived there or visited the city. But even an optimist like Leibniz had to recognise that the prospects of paid employment that would have enabled him to stay were slight. So he reluctantly accepted the Duke's offer out of necessity, while continuing to hope for a post in the Academy, which no doubt he believed he would be able to accept without losing the Duke's goodwill. Despite the uncertainty concerning his future and the anxieties that must have been generated by his personal situation during his last months in Paris, Leibniz continued his own researches with undiminished enthusiasm.

Early in the year, as already remarked, he made an examination of Pascal's papers with Tschirnhaus. Then in April he met the Danish mathematician Georg Mohr, who came to Paris from London and reported on the conversations he had there with Collins. When Mohr passed on to Leibniz the sine and inverse-sine series he had received from Collins, Leibniz wrote to Oldenburg (*A* III **1**, pp 374–81), remarking particularly on the elegance of the sine series,

$$\sin z = z - \frac{1}{6}z^3 + \frac{1}{120}z^5 - \frac{1}{5040}z^7 + \frac{1}{362880}z^9 - \ldots$$

as if he had just seen it for the first time. In fact, Oldenburg had communicated both series to him a year earlier (*A* III **1**, p 233). Many years later, Leibniz attributed his lapse of memory to the pressure of his affairs in Paris at that time. Through Oldenburg, Leibniz requested from Collins a proof of the sine series and promised in return a proof of his own arithmetical quadrature of the circle, to which he was then putting the finishing touches. After mentioning briefly some other problems on which he was working, Leibniz ended his letter with a plea that the papers of Gregory (who had died the previous year) should be adequately preserved and he expressed his interest in their contents.

Two days before writing the letter to Oldenburg, Leibniz addressed another letter to the Hamburg jurist Vincent Placcius, who had referred to Leibniz's legal works in a publication of his own (*A* II **1**, pp 259–60). Since moving to Paris, Leibniz explained, he had devoted his attention to mathematics rather than law, adding that he hoped to take up the study of law again, in which he had so far started much but completed nothing. Also at this difficult time Leibniz kept up his correspondence with mathematicians and philosophers on a variety of subjects. For example, he sent to Claude Perrault a detailed criticism of the theory of gravity Perrault had proposed in the Academy during a debate on the subject in 1669 (*A* II **1**, pp 262–8). Leibniz himself later adopted a modification of the theory Huygens had introduced on this occasion. To the experimentalist natural philosopher Edme Mariotte, Leibniz addressed letters stressing the value of the search for causes and logical demonstration in physics (*A* II **1**, pp 268–71).

In mathematics, Leibniz continued and brought to completion his investigation of the problem of Debeaune already mentioned. As stated by Debeaune, the problem is equivalent to finding the curve through the origin satisfying (in rectangular coordinates) the differential equation $\mathrm{d}y/\mathrm{d}x=(x-y)/a$, where a is a constant. Leibniz made little progress until he followed the approach adopted by Descartes of referring the curve to oblique axes, one of which was the asymptote, inclined at 45°. In this case the curve (figure 3.8) no longer passes through the origin B but through another point A$(0, c)$. Also the subtangent $t=\mathrm{RN}$ is constant, so that $t=\mathrm{BC}=\sqrt{2}c$. Taking $\mathrm{SX}=\mathrm{d}x$ and $\mathrm{SV}=\mathrm{d}y$, it follows from the similar triangles SVX and RXN that $\mathrm{d}y/\mathrm{d}x=y/t$. Hence $\mathrm{d}y/y=\mathrm{d}x/t$. At first Leibniz did not recognise the integral but then remembered that, some months before, he had shown it to be a logarithm (Scriba 1964). Inserting the minus sign omitted by Leibniz (for his $\mathrm{d}x$ and $\mathrm{d}y$ are of opposite sign), putting $t=\sqrt{2}c$ and using the condition that the curve must pass through A$(0, c)$, the equation of the curve becomes $x=-\sqrt{2}c\log(y/c)$.

At the beginning of July, Kahm again wrote to Leibniz on behalf of the Duke, pressing him to proceed to Hanover as quickly as possible (*A* I **1**, pp 515–16). Reminding Leibniz that half a year had gone by since his official appointment as Counsellor, Kahm expressed surprise that he had delayed for so long his departure from Paris. Besides the office of Counsellor, Kahm added, Leibniz would be in charge of the ducal library, following the departure in September of the first librarian, Tobias Fleischer, for a post as

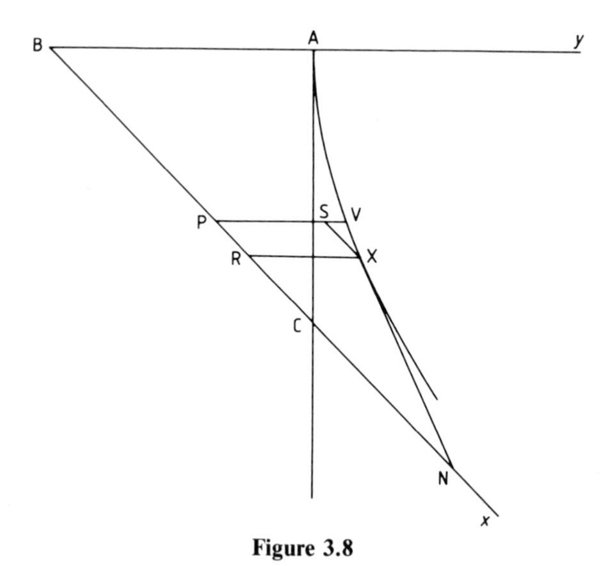

Figure 3.8

Counsellor to the Danish king in Copenhagen. When Christophe Brosseau,[4] the Hanoverian ambassador in Paris, paid Leibniz his travel expenses at the end of July and pressed for his departure, it must have been difficult for him to stay longer. Yet it was just at this time that he had his first full exchange of views with Newton.

Oldenburg formed the intention to respond to Leibniz's request for information concerning important items in Gregory's correspondence and also to pass on to him as many of Newton's latest results as he could persuade Newton to release for the purpose. The need to place on record for the benefit of Leibniz and his friends in the Paris Academy the achievements of the English mathematicians became evident to Oldenburg when, in the middle of June, he received from Tschirnhaus a letter in which he repeated the claim that Descartes was the true founder of the new mathematical method and that the contributions of his successors were only a continuation and elaboration of Descartes' ideas (*A* III **1**, pp 408–24). In Oldenburg's view, there had to be a clear demonstration that the English had made significant strides beyond Descartes. From Collins, Oldenburg obtained for this purpose an elaborate, fifty-page document now known as the *Historiola*. This he considered to be too long, whereupon Collins produced an 'Abridgement' (*A* III **1**, pp 504–16), which Oldenburg translated into Latin to form the basis of the reply he addressed jointly to Leibniz and Tschirnhaus on 5 August 1676 (*A* III **1**, pp 517–33).

First, Collins gave the rule for forming the coefficients of the inverse-sine series, which he supposed to be not inferior in elegance to the sine series, on

[4]On Leibniz and Brosseau, see Jurgens and Orzschig (1984).

which Leibniz had particularly remarked. Thus

$$\frac{1}{6}=\frac{1\times1}{2\times3},\ \frac{3}{40}=\frac{1\times3\times3}{6\times4\times5},\ \frac{5}{112}=\frac{3\times5\times5}{40\times6\times7},\ \frac{35}{1152}=\frac{5\times7\times7}{112\times8\times9}$$

and so on to infinity. Gregory was of the opinion, Collins added, that the method of series swept away all difficulties, so that all that was known before it was as the dawn to the noon-day. To illustrate the power of Gregory's method, Collins quoted his treatment of Kepler's problem; that is, the division of the area of a semicircle in a given ratio by a line drawn from a point on the diameter to the circumference. Only the result was given without any indication of the rule of formation of the terms of the series. Collins then gave two more series, inaccurately transcribed, so that each contains a mistake:

$$\tan x = x+\frac{x^3}{3}+\frac{2x^5}{15}+\frac{17x^7}{315}+\frac{3233x^9}{181440}+\ \ldots$$

$$\log\tan\left(\frac{\pi}{4}+\frac{x}{2}\right)=x+\frac{x^3}{6}+\frac{5x^5}{24}+\frac{61x^7}{5040}+\frac{277x^9}{72576}+\ \ldots .$$

Leibniz had already received the tangent series (with the same mistake) in Oldenburg's letter of 22 April 1675. Collins gave no indication of the method of repeated differentiation employed by Gregory to obtain the series. In fact Collins himself had never received from Gregory any explanation of his method, so that he could not have passed this on to Leibniz even if he had wished to do so. Finally, Collins expressed the hope that the results he had quoted would satisfy the curiosity of the Paris mathematicians and that they would in return communicate an account of their own work.

With his own letter Oldenburg enclosed a copy of the letter Newton had sent him for communication to Leibniz and Tschirnhaus. Before describing the contents of the letter, a word may be said about the purpose of such exchanges between scholars and also about the state of affairs in the Royal Society at this time. Owing to the reluctance of printers to accept books on mathematics, because of the difficulties of typesetting and the small number of potential readers, the statement of results in letters, especially when these were registered in the Royal Society or the Paris Academy, provided a means of establishing a claim to invention, pending possible publication at a later date. The most precious possessions of a mathematician were, of course, the original methods by which new results could be obtained. While communicating results, in order to establish his possession of a general method, to which he might refer in impenetrably opaque terms, he took pains to eliminate any clues that would enable his correspondent to guess the method and use it himself to make further discoveries. As we have seen, Leibniz expressed to Oldenburg a willingness to reveal a method in exchange for another that he desired, but both he and Newton were very cautious, fearing that they might give away something valuable without receiving anything in return.

Since the autumn of 1675 Newton had been embroiled in disputes concerning his theory of colours; Hooke, for example, had declared that the essence of Newton's 'Hypothesis of light' was taken from his own *Micrographia*. Even after the successful demonstration of Newton's experiments in the Royal Society, the disputes continued. In the atmosphere created by these disputes, Newton may well have suspected that Leibniz, who had compromised his reputation during his first visit to London, was trying to obtain the best results of the English by pretending to possess discoveries of his own.

At the beginning of his letter to Oldenburg (*A* III **1**, pp 533–54) Newton made some polite remarks concerning Leibniz. Though Leibniz had paid great tribute to the English for a certain theory of infinite series, Newton had no doubt that Leibniz had discovered not only a method for reducing any quantities whatever to such series, as he asserted, but also various shortened forms, perhaps like those of the English if not even better. Newton then explains that he fell upon the theory of infinite series some years before and was sending some of the things which had occurred to him in order to satisfy Leibniz's wishes. First he gave the binomial theorem, by means of which the extraction of roots as infinite series was much shortened, and, moreover, simple or repeated division could be carried out. There followed several examples of varying difficulty. Next he gave an example of his scheme for solving equations in series, omitting an explanation of vital steps 'for the sake of brevity'. Again he remarked that the explanation of the derivation of areas and lengths of curves, volumes and surfaces of solids or of any segments of such figures, and centres of gravity from equations thus reduced to series, would take too long to describe. Newton continued his catalogue of results with his solution of Kepler's problem, evidently unaware that Gregory's solution of this problem had already been supplied by Collins, and series for the arc of an ellipse and area of the hyperbola. Some other things he had devised, Newton added, there was then no time to explain.

All the theorems or examples cited by Newton were already known in one form or another. And of course he gave only the results but not the methods of derivation. In other words, Newton's letter contained nothing that Leibniz could not have learned from elsewhere. At the end of the letter, Newton added a comment in English for Oldenburg, concerning his theory of colour, thus forcing Oldenburg to send a copy to Leibniz rather than the autograph.

In a postscript to the copy of Newton's letter, Oldenburg again reminded Leibniz of his promise to send his calculating machine to the Royal Society. Oldenburg pointed out that the failure of Leibniz, a German and a member of the Royal Society, to fulfil the promise he had given before the members of the Society had been a great embarrassment to him.

The letters of Oldenburg and Newton were despatched to Leibniz on 5 August 1676. To ensure safe delivery, they were entrusted to a German mathematician, Samuel König, about to leave for Paris, instead of being sent by the ordinary post. When he failed to find Leibniz at home, König left the package at the local apothecary's shop, where Leibniz received it when he chanced to call on 24 August. Although Leibniz in his reply remarked on the circumstances which had delayed his receipt of the letters, this passage was omitted as unimportant by Oldenburg when he transcribed Leibniz's letter into the Royal Society Letter Book. As a consequence, Newton was able to claim later, in the context of the priority dispute, that Leibniz had six weeks to study his letter before composing his reply on 27 August 1676. In fact he had only three days.

Leibniz immediately recognised the importance of the contents of Newton's letter and he knew that he must communicate his own discoveries quickly, to avoid any suspicion that these were based on a careful study of Newton's results. The reply was written in haste, contained mistakes in the writing of formulae, and the handwriting was even more difficult to decipher than usual, so that the copy transcribed by Collins for Newton included some serious errors (*A* III **1**, pp 558–86). For example, Leibniz is interpreted as calling the curve defined by Debeaune a 'sport of nature', whereas he referred to a curve of 'this' nature, that the subtangent is constant.

After commenting that Oldenburg's letter contained 'more numerous and more remarkable ideas about analysis than many thick volumes published on these matters', Leibniz praised Newton's discoveries. But, he added, Newton's method of obtaining roots of equations and areas of figures by infinite series was quite different from his own, which he characterised as a corollary of a general method of transformations by the help of which any given figure is reduced to another analytically equivalent figure in which the ordinate can be expressed as an infinite series by Mercator division. His arithmetical quadrature of the circle, communicated to his friends in Paris, would establish his possession of a powerful method. To Oldenburg he presents the basis of this method, so that what are held by the English to be brilliant discoveries will not be denied to himself also. Essentially, his method of transformation consisted of resolving a figure into parts and then reassembling these—or other parts equal to them—in another position or another form to compose another figure, equivalent to the former in area, even if the shape is quite different. As an example, he presents his arithmetical quadrature of the circle, starting with the equation $y^2 = 2ax - x^2$ and substituting $ay = xz$. By suppressing the connection between z and the tangent, however, he conceals the essential element in his general transmutation theorem. Since Collins and Newton had communicated only results, Leibniz was justified in withholding the key to his own method.

As further evidence of his possession of important discoveries, Leibniz informed Oldenburg that he had solved Debeaune's problem; indeed he had solved it by a sure analysis on the day that he first put his mind to it, though he indicated that he was not yet quite satisfied with his solution. It is curious that Leibniz questioned the power of the method of infinite series by claiming, in opposition to Newton, that there were many problems which depended neither on equations nor quadratures, such as problems of the inverse method of tangents like that of Debeaune. For he had previously emphasised the close connection between inverse-tangent problems and quadratures.

Lastly, Leibniz informed Oldenburg that when he had leisure, he would take up the reduction of mechanics to geometry, problems of elasticity, fluids, pendulums, projectiles, resistance of solids and friction; manuscripts from the Paris period in fact exist in which studies of such themes are undertaken (Hess 1978). Concerning his unfulfilled promise in relation to the calculating machine, over which Oldenburg had expressed his embarrassment, Leibniz maintained a complete silence.

Leibniz was still in Paris on 13 September when Brosseau informed him that his plan to travel by way of the Spanish Netherlands was not possible, since passport formalities would entail a further delay of several months (*A* I **1**, pp 516–17). At the end of September Brosseau informed him of the Duke's impatience and it was clear that his departure could not be further delayed. So, having taken leave of the ambassador on 2 October, he left Paris on Sunday morning, 4 October 1676, travelling with the mail-coach via Abbeville to Calais, never to set foot in his favourite city again (*A* I **2**, p 3).

The second visit to London

The decision to travel to Hanover via London enabled Leibniz to follow up his recent exchanges by a personal meeting with Oldenburg and Collins. Having arrived in Calais on 10 October, he was held up for five days by storms and unfavourable winds in the English Channel. After crossing to Dover, where he stayed one night, he arrived in London late in the evening of 18 October. Leibniz remained in London for just over a week, during which he at last showed his calculating machine to Oldenburg. A demonstration in the Royal Society was not possible, for the meetings had not yet started after the summer vacation. The other highlight of his visit was the meeting with Collins, who formed a very favourable impression of the young German, despite being unwell and having to contend with difficulties of communication, for Collins' Latin was limited while Leibniz could read English but probably could not speak it very well (*NC* **2**, p 109). Apart from Oldenburg and Collins, there is no record of Leibniz having met any other member of the Royal Society, though he would have been delighted to meet Wallis or Newton, if an opportunity had presented itself.

During his stay in London, Leibniz was allowed by Collins, who was the librarian of the Royal Society, to make excerpts from the manuscript of Newton's *De analysi* (*A* III **1**, pp 664–77) and also from his own 'Historiola', which had been composed with the intention of sending it to Leibniz (*LH* **XXXV**, 8, 23, f 1–2). From Newton's work, Leibniz copied only expansions in series and completely disregarded sections relating to infinitesimals, presumably because they offered nothing new to him. The 'Historiola' contains a transcription of the tangent method Newton had given to Collins in a letter of 1672 (*NC* **1**, pp 247–55). Leibniz just copied Newton's example and then added the general rule that he had already received from Oldenburg. In the course of the later priority dispute, Newton claimed that Leibniz had appropriated this tangent rule as if it were his own. But Leibniz was able to point out that his differential method was something quite different, and that, in any case, his source of inspiration had been not Newton but Sluse. The mere fact that Leibniz made excerpts from the 'Historiola' in London establishes that, contrary to Newton's belief, this document had not been sent to him in Paris. On the cover there is a note asking him to return the manuscript after he had inspected it but this, of course, refers to his study of it in London.

Although his second and last visit to London had been short, Leibniz could feel well satisfied with the friendly reception he had received. Having fulfilled his promise to Oldenburg in respect of the calculating machine, besides making a good impression on Collins, he had every reason to hope for a continuance of his good relations with the Royal Society, so that, despite the isolation he was facing in Hanover, he would be kept informed of the latest developments of mathematics in England.

The journey to Hanover

In a letter to Kahm (*A* I **2**, pp 3–4), sent from Holland towards the end of November, Leibniz reported on the details of his journey and activities since leaving Paris. Having completed his business at the Royal Society, he had called on Prince Ruprecht von der Pfalz, whose sister Sophie later became Duchess of Hanover and his intimate friend. The Prince offered Leibniz the opportunity to sail with a yacht he was sending to Germany in a few days' time for a cargo of wine. Leibniz boarded the yacht on 29 October and two days later it left harbour in the morning, reaching Gravesend the same day, where it waited four days for the loading of cargo. Then it was held up for six days at Sheerness in the mouth of the Thames by a strong head wind. At last the wind changed to northwest and in twenty-four hours Leibniz arrived in Rotterdam. Altogether he had waited fourteen days but when the conditions were right, the crossing had been fast.

During the uncomfortable waiting time on board the yacht at Sheerness, Leibniz composed a dialogue 'Pacidius Philalethi prima de motu philosophia' and reflected on his ideas of a real characteristic or universal language that would speak to the understanding rather than the eyes and the more powerful universal characteristic that would permit reasoning and demonstration in metaphysics and ethics by a calculus analogous to arithmetic and algebra. These thoughts on the wonderful idea he had first advanced in *De arte combinatoria* occupied his mind because, as he remarked in a letter to Gallois (*GP* **7**, pp 21–2), he had no-one to talk to except the sailors. The dialogue that Leibniz composed on the philosophical principles of motion was just one of a number of works that he wrote in this form. Another dialogue, probably written in the same year, takes for its underlying theme Plato's 'recollection' theory of learning (Knobloch 1976). The son of Aretaeus, in whose garden the dialogue is conducted, takes the place of the slave boy in Plato's *Meno*,[5] and Charinus (a pseudonym of Leibniz) demonstrates the theory in terms of a complete introduction to arithmetic and algebra rather than the special geometrical problem employed by Socrates. Charinus appears again as one of the characters in the dialogue composed on the yacht, but here Leibniz himself is represented by the character Pacidius. Guilielmus Pacidius, evidently a conciliator to unite all the scholars in a common task, was the pseudonym under which he intended to publish his 'Encyclopedia', the achievement of which would need the cooperation of many collaborators.

In the dialogue, Leibniz returns to one of his favourite themes, already mentioned in his letter to Oldenburg of 11 March 1671 (*A* II **1**, pp 88–91) and later to lead him to his metaphysical theory of monads, namely the labyrinth of the continuum. Unless we enter into this labyrinth, Leibniz declares, we cannot penetrate into the nature of motion itself (Couturat 1903, pp 609–10). Since space cannot be simply an aggregate of points nor time an aggregate of instants, the composition of the continuum is revealed to be the most basic problem that needs to be resolved before a rational theory of motion can be constructed.

From Rotterdam Leibniz travelled to Amsterdam—arriving on 13 November—where he met the microscopist Jan Swammerdam, famous for his investigations of insects, and the mathematician Jan Hudde, whose papers, Leibniz reported to Oldenburg, contained quite remarkable discoveries. He had long known Sluse's method of tangents, Leibniz added, and had improved upon it, while Mercator's quadrature of the hyperbola was already known to him in 1662.

After a few days in Amsterdam, Leibniz made a small circular tour to Haarlem, Leiden, Delft and The Hague, sleeping on the boat, which would

[5]In Plato's *Meno* Socrates teaches a slave boy a special case of the theorem of Pythagoras by asking questions simply requiring an answer in the affirmative.

take off in the middle of the night when a good wind blew up. The damp, cold air of winter in Holland and the rigours of about ten nights afloat adversely affected his health, so that on returning to Amsterdam, he had to spend several days in warm surroundings to recover his strength and appetite. However, he assured Kahm that he would be able to continue his journey to Hanover by the end of the week; that is, the last week of November.

In Delft Leibniz met Antoni van Leeuwenhoek, whose study of microorganisms, revealing a microscopic world teeming with life, may well have had some influence in the development of his theory of monads. Leeuwenhoek's researches certainly seemed to confirm the concept of preformation that would be demanded by this theory. It seems that Leibniz had more time in The Hague, for he was able to hold many long conversations with the great Benedict Spinoza (*A* II **1**, pp 378–81). Among the subjects of discussion, the most important concerned ethics and theology. Already, on board the yacht at Sheerness, Leibniz had reflected on one of the two famous labyrinths in which he believed our reason goes astray; namely the labyrinth of the continuum, relating to the problem of continuity and the antinomies of the infinite. Now his conversations with Spinoza brought him into the other labyrinth; that is, the problem of the freedom of the will.

Spinoza's most important work, the *Ethics*, appeared only after his death in 1677, though the manuscript had been circulated among his friends and Leibniz may have gleaned something of its contents from Tschirnhaus. With Leibniz Spinoza discussed the basic ideas of this work. They constituted a metaphysics which Leibniz found strange and full of paradoxes, as he remarked soon afterwards in a letter to Gallois (*A* II **1**, pp 378–81). In particular, Spinoza's pantheistic identification of God with the universe and his strict determinism, which deprived God and man of any freedom of action, were unacceptable to Leibniz.

One of the topics Leibniz discussed with Spinoza was the ontological argument—that is, the demonstration of the necessary existence of God—and when Spinoza was not at first convinced by his reasoning he put it in writing and gave Spinoza the paper, which is entitled 'Quod ens perfectissimum existit' (*A* II **1**, pp 271–3). First, Leibniz defines a perfection as a simple quality, positive and absolute. Such qualities express whatever they express without limits. Perfections are therefore indefinable, for definition would impose limits. Consider the proposition that two perfections A and B are incompatible. This cannot be demonstrated. For a demonstration would need definitions of A and B; otherwise the nature of A and B would not enter into the argument, so that the demonstration would apply to any qualities whatsoever, which would clearly be an absurd conclusion. Since all propositions that are necessarily true can be demonstrated, it follows that this proposition is not necessarily true. In other words, it is possible for A and B to exist in the same subject. More generally, a being possessing all perfections can exist, and since existence is contained in the number of perfections, such a being (God) necessarily exists. On the day following his conversation with Spinoza, Leibniz added to his paper the remark that not all possibles can exist along with others, for such existence would lead to absurdities (Couturat 1903, pp 529–30). For example, although God would exist in so far as he is possible, he would have no reason to exist; in particular he would not have the freedom of action ascribed to the Jewish and Christian God. The principle adopted by Leibniz is that whatever is possible within itself and compatible with other things necessarily exists. By substituting compossibility (as he called it) for possibility as the condition of existence, Leibniz opened a path for God's freedom of action in deciding from among the compossibles those to be brought into existence. But at the same time he weakened the ontological argument. For it was not now sufficient to demonstrate the possibility of God in order to infer his existence. Some other criterion was needed. This additional principle Leibniz found in the idea of the best or most perfect. Since God possessed all perfections, he had the greatest claim to existence and in accordance with his own perfection, chose to bring into existence the best of all possible worlds.

The circumstances of his delay at Sheerness and his conversations with Spinoza had been the occasions for Leibniz to concentrate his attention on the two basic issues whose resolution would lead him to his definitive metaphysics of monads. In the four years that he spent in Paris, philosophical problems were never far from his mind. For although his efforts were mainly centred on mathematics, there exist many manuscript notes from this period dealing with the entire range of fundamental metaphysical issues.[6]

At last, towards the end of December 1676, almost exactly seven years after Habbeus von Lichtenstern had recommended him to the Duke, Leibniz arrived in Hanover to take up the offices of Counsellor and Librarian that he would have preferred to be in a position to decline.

[6]A number of these are discussed in *SL Supplementa* **18** (1978).

4
Hanover under Duke Johann Friedrich
(1676–1679)

Having taken up his residence in the Library, at that time housed in the Palace, Leibniz found an early opportunity to place his plan of work before the Duke (*A* I **2**, pp 15–18). With regard to the Library, he proposed further acquisitions to transform the existing 3310 volumes and 158 manuscripts into a comprehensive collection covering all the major fields of knowledge. The quality of the books was more important than quantity, so that careful choice would be needed, and he was in an excellent position to make the selections, Leibniz told the Duke, owing to his wide knowledge of recent publications, catalogues and libraries, and his many contacts in the world of learning. Provided he could have the services of a copyist, Leibniz also planned to provide indexes of a new kind that would facilitate more speedy reference. In his role as Counsellor, Leibniz proposed that, besides carrying out the direct commissions of the Duke, he would collect through correspondence with the many scholars of Italy, France, England, Holland and Germany known to him all results of scientific importance. After listing some useful inventions, Leibniz mentioned his own calculating machine and his researches in natural theology, law, physics, geometry and mechanics, which he hoped the Duke would support.

A month later Leibniz reminded the Duke (*A* I **2**, pp 19–21) of his qualifications and experience, mentioning especially his appointment as a judge in the High Court of Appeal in Mainz at such an early age, and asked for an appointment as Privy Counsellor. This, he believed, would be justified by his record and would provide a firmer ground for both his fortune and his reputation. Following a further request to the Duke in October (*A* I **2**, pp 35–6), Leibniz was at last installed as a Privy Counsellor towards the end of 1677, when he engaged Jobst Dietrich Brandshagen as his servant and secretary.

At the Christmas Day service in the Church of the New Town, Leibniz took his place among those reserved for Privy Counsellors, only to find that he had offended the Duke's physician Jakob Franz Kotzebue who, though not a Privy Counsellor, resented Leibniz taking what he regarded as his place. Liebniz wrote a long explanation to the Duke about this embarrassing incident (*A* I **2**, pp 43–5). Although Leibniz avoided the occasions of further dispute by staying away from the Church for a whole year, Kotzebue not only remained hostile but publicly expressed the view that Leibniz, by his absences, had admitted he was in the wrong. Since Leibniz had in fact committed no offence and recognised moreover that Kotzebue's claim to precedence was likely to cause more trouble when new Privy Counsellors were appointed, he suggested to the Duke, who had thus far taken no action, a compromise that should settle the affair to the satisfaction of all (*A* I **2**, pp 107–9).

Besides being deprived of the place in Church to which his rank entitled him, Leibniz suffered another indignity when he was paid only 500 taler[1] instead of the 600 taler received by other Privy Counsellors. But this was probably just an oversight, to which he drew the Duke's attention in a letter of November 1678.

By the beginning of 1678, after he had been appointed a Privy Counsellor, Leibniz had come to the conclusion that his acceptance of the Duke's invitation to Hanover had been a wise decision after all. In a letter to Gallois he remarked that his salary was higher than the one offered to him initially and he was pleased to be in the service of a prince with an extraordinary degree of discernment, who showed him much favour (*MK,* pp 52–3). Writing to Conring in June (*A* II **1**, pp 418–20) he again expressed satisfaction with his position in Hanover, for the Duke allowed him freedom from routine duties, such as his presence at meetings, whenever this appeared to be necessary for the performance of his personal duties, which included running the library, correspondence with scholars and private commissions for the Duke himself. To Martin Geier in Leipzig, Leibniz wrote that he preferred the service of a prince whose virtues were so great to every kind of freedom (*A* I **2**, pp 398–9).

The Duke was a pious convert to Catholicism, whom Leibniz praised for his moderation, in that he never gave the least cause of complaint to Protestants (*GP* **3**, p 212). His generous treatment of Leibniz, especially in allowing him ample freedom to pursue his own studies, may have been to some extent the result of a communication from Antoine Arnauld to one of the Franciscans at the Court, in which Arnauld remarked that Leibniz lacked only the true religion (that is, Catholicism) in order to be one of the greatest

[1]The 'taler' was the standard unit of currency (a silver coin) in the Empire, named after the town in Bohemia, St Joachimsthal, where it was minted. In 1704, 1 corn measure (about 100 kg) of rye cost 3 taler. The 6000 inhabitants of the mining town of Clausthal consumed 300 corn measures of rye each week (Schnath 1978, pp 284–5).

Plate 2 The Palace in the Leinstrasse in the seventeenth century. Courtesy of the Historisches Museum, Hanover.

men of the century. The Duke was greatly impressed by this testimony from the great theologian. Leibniz had himself brought the letter from Paris, without knowing anything of its contents (*DS* **2**, app, pp 66–7).

First contacts in Hanover

At the end of February 1677, the Pastor in Hattorf, Jakob Schwachheim, wrote to Leibniz of his intention to visit him soon to request an exchange of some duplicate copies he had found in the Duke's library for some books of his own. His description of the location of the books he wanted as 'by the window' and 'on the left hand by the entrance' perhaps indicated the need for some of the new indexes Leibniz intended to prepare. Also early in the year, Leibniz met Gerhard Wolter Molanus (van der Muelen), the President of the Ecclesiastical Court in Hanover and Abbot of Loccum, who had been formerly Professor of Theology and Mathematics in Helmstedt. Molanus had a friend, Arnold Eckhard, at that time Professor of Mathematics in

Rinteln, who wished to meet Leibniz in order to discuss the Cartesian philosophy, of which he was an ardent admirer. Molanus provided the opportunity by arranging a debate between Eckhard and Leibniz on the Cartesian demonstration of the existence of God. This debate took place in the presence of Molanus and his brother monks on 15 April 1677 (*A* II **1**, pp 311–14).

In the following month, Johann Daniel Crafft,[2] an expert on the manufacture of wool and an old friend from his days in Mainz, visited Leibniz in Hanover. Employed in the service of the Elector of Saxony, Crafft had, in the course of his visits to England and Holland, acquired considerable knowledge of processes, including the preparation of dyes, which he had successfully applied in promoting the manufacture of wool in Saxony. Having learned of the Duke's interest in the manufacture of wool, following his conversation with Crafft, Leibniz suggested to the Duke that it might be advantageous to retain the services of Crafft as an adviser, if this would not be incompatible with his obligations in Saxony. Indeed it might serve as a pretext for more important negotiations for cooperation between Hanover and Dresden (*A* I **2**, pp 23–4).

Besides the manufacture of wool, Crafft also discussed with Leibniz the production of phosphorus, which had been discovered by Heinrich Brand of Hamburg in about 1674. The unexpected discovery came about by following the procedure described in an alchemical book for the extraction from urine of a fluid that would change silver into gold. Brand had sold the secret of the manufacture to Crafft. Fascinated by the properties of the strange substance, Leibniz published a description in the *Journal des Sçavans* in August 1677 and later, at the request of Tschirnhaus, sent him a report to be read to the Royal Academy of Sciences on one of his visits to Paris (Ravier 1937, p 47).

Through Crafft Leibniz saw an opportunity to make contact with Bishop Cristobal de Rojas y Spinola, the chief irenical negotiator, and through him to become known to Emperor Leopold I (Miller and Spielman 1962). Writing from the hunting lodge in Linsburg on 3 June 1677, where he was staying with the Court, Leibniz asked Crafft to recommend him to Rojas (*A* I **2**, pp 272–5). At the beginning of 1679, Leibniz had an opportunity to meet the Bishop when he visited Hanover for political discussions on behalf of the Emperor (*A* I **2**, pp 408–9). These discussions, Leibniz remarked, had hardly any relation to the outcome of the Peace Congress of Nijmegen,[3] for which he had been preparing a political document at the time of his request

[2]On Crafft, see Forberger (1964).

[3]The Treaty of Nijmegen (1678) brought the conflict between France and Holland to an end. Although Holland remained intact, France kept Lorraine and 'Louis le Grand', as he was officially styled from 1680, used his position of strength to continue his territorial expansion by means of bribery and diplomacy, claiming 'dependent territories' on his eastern frontier.

to Crafft for the introduction to Rojas. This document, entitled *De jure suprematus ac legationis principum Germaniae* and published in Amsterdam in 1677 under the pseudonym Caesarinus Fürstenerius (*A* IV **2**, pp 13–270), reveals remarkable insight into the nature of government and the problems of applying political and legal theories in a complex situation. The primary concern was the question whether the representatives of the German Electors and other Princes at the Congress should be regarded as representatives of sovereign states; for although these states had armies and many of the outward marks of sovereignty, they were subject to the Imperial Court. Leibniz also composed a dialogue on the subject, *Entretien de Philarete et d'Eugene sur la question du temps agitée à Nimwegue touchant le droit d'ambassade des électeurs et princes de l'Empire* (*A* IV **2**, pp 293–338), which was distributed in the form of a pamphlet to the delegates at the Peace Congress.

At the time Leibniz was engaged in the writing of his political document, he received Newton's second letter, containing the key to his infinitesimal calculus in the form of two indecipherable anagrams. Leibniz replied immediately in a letter setting out clearly the principles of his own differential calculus.

Towards the end of 1677, the Duke invited to Hanover as apostolic vicar the Danish theologian Nicolaus Steno. Formerly a doctor and a convert to Catholicism, Steno abandoned anatomy and natural science to become, in Leibniz's words, a mediocre theologian (*GP* **6**, p 158). To Conring, Leibniz expressed his regret that Steno was disinclined to continue the scientific studies in which he was excellently skilled (*A* II **1**, p 385). His studies in palaeontology inspired the similar investigations Leibniz conducted in the Harz mountains and reported in his *Protogaea* (*D* **2**, 2, p 181). It seems that Leibniz formed his impression of Steno as a poor theologian following a discussion with him on the freedom of the will. This conversation was given literary form by Leibniz in his unpublished *Dialogue entre Poliandre et Théophile* (Baruzi 1905). After a long discussion, Théophile (representing Leibniz) brings Poliandre to assent to the principle that existence is determined by God's choice of the best among the compossibles.

Leibniz kept in touch with Paris through the ambassador Christophe Brosseau, who sent him information concerning new books, Friedrich Adolf Hansen, who acted as his general agent, and Henri Justel, the king's secretary, who later found a new home in England. For Justel, Leibniz conducted a small genealogical research on the descent of the Counts of Löwenstein (*A* I **2**, pp 335–7). This rather hastily written piece was his first purely historical work.

In Amsterdam Leibniz had met Spinoza's friend Georg Hermann Schuller, who corresponded with him in Hanover. It was Schuller who distributed his pamphlet at the Peace Congress in Nijmegen. When, in January 1678, Schuller sent Leibniz a copy of Spinoza's *Opera posthuma,*

Leibniz informed Justel that Spinoza's works had at last been published, the most important part being the *Ethics,* in which he found many ideas in agreement with his own, but also some paradoxes.

Molanus and Eckhard

In the debate that Molanus arranged between Eckhard and Leibniz on the Cartesian demonstration of the existence of God, Eckhard defended the Cartesian position, while Leibniz pointed to flaws in the demonstration without explicitly stating his own version of the ontological argument. Following the public debate, Eckhard wrote a letter to Leibniz on 19 April 1677 (*A* II **1**, pp 317–21), to which Leibniz replied the following week (*A* II **1**, pp 321–4). Then, in May, Eckhard wrote a very long letter on the subject (*A* II **1**, pp 326–62). Besides sending a brief reply (*A* II **1**, pp 362–6), Leibniz added his own notes to the manuscript of Eckhard's letter. One of these (*A* II **1**, p 358, n 103) is especially interesting for its implication of the existence of the unconscious; Leibniz doubted Eckhard's assertion that there was nothing in the mind of which we are not conscious.

Despite his many criticisms, Leibniz was sympathetic to Eckhard, for his real target was Descartes, and in particular the highly regarded but in his view greatly overrated Cartesian method. What was this method, Leibniz asked, of which Archimedes and Galileo had been ignorant? Moreover, the advances made since Descartes were not the result of an application of the Cartesian method, for they seemed to have been achieved by anyone but Cartesians. What Eckhard had done, Leibniz told him, was his own and not Cartesian, for Descartes had not even tried to prove the possibility of a most perfect being, from which alone the existence of God could be inferred. Leibniz commended Eckhard for setting about the demonstration in the right way, seeking to show that the concept of the most perfect was not composite and hence not contradictory, but there were many flaws in the argument, which Leibniz pointed out to him. To many of the Cartesian ideas Eckhard had mixed in throughout, Leibniz could not give his assent. For example, the Cartesian idea that all truths depended on God's will seemed to him to imply a contradiction. In one of his notes on Eckhard's manuscript, Leibniz elaborated this point (*A* II **1**, p 351, n 74). According to the Cartesian idea, God's will was both prior and posterior to itself, since the necessity of God's existence, and hence of God's will, would depend on God's will. Again there were clearly truths that did not depend on God's will, for who could claim, Leibniz asked, that *A* was not non-*A* because God had decreed it? God may not will to determine those things which are already determined by themselves, such as the truths of reason and the free choices of individual minds. There is here at least an indication of Leibniz's fundamental distinction between possibilities determined by God's understanding and existence determined by God's will.

Through the intercession of Molanus and Leibniz, Eckhard in 1678 received a parish near Hanover, which no doubt enabled him to meet his friends more frequently, though there is no record of further meetings (*GP* **1**, pp 209–10). To test Eckhard's mathematical skill and also the Cartesian method which he prized so highly, his friends put to him a problem in indeterminate analysis, supposedly formulated by a nobleman at the Court but which had in fact been proposed to Leibniz by Mariotte (Hofmann 1974, p 201). Following his introduction to indeterminate analysis by Ozanam, Leibniz had discussed such problems with Arnauld and Mariotte in Paris (*A* III **1**, pp 311–26). Molanus acted throughout simply as an intermediary to whom both Eckhard and Leibniz addressed their contributions to the discussion. The problem which Eckhard received from Molanus required him to find a right-angled triangle whose sides were rational numbers and whose area was the square of a rational number (*GP* **1**, p 272). Eckhard addressed a solution to Molanus at the beginning of 1679; he made some elementary mistakes but Leibniz pointed them out politely in the reply he sent to Molanus (*GP* **1**, pp 276–8), for he wished to locate the causes of failure to solve the problem not in Eckhard's lack of skill but in the Cartesian method that Eckhard claimed to follow.

On receiving Eckhard's next attempt, Leibniz remarked to Molanus that his friend had shown himself to be an able man who was capable of going as far as the Cartesian method permitted (*GP* **1**, pp 283–6). Eckhard had reached the correct conclusion that a solution was impossible but he had not achieved a demonstration. In a postscript to his letter to Molanus, Leibniz excused himself for writing in French (which he believed Eckhard could understand), because to write a whole letter in Latin would give him more trouble than the solution of the problem would give to his friend. Since Leibniz wrote Latin very fluently, this remark must be interpreted as a joke.

In further letters Leibniz returned to the question of method. Thus far, he remarked, they had only turned the wheel of Ixion in their letters; a classical allusion meaning to go round in circles (*GP* **1**, p 293). It was very easy to fail in algebra, Leibniz remarked, when one did not reason with rigour in the way of the ancient geometers (*GP* **1**, p 300). Leibniz generally demolished Eckhard's false assertions by counter-examples and syllogistic reasoning. Recognising that Eckhard was not going to be convinced of his failure to demonstrate the solution to the problem, Leibniz finally suggested to Molanus on 8 June 1679 that, if Eckhard consented, he would send the correspondence to the Royal Academy of Sciences in Paris for the judgment of Huygens and others there who were well versed in analysis (*GP* **1**, p 303). In response to this idea, Eckhard sent to Molanus a fair copy of his solution (*GP* **1**, pp 306–14) for communication to any mathematicians he pleased to send it to. Leibniz had written out his own solution when he devised the problem at the end of 1678 (*GM* **7**, pp 120–5).

Renewed interest in chemistry

At the time of his conversations with Crafft, Leibniz composed an essay, *De modo perveniendi ad veram corporum analysin et rerum naturalium causas* (*On a method of arriving at a true analysis of bodies and the causes of natural things*) (*GP* **7**, pp 265–9) in which he applied the method of analysis and synthesis to problems of chemistry. Analysis was of two kinds; one of bodies into various qualities through experiments, the other of sensible qualities into their causes through reasoning. For an effect was understood when its cause was understood. Those experiments in which only a few ingredients were brought together were more useful because in these cases it was easier to locate the one in which the cause lay concealed. Starting with experiments involving (besides the general and necessary agents) just one single homogeneous body, as many experiments as possible should then be tried with two kinds taken together, treating them in various ways by fire, water and air, then going on to experiments with three kinds, and so on, following the pattern of combinations built up in *De arte combinatoria*.

In the summer of 1678, Liebniz visited Hamburg in order to purchase the library of the late Martin Fogel on behalf of the Duke. There he met Heinrich Brand, the discoverer of phosphorus, with whom he made an agreement, after consultation with the Duke, whereby Brand undertook to communicate the results of his investigations on the manufacture of phosphorus and other chemical experiments in return for an annual payment of 120 taler (*A* I **2**, pp 60–1). The following summer, Brand arrived in Hanover and found a suitable place outside the city for his work on the manufacture of phosphorus (*A* I **2**, p 184). The urine was collected from the soldiers in a nearby camp and stored in barrels. Then Brand evaporated and distilled the urine to produce the phosphorus. Leibniz testified that Brand had honestly communicated the details of his secret, for everything that he accomplished, Leibniz had been able to repeat with his own workers in another laboratory.

On 8 September Leibniz sent a piece of phosphorus to Huygens, promising more if needed, since he could make it. A good experiment, he suggested, would be to try it in a vacuum—Boyle later concluded that phosphorus only glows in air—and he asked Huygens to demonstrate its effect to Colbert, the Duke of Chevreuse and the Academy (*GM* **2**, pp 19–20).

While in Hamburg Leibniz also met Johann Joachim Becher, the representative of chemistry in Germany, as was Boyle in England and Lémery in France. It was Becher who laid the foundation of the phlogiston theory, which as developed by Stahl held the field until the discoveries of Lavoisier. Becher was a strange character, however, full of envy and malice towards his outstanding contemporaries, which led him to write a book ridiculing the scholars of the time. This book, *Närrische Weisheit und weise Narrheit,* published posthumously in 1683, came into the hands of Duke Ernst August (Johann Friedrich's successor in Hanover), to whom Leibniz explained how Becher's attack on himself had come about (Guhrauer 1846 **1**, app, p 26). Becher had been piqued because Leibniz prevented an alchemical deception he had contemplated. This was Becher's proposal to extract gold from sand, concerning which Leibniz had expressed his doubts in the letter he sent to Huygens with the piece of phosphorus. In searching for a means of revenge, Becher had taken as a pretext one of their conversations in Hamburg, when Leibniz made some passing remarks on the improvement of carriages. Becher distorted these comments so as to attribute to Leibniz the ridiculous claim that he could devise a carriage capable of completing the journey from Hanover to Amsterdam in six hours. Leibniz remarked to the Duke that he was in good company, for Becher had also ridiculed others of merit and reputation, such as Huygens and the King of France.

Newton's second letter received

Newton believed that Leibniz had six weeks in which to study his first letter (*epistola prior*) before composing his reply, whereas he had replied immediately. In Newton's view, Leibniz had not produced anything essentially new, but in the course of the mythical six weeks had been able to recast the contents of Newton's letter so as to claim that he had himself discovered the results by a different method. Leibniz had indeed discovered the results himself by a different method, but in his reply to Newton he had effectively concealed the most original element of this method, his general transmutation theorem. The significance of this result was never in fact appreciated by the Newtonians in the course of the subsequent priority dispute. So Newton firmly believed in Leibniz's plagiarism when he wrote his second letter (*epistola posterior*) on 24 October (3 November) 1676 (*NC* **2**, pp 110–61). An incompetently prepared copy (*GBM,* pp 203–25) was sent to Hanover by Oldenburg on 2(12) May 1677. This reached Leibniz on 21 June (1 July) 1677 and he replied on the same day (*NC* **2**, pp 212–31). The Julian Calendar, it may be noted, was in use in Hanover until 1700.

Although Newton's letter is politely phrased, there is an evident lack of warmth or enthusiasm for the correspondence. To Oldenburg (*NC* **2**, p 110), Newton expressed the hope that Leibniz would be satisfied, so that it would not be necessary for him to write more on the subject, since he was then more interested in other subjects. It was evidently Newton's intention, by his long and carefully prepared letter, to bring the dialogue with one he regarded as an unworthy opponent (a plagiarist, no less) to an end.

At the beginning of his letter, Newton wrote vaguely concerning Leibniz's method of series. The bare comment that he knew three methods of arriving at series and could hardly have expected a new one to be communicated to him could be described as damning with faint praise. Having referred briefly to his three methods, Newton described in detail his path to the binomial theorem, as Leibniz had requested.

Turning to his method of finding tangents, Newton claimed that if anyone knew the basis of this method, he could determine tangents in no other way, unless he were deliberately wandering from a straight path. The foundation of this method, however, he concealed in an anagram *6accdæ13eff7i3l9n4o4qrr4s8t12vx*. Even when deciphered (*NP* **2**, p 191, n25), 'from an equation containing fluents to find the fluxions, and vice versa', it would reveal little to anyone not already familiar with the method. On the basis of this principle, Newton devised certain general theorems for finding quadratures, one of which he enunciated for Leibniz and illustrated by particular examples. The theorem evaluates the integral of $z^{\theta}(e+fz^{\eta})^{\lambda}$ as a series. No explanation was given by Newton but in a marginal note (*GBM*, pp 208–9) Leibniz set down how he thought the series had been worked out, using integration by parts to lead to a reduction formula. Among examples given by Newton were the integral of $\surd(az)$ and $a^4z/(c^2-z^2)^2$. The general theorem described by Newton in the letter is a special case of the more general multinomial integral of $z^{\theta}(e+fz^{\eta}+gz^{2\eta}+\ldots)^{\lambda}$ given in his *De quadratura curvarum*.

Following remarks on the combination of series to obtain more rapid convergence and on the construction of tables of logarithms and sines, Newton explained his method of inverting series. Finally, he explained that for inverse-tangent problems and other still more difficult questions he used two methods, one particular and one general, which he concealed in a long anagram, inaccurately transcribed in the copy received by Leibniz. The two methods relate to finding the fluent from an equation involving the fluxion (that is, the general integration of differential equations) and to deriving infinite series with indeterminate coefficients.

The problems of Debeaune solved by Leibniz did not need Newton's general methods, and he would not call it a 'sport of nature', as he supposed Leibniz had done, unaware that this expression had arisen as the result of an error in the transcription of Leibniz's letter at the Royal Society. Having failed to recognise the true nature of Leibniz's general theory of transmutation, it might have seemed to Newton that Leibniz had been concerned only with some peripheral problems rather than the central ideas concealed in the anagrams.

In return for Newton's explanation of the binomial theorem, Leibniz gave a detailed account of his general method of determining the tangent to a curve of the form $f(x,y) = a + by + cx + dyx + ey^2 + fx^2 + gy^2x + hyx^2 + \ldots = 0$, thus introducing the principles and notation of his differential calculus. By means of particular examples, he demonstrated how the procedure could be extended to irrational quantities using the function of a function rule for differentiation. Three weeks after sending his reply to Newton's second letter, Leibniz posted another short letter to Oldenburg explaining that the clarification he had requested concerning the inversion of series was no longer needed, since, on a second reading, he had been able to understand the derivation (*NC* **2**, pp 231–4). At the end of his letter, Leibniz asked Oldenburg to send him the *Transactions* punctually each month, and also to send copies of interesting items that came his way, adding that he was prepared to pay the cost of copying. Leibniz's hopes of keeping in touch with London were soon dashed, for a month after Oldenburg had acknowledged receipt of Leibniz's two letters, telling him not to expect a quick reply from Newton or Collins, since they were out of town and occupied with other things, Oldenburg was himself dead.

In the May of the following year, Leibniz sent for publication in the *Journal des Sçavans* (*GM* **5**, pp 116–17) his quadrature of a cycloidal segment, his first result in the new geometry of infinitesimals, that he had communicated to Oldenburg in 1674 and to several of his friends in Paris. Although he mentioned the general transmutation theorem on which his discovery of the quadrature of the cycloidal segment was based, he still reserved a detailed explanation for another occasion. This was to be the publication of his *De quadratura arithmetica circuli ellipseos et hyperbolae cuius corollarium est trigonometria sine tabulis,* which he had completed before leaving Paris. In the period between 1677 and 1680 Leibniz made several attempts to have it published in Paris or Amsterdam but these were all unsuccessful (Hofmann 1974, p245, n78). A belated request for publication came from Johann Bernoulli in 1698, but by then Leibniz considered publication to be no longer useful, as the work had been overtaken by newer developments (*GM* **3**, p537). Even today, only an abbreviated version exists in print (Ravier 1937, p530). Yet this work contains a rigorous and unexceptionable proof of what was probably Leibniz's most important step in his path to the new calculus; that is, his general transmutation theorem.

Reaction to Spinoza's *Ethics*

Although, in his letter to Justel of 14 February 1678 (*A* I **2**, pp317–18) Leibniz remarked that he had found in Spinoza's *Ethics* many ideas in agreement with his own, he left his friend in no doubt of his fundamental disagreements with Spinoza and even warned that the *Ethics* was a dangerous book for those who would take the pains to master it. The grounds of his disagreement are neatly summarised in a list of paradoxes he had found. These were statements which he judged to be neither true nor even plausible. They were:

1 that there is only one substance, namely God
2 that creatures are modes or accidents of God
3 that our mind perceives nothing after this life
4 that God thinks but neither understands nor wills
5 that all things happen by a kind of fatal necessity
6 that God does not act for a purpose but only from a certain necessity of nature.

Spinoza, he added, had retained Providence and immortality in words but denied them in fact. In a note on a letter of Spinoza to Oldenburg written in 1671, Leibniz made the more forceful comment that Spinoza had 'destroyed the principles of ethics' (*GP* **1**, p 124, n 3).

Leibniz was first of all opposed to Spinoza's monism or pantheism, the idea that God was the only substance and that creatures were just modes or accidents of God. In one of the extensive notes he wrote on the *Ethics* (*GP* **1**, pp 139–52), Leibniz commented that, although it did not yet seem certain to him that bodies were substances,[4] the case with minds was different. From his earliest youth he had believed in a plurality of substances, taking for the subject of his first dissertation the demonstration of the principle of individuation. On this issue Spinoza was easily refuted; his definition of substance was obscure and his reasoning unsound. Leibniz repeated these remarks in a letter of 10 December 1679 (*GM* **2**, p 34) in which he asked Huygens whether he had read the works of Spinoza, adding that Spinoza had not really defined a substance.

Another of Spinoza's propositions rejected by Leibniz was the assertion that existence necessarily belongs to substance. For not everything that can be conceived, he argued, can be produced, since it may be incompatible with more important things. In other words, using an expression introduced earlier, not all possibles can exist but only those which are compossible. To establish existence, Leibniz in general appeals to experience, though he still believes that the necessary existence of God can be demonstrated. But in Leibniz's view Spinoza failed here also. For he made no attempt to demonstrate that the concept of God is possible. As Leibniz points out, having defined God as a substance which consists of infinite attributes, he should have demonstrated that they are compatible. Concerning Spinoza's proposition that the world could not have been produced by God in any other way than it has been produced, Leibniz agreed that, on the hypothesis that God chooses the best, only this world could have been produced, but if the nature of the world is considered in itself, a different world could have

[4]That is, real (metaphysical) unities.

been produced. In opposition to Spinoza's necessitarianism Leibniz thus sets his belief in the determination of existence by God's free choice of the best among the compossibles.

A corollary to one of Spinoza's propositions provided for Leibniz the occasion for the development of one of his own philosophical theories, an epistemology founded on symbolic representation. Spinoza had claimed particular things to be modes or attributes of God, which express these attributes in a particular way. According to Leibniz, particular things in themselves were not such modes, but the ways of conceiving particular things corresponded to the ways of conceiving the divine attributes. This principle is developed further in a short paper, *Quid sit idea* (*What is an idea?*) (*GP* **7**, pp 263–4), which he wrote in response to a definition given by Spinoza in the section of the *Ethics* concerned with the nature and origin of the mind. An idea, Leibniz supposed, was a faculty of thinking (or calling to mind) which leads the mind to the thing and also expresses it. The relations in the symbol corresponded to the relations in the thing expressed. This means, as Leibniz explains, that God, the creator alike of things and the mind, has impressed upon the mind a power of thinking so that by its own operations it can derive what corresponds perfectly to the nature of things. For example, the idea of a circle expresses the circle in the sense that truths can be derived from the idea which investigation of the circle itself would confirm.

The library of Martin Fogel

Leibniz heard about the impending sale of the library of Martin Fogel, who had died in 1675, from two former students of Joachim Jungius in Hamburg: Vincent Placcius, who wrote on 23 April 1678 (*A* II **1**, pp 407–9) telling him that the catalogue was in the press, and Heinrich Siver, writing on 16 June (*A* II **1**, pp 415–17) to inform him that the sale would take place in the autumn. Having received the sale catalogue from Placcius (*A* I **2**, p 71) and shown it to the Duke (*A* I **2**, p 55), to whom he described the library as small but one of the most select in Germany, Leibniz set out for Hamburg early in July with the Duke's commission to acquire the collection. Negotiations were completed by the middle of August when Leibniz purchased the 3600 books for 2000 taler. To the Duke he explained that he had to bargain for this price (*A* I **2**, pp 60, 70–1) but the new acquisitions would complement the Duke's existing collection, especially in natural sciences, which were important for medicine and economics.

When he left Hamburg on 2 September, Leibniz borrowed 86 of Fogel's manuscripts, which he later returned (*A* I **3**, p 391). Expenses in Hamburg and on the journey had been high, especially as he needed to have a servant and secretary with him, so, on his arrival in Hanover again, he requested from the Duke an extra 5 taler per week to cover these expenses (*A* I **2**, p 68).

The visit to Hamburg provided an opportunity for Leibniz to meet several former students of Jungius, such as Siver, Placcius and Johann Vagetius, and also to inspect the unpublished manuscripts of Jungius, whom he had long admired (Kangro 1969). In his letter to Thomasius of 30 April 1669, Leibniz had mentioned Jungius as a defender of the logic of Aristotle against that of the Scholastics. When in August of that year, Leibniz visited Bad Schwalbach with Boineburg, the jurist Erich Mauritius, to whom it will be recalled he was indebted for information concerning the experiments of Huygens, Wren and Wallis on elastic collision, told him that Jungius was the author of the Rosicrucian manifesto, the *Fama*. It seems likely that in the course of conversation Mauritius also mentioned the name of Martin Fogel, another student of Jungius in Hamburg, from whom an edition of Jungius's posthumous works might be expected. For in the preface to his edition of Nizolius, printed in the following spring, Leibniz expressed the hope that Jungius's works would soon be edited by the enlightened Fogel. Then at the beginning of 1671 he had written to Fogel enquiring about the publication of Jungius's works, mentioning especially that on the species of insects (*A* II **1**, pp 77–8). Nothing came of the repeated entreaties to his friends in Hamburg over a number of years for the publication of Jungius's works and in 1691 a significant part of this treasure, as he described it, was irretrievably lost in a fire at the house of Vagetius.[5] At the time of his visit to Hamburg, when Siver told him that it was not known whether Jungius was the author of the *Fama,* Leibniz published a piece in the *Journal des Sçavans* (Ravier 1937, p 93) in which he described Jungius as one of the greatest mathematicians and philosophers of his time and one of the ablest men Germany had ever possessed.

Besides the meetings with the students of Jungius and those with chemists which have already been described, Leibniz also had several meetings during his stay in Hamburg with Christian Philipp, the emissary of the Elector of Saxony. Together they visited the Swedish diplomat Esaias von Pufendorf, when the subject of conversation was the philosophy of Descartes. A year later, Philipp recalled the conversation and asked Leibniz to refresh his memory with the details. In reply, Leibniz summarised the argument as follows (*A* II **1**, p 495). In opposition to Descartes, he believed: (i) that the laws of mechanics, which are the foundation of the whole system, depend on final causes, that is to say, on God's will determined to make the most perfect; (ii) that matter does not take all possible forms but only the most perfect. He then added that, if Descartes had given greater attention to experiment and less to his imaginary hypotheses, his physics would have been more worthy of following. Moreover, there were several errors in his metaphysics. He had not known the veritable source of truths nor the general

[5]The 25 000 sheets of manuscript which survived the fire have been little studied. The main collection is now in the Staats- und Universitätsbibliothek, Hamburg.

analysis of notions, which, in Leibniz's view, Jungius had understood better. Finally, Leibniz remarked that the reading of Descartes was nevertheless very useful, for his philosophy was preferable to those of other schools and should be regarded as the antechamber to the true. In another place (*GP* **7**, p 186), Leibniz expressed the view that, if Jungius had been better known and supported, he might have taken the reform of the sciences further than had Descartes.

Memoranda for the Duke

On his return from Hamburg, Leibniz composed three memoranda for the Duke, setting down his thoughts on ways of improving public administration, the organisation of archives and the practice of farming and agriculture (*A* I **2**, pp 74–9).

Leibniz evidently believed that, for the conduct of affairs in such a way as to promote the harmony and prosperity of the people, the administrators and rulers needed to have for their information a comprehensive survey of the present state of the economy. If in addition similar surveys could be produced for 1618 and 1648 (which would require historical research), these would be valuable in indicating the long term trends over two generations. Among the aspects to be covered in the survey Leibniz included natural resources, such as materials, woods and streams, as well as human resources in terms of population, craftsmen and merchants. The balance of imports and exports and the level of the money supply he regarded as especially important.

To promote trade, Leibniz suggested the setting up of an Academy of Commerce and Languages, so that the young could be instructed in the current practices and acquire the necessary skills. This idea was inspired by the existence of such an Academy in Turin. He also recommended the establishment of a Bureau of Information from which people throughout the land could find particulars of goods needed and for sale, things for hire and what there was to learn and see. The Bureau, he added, could also publish a magazine. In addition, he made a few proposals for social reform, such as the provision of Department Stores, where everything could be bought cheaply, and a contributory pension scheme for widows and orphans. Among other proposals may be mentioned the adoption of the English practice of entering the cause of death in the official register.

The most extraordinary proposal in the first memorandum provided for the setting up of L'Ordre de la Charité, Societas Theophilorum, which would start where the Jesuits left off, studying the secrets of nature and giving free medical treatment to the poor, instructing the youth in classical studies, especially mystical theology and chemistry. Scholastic theology and philosophy would be left to the Jesuits. Leibniz's idea has almost a

Neoplatonic or Hermetic flavour. Such an order he envisaged spreading throughout the world, but in accordance with the ideal of religious harmony it would maintain good relations with the Jesuits and other orders.[6]

In his second memorandum Leibniz explained the need for the establishment of an organisation for Archives, consisting of departments and secretaries working under the general guidance of a director. This would serve the same role for the state that his new indexes served for the library, namely to make information easily available. In the Chancellery, for example, one hardly knew which office to consult concerning particular problems and the secretaries were often unwilling to search for material they believed to be in another section. To remedy this state of affairs, the new organisation would compile a number of handbooks for immediate reference. One such volume, containing the regulations of all committees and offices, as well as the important forestry regulations, should lie on the table of the Chancellery, like the Brunswick legal code (*Corpus juris Brunsvicense*). Other volumes would contain a general index of all correspondence and papers in the Chancellery, legal proceedings and things like settlements and privileges granted to communities such as towns, families, crafts and professions.

In the third memorandum, Leibniz noted the need for incentives to promote good farming and pointed to the importance of the cultural and social aspects, such as rural music and dancing and the introduction into the country of a good beer.

Leibniz asked the Duke to treat his memoranda and proposals as confidential but entreated him to make decisions concerning them (*A* I **2**, pp 79–89). All new proposals, he remarked, were naturally suspect and he always hesitated before making suggestions because there was always a danger that a poor initial presentation could effectively prevent a good idea from ever again being taken seriously. It was clear that the proposals for reform should be based on detailed and accurate knowledge of all the facts but access to such information was often lacking. This was just one of the problems that would be solved by appointing a Director of Archives, and although Leibniz remarked with mock modesty that he did not doubt there were others more qualified than himself to occupy that position, there was no-one, he assured the Duke, who would carry out the duties of the office more conscientiously.

The principal proposal Leibniz had in mind when writing his letter to the Duke was in fact an entirely new project not mentioned in the memoranda. This proposal, he remarked rather mysteriously, could eventually lead to the setting up of an Academy of Sciences that would surpass those of London and Paris in promoting useful applications of science besides satisfying

[6]Frances A Yates (1972, p 154) has remarked that the rules for the society are practically a quotation from the Rosicrucian *Fama*.

simple curiosity. The proposal itself, however, concerned a scheme to increase dramatically the output of the mines in the Harz mountains by using his new invention for draining the water. To dispel any fear that the more intensive mining would deprive future generations of resources, Leibniz reminded the Duke of the general consensus of geologists, that the minerals rise and fall, so that, in a hundred years' time, it was likely that nothing would remain in the Harz, even if all mining were to cease immediately.[7] In return for the large profits that would accrue to the state from the use of his machinery, Leibniz would naturally expect some recompense. It was the most important invention he knew, Leibniz confided to the Duke, and boldly expressed the wish that the Duke (perhaps as a mark of esteem) would henceforth take him a little more into his confidence in serious matters, showing him more favour in this respect than he was wont to grant to others.

The Harz project

Leibniz had already planned a journey to the Harz mountains in May 1678, before the visit to Hamburg. He had received from Paris letters touching the smelting of iron and suggested to the Duke that samples of ore should be sent to Paris in order to test the claims that the French were making. Now, to gain first-hand experience himself, as he explained to the Duke (*A* I **2**, pp 53–4) he wished to visit the mines in the company of Christoph Pratisius, a physician who understood metals and was leaving in a few days. If the Duke would grant his request, Leibniz added, he would also like to make excursions to see the library in Wolfenbüttel[8] (which would be of help in connection with his own office in Hanover) and perhaps to Hamburg in order to make a survey of certain books and letters. Although nothing came of this plan of a journey to the Harz, Leibniz had an opportunity to visit Hamburg soon after, when he received the Duke's commission to purchase the library of Martin Fogel.

The first general description of his project for draining the mines in the Harz was communicated by Leibniz to the Duke in the letter accompanying the memoranda which he wrote following his return from Hamburg. It seems likely that Leibniz was more informative in his personal conversations with the Duke than in his letters and memoranda. For, in the written documents, he emphasises the practical effects of the invention while withholding as far as possible the details of the operation, presumably in order to guard his valuable secret from anyone who might have access to the correspondence in the Chancellery. The details emerge gradually, as the

[7]Minerals do indeed rise and fall but only in millions of years.

[8]This famous library, containing a fine collection of books on the Reformation, was the creation of Duke August the Younger, who founded the new House of Brunswick-Wolfenbüttel in 1635. At his death in 1666, the library held 118 000 volumes (Hohnstein 1908, p 361).

negotiations proceed first with the Duke and then with the mining office in Clausthal. The general idea of Leibniz's plan was to use windmills in addition to water power for the operation of the pumps. The power generated by the windmills would be used where possible, supplemented by water power where necessary. Not only would sufficient power for the pumps be available in all seasons but there would also be an adequate surplus of water power for the operation of the stamp-mills (*A* I **2**, pp 195–6).

When he first mentioned the project to the Duke, Leibniz explained that his invention would produce a continuous stream, capable of acting in winter as in summer, that would provide power for the elevation of the water and other mining operations. Moreover, his pumps would be more efficient than those in general use, gaining three-quarters of the power usually lost through friction and needing little maintenance. Soon after Leibniz made his proposal, it appears that an old plan came up for consideration again. This involved a complete circuit of the water supplying power, first flowing from storage reservoirs to work the machinery in mine after mine and then being restored to these reservoirs again by means of pumps. In a memorandum of 9 December 1678 (*A* I **2**, pp 99–103), Leibniz dismissed this idea as worthless—especially on account of the loss of water by evaporation and other causes—and again pointed to the advantages of his own scheme. From the memorandum it may be inferred that one of the objections that had been advanced against his own plan was the expense of the new pumps that would be involved. In response, Leibniz remarked that his scheme could operate with the existing pumps, though not as well, and that his new pumps would not in any case be excessively expensive. At the end of the memorandum he summarised the principles of his own plan. Essentially these involved, first, the introduction of a new type of windmill that would work not only in all winds but in less wind and with more force than all others (Gerland 1906, pp 181–8), and second, the employment of pumps made incomparably more effective than those in operation up to that time by the elimination of friction (Gerland 1906, pp 159–69).

Early in the new year Leibniz pointed to two things which the Duke (*A* I **2**, pp 126–30) would himself have recognised from the model he had shown of the scheme for draining the mines; namely, that the cost of the plan would be as low as possible and the effect as great as could be desired. By excluding cogwheels in his design for a new windmill, he had virtually eliminated friction and resistance altogether. Again he had eliminated friction from his pumps, which could easily raise water from a depth of a thousand feet. If there were something in mechanics, he remarked to the Duke, that merited being made a secret (on account of its novelty and its potential for profitable application), this was it. The plan made maximum use of the forces of nature, namely the flow of wind and water, for these were presented directly to the object and friction was eliminated, so that nothing was lost. Leibniz was confident that the forces of water and wind employed to act together or alternately in the most advantageous way (that is to say, his own) were capable of keeping the mines clear. Part of the profits from the enterprise, he suggested, should be used for promoting research related to mining.

When the Duke gave his agreement in principle, Leibniz wanted to be quite certain that he was to be the sole director in charge of the machines for draining the mines, with a free hand to deploy them as he thought best (*A* I **2**, pp 130–2). In a letter of 8 April 1679 (*A* I **2**, pp 153–61), in which he thanked the Duke for his support concerning the Harz project, Leibniz revealed that the idea had been in his mind for some time. Indeed, he had formulated the design in Paris, when he had been called to the Duke's service, for he had recognised the importance of the problem and hoped he would have the opportunity to apply his knowledge usefully in finding a solution for the benefit of the state. Having indicated to the Duke the long and careful thought he had given to the Harz project, the moment may have seemed opportune to Leibniz for him to elaborate on what he considered to be the most important project of all and to suggest how the one could be promoted from some of the profits of the other. For he now explained that, when he suggested using some of the profits from the Harz project for research, he had in mind not just mining research but the setting up of the Academy of Sciences he had often mentioned, that would produce the universal characteristic. This Academy, he added, should be a permanent institution, continuing after his own death. For the potential of this Academy he made great claims; while the Academies of London and Paris only produced particular discoveries, his universal characteristic, when achieved, would be an organ or instrument as powerful to reason as the microscope was to the eye.

At the beginning of August (*A* I **2**, pp 188–9) Leibniz expressed to the Duke a wish to visit the Harz incognito, in order to choose a carpenter to do the building. In fact he set out on his first visit to the Harz about the middle of September and at the end of the month concluded an agreement with the Mining Office in Clausthal (*A* I **2**, pp 200–3). Following a trial period of one year, during which he would undertake to drain one mine at his own expense, the inventor would be paid 1200 taler annually for life. Then, on 2 October, Kahm wrote on behalf of the Duke, recalling Leibniz to Hanover for consultations (*A* I **2**, p 205).

Objections had been made behind the scenes and the whole project had been called into question. Leibniz only became aware of these difficulties in the course of the consultations in Hanover. A major problem concerned the expense of the additional machines that would be needed to refine the increased output of ore. This, Leibniz told the Duke, was a separate problem that had nothing to do with him, since he had only accepted responsibility for the work underground. In any case, new refining machinery would have been needed even if no-one had ever heard of his invention. Again, it had been put to him after an audience with the Duke that it would be better to communicate the invention to some experts in mechanics for their judgment. To the Duke Leibniz remarked (*A* I **2**, pp 207–12) that he had reason to be surprised by such a suggestion, which would change the face of the whole affair. In any case, he had little confidence in such experts and he insisted that the year's trial would provide a sufficient test.

Having demolished the objections in several letters written about the middle of October, immediately following the consultations in Hanover (*A* I **2**, pp 206–17), Leibniz was rewarded by the Duke's decision in his favour, formalised in the ratification (*A* I **2**, pp 218–21) of the agreement with the Mining Office on 25 October 1679.

Elizabeth and Malebranche

Soon after his visit to Hamburg and the preparation for the Duke of his memoranda on the Harz project and other topics, Leibniz had an opportunity to take up again his criticism of the philosophy of Descartes, which had been the subject of his first encounters in Hanover with Molanus and Eckhard. The occasion for this new attack was a meeting with the Palsgravine Elizabeth, Abbess of Herford, during her visit to Hanover in the winter of 1678, when she introduced Leibniz to the *Conversations Chrestiennes* of Malebranche (*A* II **1**, p 455). Having made extensive extracts with his own critical comments (*A* II **1**, pp 442–54)—again there is an allusion to the existence of the unconscious—Leibniz communicated his thoughts on Cartesianism to the Princess (*A* II **1**, pp 433–8).

Concerning the proof of the existence of God from his essence, Leibniz explains that God has an advantage over other things in this respect, that it suffices to prove that he is possible in order to prove that he exists. Also there is a presumption of possibility (in the absence of a proof of impossibility), which is sufficient for practical life but is not enough for a demonstration. The Cartesians with whom he had discussed this, Leibniz tells the Princess, had not succeeded in demonstrating the possibility of God. His further remarks provide a significant clarification of his own position, and they also emphasise the fundamental importance he attached to his favourite project of the universal characteristic. The Cartesians, he explained, could not succeed because they lacked an adequate concept of an individual substance. In order to demonstrate that a concept is free from contradiction, and therefore possible, it is first necessary to have a logically sound definition of the concept. Leibniz believed that he had such a definition in the idea of a concept as a combination of simple (that is, indefinable) forms or qualities. Such a concept of God, as a being possessing all perfections, was the basis of the demonstration Leibniz had written out for Spinoza during his visit to The Hague. In his letter to the Princess, Leibniz stated explicitly that the basis of his own demonstration—that is, the compatibility of simple forms—was the foundation of his universal characteristic, for the simple forms were the elements of this characteristic. He excused himself from giving his demonstration with the remark that this would require a lengthy explanation of the foundations of the characteristic.

When Leibniz wrote to Malebranche on 23 January 1679 (*A* II **1**, pp 455–6) telling him that Elizabeth had shown him the *Conversations Chrestiennes,* Malebranche denied that he was the author of the work, remarking that it was written by his student, the Abbé Catelan (*A* II **1**, pp 467–9). In his letter to Malebranche, Leibniz was very critical of Descartes, saying that he was very far from true analysis and the art of discovery in general. His mechanics was full of errors, his physics specious, his geometry limited and his metaphysics all of these things together. Malebranche had himself improved on Descartes, Leibniz suggested, but he had not gone far enough. For example, Malebranche's view that it was impossible for a substance possessing only extension to interact with a substance possessing only thought was undoubtedly correct, but in Leibniz's view, matter was something other than extension, as he could demonstrate. Leibniz was pleased to agree with Malebranche that God acts in the most perfect manner possible and he approved of the good use Malebranche made of final causes, having had a poor opinion of Descartes for rejecting them. Assuring Malebranche that, despite their fundamental disagreements, he had read his writings with interest and profit, he asked him to give his regards to Arnauld when the opportunity arose.

Malebranche replied in March 1679 (*A* II **1**, pp 467–8) giving Leibniz news of Paris and requesting from him the reasons for his opposition to Descartes. In his response, written on 2 July 1679 (*A* II **1**, pp 472–80), Leibniz listed a number of Cartesian propositions in philosophy, concerning which he invited Malebranche to dissipate his doubts, and other propositions in natural science, which seemed to be contradicted by the evidence. Also, he repeated the comments he had previously made to Eckhard concerning the failure of the Cartesians to make advances, despite their possession of a wonderful method bequeathed to them by their master. Finally, he explained to Malebranche the faults of Descartes' attempted demonstration of the existence of God.

The universal characteristic and related topics

The wonderful idea of an alphabet of human thought, which Leibniz hit upon in his schooldays, was the seed from which the concept of a universal characteristic grew. In the same way that words (representing sounds) were built up from letters (representing simple sounds), Leibniz believed that complex ideas were formed by combinations of a small number of simple ideas. This led him to the idea of a universal writing or language (a real characteristic) in which ideas would be represented by combinations of signs corresponding to their component parts. A language of this kind would provide a direct representation of ideas, speaking to the understanding rather than the eye. If the language also had a grammar, consisting of a set of rules for the combinations of the signs, then reasoning and demonstration could be carried out formally in a manner analogous to the calculations of arithmetic and algebra. Such was the idea of the universal characteristic. Essentially it was an instrument of discovery, which Leibniz had made a brilliant attempt to realise in his youthful essay, *De arte combinatoria.* Neither Lull nor others who had tried to reform philosophy, he maintained, had even dreamed of the true analysis of human thought (*GP* **7**, p 293).

The general principles of the universal characteristic were clearly described by Leibniz in a letter to Oldenburg (*GP* **7**, pp 11–15), written while he was still in Paris, and in a letter to Gallois (*GP* **7**, pp 22–3) written in December 1678, when he was heavily involved with the plans for the Harz project. In both letters, he mentioned the Chinese writing as an example of a real characteristic.[9]

Soon after he had proposed to the Duke the setting up of an Academy to achieve the universal characteristic, Leibniz had an opportunity to learn more about the Chinese script and language; for the physician to the Brandenburg Elector in Berlin, Johann Sigismund Elsholz, to whom Crafft had recommended Leibniz on his return journey to Dresden (*A* I **2**, pp 378–9) sent him a work of Andreas Müller, provost of the Nicolai Church in Berlin and adviser to the Elector of Brandenburg on Chinese affairs, from which he learned that Müller had written an unpublished *Clavis Sinica* (*Key to Chinese*) (*A* I **2**, p 462). In his reply to Elsholz, written on 4 July 1679 (*A* I **2**, pp 491–3), Leibniz posed fourteen questions for Müller, such as whether the script revealed the nature of things and whether Müller believed that the Chinese did not know the key to their script. In another letter, written on 15 August (*A* I **2**, pp 508–9), Leibniz asked if Müller would explain either in whole or in part a Chinese book he had of eighty pages, giving the meaning and pronunciation in Latin letters. Müller agreed (*A* I **2**, p 516) to translate the book and give the pronunciation but only if he found it worth doing. Before he could decide, he would need to know the title and contents, so Leibniz would have to send either the book or the title. It turned out that Leibniz's book had already been translated by Prosper Intorcetta (*A* I **2**, p 517).

At first Leibniz conceived his characteristic as a universal language, for which he claimed a great advantage over other projects of this kind (consisting essentially of some form of code), in that it would not only

[9]Concerning the influence of Chinese writing on Leibniz's plans, see Widmaier (1981).

present ideas directly to the understanding, like the Egyptian hieroglyphics and Chinese writing, but would also permit reasoning by a process analogous to arithmetical or algebraic calculation. In February 1678, he composed an outline for such a universal language, *Lingua generalis* (Couturat 1903, pp 277–9), based on the representation of simple ideas by prime numbers and the concepts formed from their combinations (that is, all possible concepts) by the products of the corresponding primes. In order to understand the language, the reader would need to learn a list of the essential simple ideas represented by the primes and have the ability to resolve large numbers into prime factors at sight. To convert the numerical expressions into a spoken language, he used an idea of George Dalgarno, representing 1, 10, 100, . . . by the vowels *a, e, i,* . . . (prolonging the series using diphthongs if necessary) and the numbers 1 to 9 by the first nine consonants. Thus 546 is represented by *gifeha,* or any permutation of the syllables, since the number is independent of their order; a property, Leibniz remarks, that would permit the writing of poetry in the artificial language.

Recognising that the problem was less easy than he had at first supposed, Leibniz quickly abandoned this arbitrary scheme for one based on a living language; in fact Latin, which was the international language of the scholars of the time. The new idea appears first in a fragment of April 1678 (Couturat 1903, pp 280–1) and a plan outlined in a manuscript, *Analysis linguarum,* written in September of the same year (Couturat 1903, pp 351–4). Although the goal is the analysis of concepts, this can be aided or even replaced by the analysis of language. This involves, first, resolving all the terms of discourse, by means of their definitions, into simple, irreducible terms, and, second, reduction of the grammar and syntax to their essential elements.

It was now clear to Leibniz that, in order to discover the alphabet of human thought and realise the universal characteristic, it would be necessary to analyse all concepts and reduce them to simple elements by means of definitions, then to represent the simple concepts by appropriate symbols and invent symbols for their combinations and relations, and finally, since the analysis of concepts is at the same time the analysis of truths, to demonstrate all known truths by reducing them to simple, evident principles. In other words, before the universal characteristic could be achieved, it would be necessary to construct an *Encyclopedia,* in which all existing knowledge was classified, analysed and demonstrated. For this gigantic task, Leibniz would need collaborators, organised in an Academy, such as the one he had proposed to the Duke.

Even before his days in Paris, Leibniz had considered the idea of producing an *Encyclopedia.* First he planned a compilation from the works of other authors, for which a table of contents exists (*GP* 7, pp 37–8), and later he had in mind the correction and completion of the *Encyclopedia* that Johann Heinrich Alsted had published in 1620.

Under the pseudonym Pacidius (connoting, as already mentioned, a conciliator who would unite all the scholars in a common task), Leibniz planned to publish an introduction to the proposed *Encyclopedia,* with the title *Initia et specimina scientiae generalis.* The *initia* are the principles of the general method (*scientia generalis*) that will lead to the elaboration of the *Encyclopedia,* and the *specimina* are examples to show the application of the method to particular sciences. The table of contents of the *Encyclopedia,* given in this work (*GP* **7**, pp 49–51), dates from the middle of 1679 and is a revised version of a somewhat earlier table (Couturat 1903, pp 129–33).

Leibniz's *scientia generalis* or universal method applicable to all the sciences consists of his logic in the widest sense. The theory of definition occupies a central role, which he explains to Conring (*A* II **1**, pp 385–9, 397–402) and Tschirnhaus (*GM* **4**, pp 451–63) in letters written in the early part of 1678. First, the analysis of concepts into simple elements proceeds by definition. Secondly, although the analysis of truths consists in demonstration, this analysis also rests on definition. For demonstration is effected by decomposition of the terms of the proposition, so that the analysis of truths reduces to analysis of concepts; that is, to definition. Demonstration also has, of course, a synthetic aspect involving the combination of definitions which leads to new theorems. The single *a priori* principle admitted by Leibniz in his theory of definition is the principle of identity. All truths, he holds, are reducible to definitions, identical propositions or empirical propositions. Truths of reason, which are independent of experience, can always be reduced to definitions or identical propositions.

In reply to an objection of Conring that there are indemonstrable propositions besides identical propositions, namely axioms, Leibniz explains that, for convenience and progress in science, axioms and postulates may be admitted without demonstration, but all the axioms thus admitted should be demonstrable. Although, for example, the axioms and postulates of mathematics are not reducible to definitions, they can be demonstrated by the principle of contradiction (which is equivalent to the principle of identity). In other words, they are necessary because the contrary would imply a contradiction. It follows that, for Leibniz, the truth of the propositions demonstrated is not nominal and subjective, as for Hobbes, but real and objective. A definition is only real, Leibniz explains to Tschirnhaus, if it manifests the possibility or existence of the object. Moreover, nothing can be deduced with certainty from a definition unless it is known that the object defined is possible.

Besides the general plan of the *Initia et specimina scientiae generalis,* Leibniz produced in 1679 studies for two of the examples of application of the method that would form the second part. These were attempts to formulate a logical calculus and a geometry of situation, which may be seen as contributions to the realisation of the universal characteristic.

The first attempt at a logical calculus (Couturat 1903, pp 41–9) is based on the analogy of the combination of concepts to arithmetical multiplication and the decomposition to resolution into prime factors. For Leibniz, the analogy is exact, since he takes the combination of concepts to be commutative. For example, he holds the definition of 'man' as a 'rational animal (being)' and an 'animal rational (being)' to be equivalent, the apparent distinction being purely verbal.

In a manuscript of April 1679 entitled *Elementa calculi* (Couturat 1903, pp 49–57), Leibniz explains that the way in which he regards the relation of terms is the opposite of that employed by the Scholastics. While the Scholastics suppose the species to be contained in the genus, Leibniz holds the genus to be contained in the species. For example, he interprets the proposition 'All gold is metal' to mean that the concept of metal is contained in the concept of gold, for gold contains all the properties of metal and also other properties, such as heaviest of metals. In other words, the 'species' gold contains the 'genus' metal. The Scholastics speak otherwise, he explains, because they are not considering concepts but instances subsumed under universal concepts. Today the difference would be expressed by saying that the Scholastics interpret propositions in terms of extension of classes, while Leibniz interprets them in terms of intension of qualities. He recognises that, by an inversion of his own calculus, all the laws of logic could be demonstrated from the standpoint of the Scholastics, but he prefers to consider universal concepts and their composition because these do not depend on the existence of individuals. In other words, his primary interest is in truths of reason rather than truths of fact.

Since a universal affirmative proposition, 'All S is P', is true if the subject contains the predicate (and hence if the number representing the subject is divisible by that representing the predicate), Leibniz expresses this proposition as $S=Py$, where y is a suitable number. Evidently Py is a species of P, so that $S=Py$ means that S is identical to a species of P. Similarly, the particular affirmative, 'Some S is P', can be written $Sx=Py$, but since x can always be chosen so that it is divisible by the number representing P, this would mean that all particular affirmative propositions (for example, some men are stones) would be true, which is absurd. Another defect of this system is that negative propositions (for example, no man is stone) would be excluded (Couturat 1903, pp 329–30).

In order to accommodate negative propositions, Leibniz introduced negative numbers to represent non-predicates. Each term is therefore represented by two numbers, one positive and one negative, which are co-prime. If there were a common factor, this would enter positively and negatively, so that the term would contain contradictory elements and would therefore itself be contradictory. Care has to be taken in assigning numbers to concepts to ensure that the numbers representing their combinations are also co-prime. For example, taking animal $= +13 \; -5$ and rational $= +8 \; -7$, then man (rational animal) $= +104 \; -35$. But if we take rational $= +10 \; -7$, then man = +130 −35, which contains a contradiction, since 130 and 35 have a common factor.

A universal negative proposition will be true if two numbers of opposite sign appertaining to the two terms have a common factor. For example, taking man = +10 −3 and unhappy = +5 −14, the proposition 'No man is unhappy' is demonstrated, because 10 and 14 have a common factor 2, signifying that the terms 'man' and 'unhappy' are incompatible. Now the equivalent universal affirmative proposition is 'All men are happy'. But man = +10 −3 and happy = +14 −5 implies that the proposition is false since 14 is not a divisor of 10 and 5 is not a divisor of 3. Thus the scheme leads to a contradiction and is therefore not valid.

Leibniz mentioned the project of a geometry of situation in a letter to Gallois at the end of 1678 (*GM* **1**, pp 182–90). He had reached the conclusion that the application of algebra was not the natural way to solve geometrical problems and he had found a better characteristic for geometry. This project of a geometrical characteristic he developed in two essays, *Characteristica geometrica* (*GM* **5**, pp 141–71), dated 10(20) August 1679, and the undated *De analysi situs* (*GM* **5**, pp 178–83), which probably belongs to 1693. Besides these essays on the geometrical characteristic, Leibniz also contributed early in 1679 another study of the foundations of geometry, *Demonstratio axiomatum Euclidis* (Couturat 1903, p 539) in which he demonstrated the axioms of Euclid from two definitions of his own.

By the beginning of September 1679, Leibniz was ready to communicate some details of his new project; for, in the packet to Huygens containing the specimen of phosphorus already mentioned, he also enclosed an essay based on the *Characteristica geometrica,* asking for Huygens' opinion of it. No doubt he had reason to expect that his new work, helped by the curiosity value of the phosphorus, would make a good impression on the circle of members of the Paris Academy of Sciences.

In his letter to Huygens, Leibniz expressed his dissatisfaction with the application of algebra to geometry, in that it gave neither the shortest ways nor the best constructions in geometry. He neatly illustrated this point in an appendix to the *Characteristica geometrica* by comparing an algebraic construction of a triangle on a given base and of a given height and apex angle with a geometrical construction based on intersection of loci. Another fault of algebraic geometry was that it presupposed the theorems of elementary geometry, such as those on similarity and the theorem of Pythagoras, and so did not really rest on primary axioms. Mathematical analysis as generally understood was concerned with analysis of magnitude and hence, in Leibniz's view, was only indirectly applicable to geometry, in which science the fundamental concept was that of situation rather than magnitude. His geometrical characteristic was therefore designed to express directly situation, angles and movements by symbols. This new calculus, Leibniz claimed, would not only allow solutions, constructions and geometrical demonstration to be achieved in a way that was both natural and analytical but would also have applications hitherto unknown in the invention of machines and the description of the mechanism of nature as well as geometry.

Leibniz took similarity and congruence to be the basic relations between figures in his new geometry. Similar figures were defined as figures which are indiscernible when viewed separately. In *De analysi situs,* he showed how this definition could be used to demonstrate simply theorems that Euclid had only proved in a roundabout way. For example, the theorem that circles are to each other as the squares of their diameters follows immediately from the definition of similarity. For otherwise, there would be a difference in the relation between the square and the circle in the two cases (figure 4.1) that would enable them to be identified when viewed separately; in other words they would be discernible in form. Euclid demonstrated this theorem, Leibniz noted, only in Book X of the *Elements* and then by using inscribed and circumscribed figures and *reductio ad absurdum*.

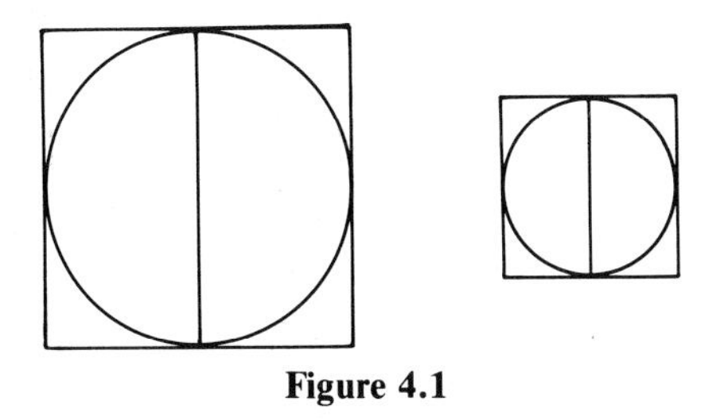

Figure 4.1

Having recognised that, with the designation of the points of a figure, many of its properties could be made clear, the problem for Leibniz was to find a suitable situation characteristic. This characteristic should enable true definitions of the elements of the geometry to be formulated and everything that Euclid had accomplished with his definitions, axioms and postulates would have to be derived before the calculus could in a proper sense be applied to the solution of geometrical problems. Euclid's *Data* provided him with the inspiration. For the trend of this work seemed to be a kind of analysis concerning situation, dealing with data and the positions of unknown entities or their loci. In the essay sent to Huygens, Leibniz took congruence to be the basic relation and set up symbolic expressions for loci. Fixed points were represented by letters at the beginning of the alphabet, variable points by letters at the end and congruence by ♉ . Thus A♉Y represented the locus of a point congruent with A; that is, the whole of space, since all points were taken to be congruent. As another example, consider the locus ABC♉ABY (figure 4.2). Clearly this is a circle.

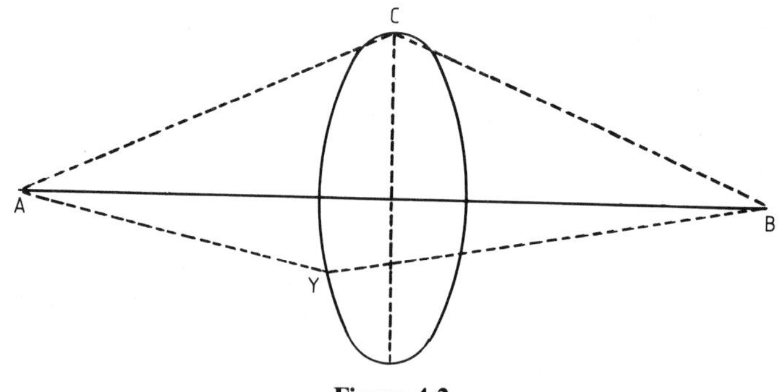

Figure 4.2

As illustrations of the use of the characteristic in reasoning, Leibniz demonstrated the nature of the intersections of some loci. For example, from the expression ABC♉ABY, which represents a circle, it follows that AC♉AY and BC♉BY. But these expressions represent spherical surfaces with centres A and B. Consequently, the intersection of the two spherical surfaces is demonstrated to be a circle.

The greatest defect of the system is that, while the equivalence relation of congruence (Leibniz clearly recognised the properties of symmetry and transitivity), by means of which he defined the loci and which was the only relation he admitted, can determine, for example, a line or a plane, it cannot describe the concepts of line and plane in themselves. In other words, Leibniz's characteristic does not provide definitions of the elements of his geometry, nor does it contain relations of incidence, which are needed to express in symbols such things as the conditions for a given point to lie on a given line (Münzenmayer 1979).

Huygens received Leibniz's essay without enthusiasm and had to be pressed to give an opinion. Leibniz must have been disappointed, for Huygens, from whom he expected a favourable response to what he considered to be a significant achievement, had misunderstood the intentions of the geometrical characteristic. First he remarked that he did not see how the characteristic could be applied to things like quadratures and tangents, showing that he had failed to recognise the complementary natures of the geometrical calculus and infinitesimal analysis. Secondly, he complained that the results given by Leibniz were already very well known, thus missing the point that Leibniz was offering what he considered to be a powerful general method, from which significant results could be expected in the future (*GM* **2**, pp 29–32). Alas, these did not materialise. Many years later, in 1695, Leibniz wrote to the Marquis de L'Hospital that he would not yet venture to publish his project concerning the *characteristica situs,* for unless he could render it plausible by examples of consequence, it would be taken for a phantasy (*GM* **2**, p 258).

The project of the *Catholic Demonstrations* revived

After the difficulties and frustrations concerning the Harz project had been resolved with the ratification by the Duke of his agreement with the Mining Office, Leibniz tried to turn the Duke's attention to another matter and enlist his help in a plan he had for promoting the reunification of the Churches, which they both desired. Reminding the Duke (*A* II **1**, pp 487–9) that he had often discussed controversies of theology with Boineburg, Leibniz recalled their conclusion that the decisions of the Council of Trent[10] could be approved apart from a few passages. An interpretation of these passages could be given, Leibniz explained, that was neither contrary to the words nor the teachings of the Catholic Church, though far removed from the common opinions of certain powerful Scholastic theologians in the Church. At the time, he had agreed with Boineburg, that if a declaration could be obtained from Rome to the effect that these interpretations, which seemed to him to be true, were tolerable and contained nothing heretical or contrary to faith, he would compose an apologetic work, *Demonstrationes Catholicae,* which might serve the cause of reconciliation between Catholics and Protestants. When the Baron died unexpectedly, he did not think the time opportune to enlist the support of Arnauld in Paris, to whom Boineburg had given him a letter of recommendation, especially as he knew that Boineburg had intended to speak to the Duke on the matter, and so the plan had come to a stop.

Earlier in the year, Leibniz had entered into correspondence with Jacques Bénigne Bossuet (*A* I **2**, pp 428–9) whose *Exposition de la foi de l'église catholique* had been approved by the Pope and well received by Duke Johann Friedrich. If he had known that his work would be welcomed in Germany, Bossuet told Leibniz (*A* I **2**, pp 468–9) he would have added some articles for the Lutherans. As Leibniz also believed that the Pope was not only a good man but enlightened and fair minded as well, the time had evidently come to revive his earlier project, if the desired declaration could be obtained from Rome. This would have to be sought with circumspection, in order to avoid misunderstanding. If the request were to come from such a respected source as the Duke himself, Leibniz suggested, there could be no suspicion that it was not prompted by the highest motives.

The time was opportune for Leibniz to make his request, as the Duke was about to make a visit to Italy before withdrawing from the business of government in order to improve his health and follow the more ascetic, religious life he desired.

The *Demonstrationes Catholicae* was to consist of three parts. The first would cover demonstration of the existence of God and natural theology,

[10]The Council of Trent lasted from 1545 to 1563 and was dominated by Spanish Jesuits who led the Counter-Reformation.

while the second would defend revealed theology and the third would expound the relation between Church and State. The truths of revealed theology, Leibniz noted, as truths of fact, could only be established with moral certainty. In addition to the main work, he planned to write a philosophical introduction and in order to render the demonstrations incontestable, he would offer an essay on his universal characteristic or language. The rather optimistic claims he made for his universal characteristic (still only a project in its infancy) were perhaps designed to persuade the Duke to establish the Academy of Sciences he had proposed for its realisation.

Johann Friedrich set out for Italy on 26 December, having sent his wife and the princesses to their relations in Paris. Meanwhile Leibniz had gone to Herford where he visited the Palsgravine Elizabeth, who was then very ill. There he met her sister Sophie and also the physician François Mercure van Helmont, an acquaintance from his days in Mainz, who attended the Princess.

On 4 January 1680, news reached Leibniz of the Duke's death in Augsburg. Abandoning his original plan to journey on to Osnabrück and Paderborn to visit the bishop and his secretary, whom he had met in Hanover, Leibniz immediately returned to Hanover (*A* I **3**, pp 12–13). Again everything was in the balance. For the second time, his plan for the *Demonstrationes Catholicae* had been halted by the death of a patron when it seemed to be on the verge of success, and without the support of the Duke, he would have reason to fear renewed opposition to his Harz project and in consequence a setback to his hopes for the establishment of an Academy of Sciences and the realisation of his universal characteristic.

5
Hanover under Duke Ernst August
(1680–1687)

On his return journey to Hanover, Leibniz composed a letter of condolence for Johann Friedrich's widow Benedicte (*A* I **3**, pp 6–8) and also a poem of consolation for his friend, the Duchess Sophie, who remained in Herford (*A* I **3**, pp 8–11). Her husband, the new Duke Ernst August, who had been awaiting his brother in Venice, hurried to Augsburg to make arrangements for the lying-in-state until the funeral, which took place in Hanover on 1 May. Before the end of January, Leibniz had already received an invitation to become a Counsellor to Count Ahlefeldt in Oldenburg, but he was able to decline this offer on 8 February (*A* I **3**, p 350), having been confirmed in his office by the new Duke.

With his official position at the Court thus secured, Leibniz lost no time in presenting to the new Prime Minister and Grand Marshall Count Franz Ernst von Platen and also to the Duke himself details of his career and the many projects for which he hoped to gain their support. In a memorandum to Platen, he suggested the writing of a short history of the House of Brunswick-Lüneburg[1] and explained his ideas for the extension of the Library, including the provision of a laboratory and a museum, and proposed the establishment of a Ducal Printing Press (*A* I **3**, pp 16–21). To the Duke he suggested that, if the care of the Archives were in some way assimilated to his duties as Librarian, he could help the Duke in the conduct of affairs by preparing summaries and memoranda that would convey the essential information concerning a given item at a glance (*A* I **3**, pp 23–5). At the same time, he informed the Duke of his contract with the Mining Office

[1]At this time, the Dukedom of Brunswick-Lüneburg was divided between the new House of Lüneburg, ruling in Celle and Hanover, and the new House of Brunswick, ruling in Brunswick and Wolfenbüttel.

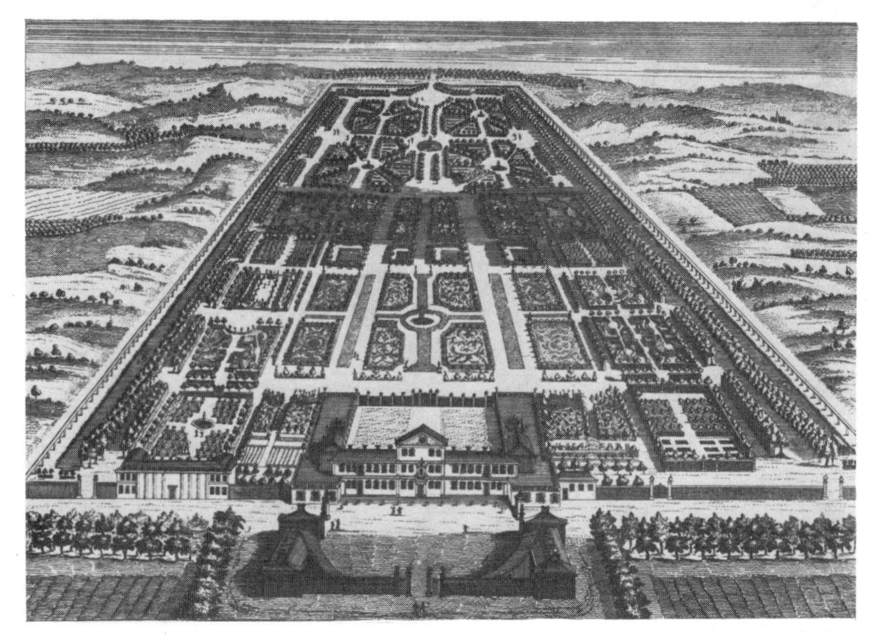

Plate 3 The Palace and Park at Herrenhausen, about 1710. Courtesy of the Historisches Museum, Hanover.

for the draining of the mines in the Harz and asked for his support with the project.

Before taking up residence in Hanover on 13 March, Ernst August had returned to Osnabrück where up to that time he had been the Prince Bishop. Leibniz visited Osnabrück in order to introduce himself in person to the Duke. During his stay, which lasted about a month, he no doubt had several meetings with the Duchess Sophie. Over the following years, Leibniz developed a close friendship with Sophie. When both he and the Duchess were in Hanover, he was a guest in Herrenhausen[2] almost every day and accompanied her on the walks that she liked to take in the park (*K* **7**, p xiv). The closeness of the relationship may be gleaned from Sophie's remark in 1690, that she valued Leibniz's New Year greeting more than those she received from kings and princes (*A* I **5**, p 519) and her confession in 1701, that she wrote only in the hope of receiving his letters (*K* **8**, p 295).Their extensive correspondence, in French, concerned questions of politics, religion and philosophy.

Leibniz found in Sophie a companion receptive to the scholarly subjects of common interest he had previously shared with Duke Johann Friedrich. Duke Ernst August seems to have had little interest in purely academic

[2]The Herrenhausen Palace with its adjoining park was the summer residence. Normally the Duke and his Court resided at the Palace in the Leinstrasse.

Plate 4 Electress Sophie of Hanover. Painting by Andreas Scheits, about 1710. Courtesy of the Historisches Museum, Hanover.

questions of theology and philosophy. He was interested in the reunion of the Churches, for example, but only on a political level, while his primary concern was the enhancement of the power and lustre of his House. In his correspondence with the Duke, Leibniz confines his attention to matters of practical advantage to the State, avoiding fundamental issues of religion and philosophy. There is no mention, for example, of his universal characteristic and the Academy of Sciences that would be needed to achieve it. It may be a reflection of the differing interests of the two Dukes that, in three years, Johann Friedrich spent 4500 taler on the Library, whereas in seven years, Ernst August spent only 700 taler, of which 440 taler was for bills relating to 1679, the year before his accession (*A* I **3**, p xxxiii).

From the beginning, Leibniz's talents were recognised by Duke Ernst August, who evidently promised to assist him in the perfection of some useful inventions. For in thanking the Duke for his generosity in this respect, Leibniz explained that the arsenal, especially if directed by an officer of artillery well versed in mechanics, could provide the workmen needed to construct models and then full-scale prototypes of such things as pumps and windmills (*A* I **3**, pp 30–3). Among other inventions he wished to develop Leibniz mentioned some for the more efficient conveyance of canons and other heavy loads and also his calculating machine, the model of which had been so greatly admired in Paris. About this time Leibniz also outlined a design for a calculating machine to operate the four rules in binary arithmetic, though he recognised that the development of such a machine would not be easy. Owing to the great number of wheels needed, the problems related to friction and smooth movement already encountered with the ordinary calculating machine would be more serious, while the greatest difficulty would be the mechanical conversion of ordinary numbers into binary and the binary answers into ordinary numbers. Perhaps it was on account of these seemingly insuperable obstacles that Leibniz failed to mention the binary calculating machine in his correspondence (Mackensen 1974). Concerning the 'binary progression' itself, he remarked to Tschirnhaus in 1682 that he anticipated from its use discoveries in number theory that other progressions could not reveal (*GM* **4**, pp 491–8).

Following a memorandum to the Duke in which he answered the objections that had been made to his project for draining the mines in the Harz (*A* I **3**, pp 35–45), Leibniz was rewarded on 24 April 1680 by the Duke's decision in his favour (*A* I **3**, pp 47–8). The cost of the test, to be carried out on the Catharina mine with three windmills, was to be divided equally between the Duke, the Mining Office in Clausthal and Leibniz himself. Subject only to the completion of a successful test, the financial provisions of the agreement of 25 October 1679 were confirmed.

For Bishop Ferdinand of Paderborn, Leibniz composed a long Latin poem in commemoration of Duke Johann Friedrich (*A* I **3**, pp 374–84) and also contributed an appreciation in German, which was read out in the

Chapel at the Palace (though without mention of the author's name) on the Sunday following the funeral on 1 May (Guhrauer 1846 **1**, p 367). Leibniz sent copies of his French and Latin poems in memory of Johann Friedrich to his friend Brosseau, the ambassador in Paris, for distribution to people whose names he supplied (*A* I **3**, p 440). The poem composed for the Bishop of Paderborn was later described by Fontenelle as one of the finest accomplishments of neo-Latin poetry (Hankins 1972, p 3).

Besides Brosseau, Justel also continued to write to Leibniz from Paris, sending him in July 1680, for example, data on the magnetic variation at St Helena and the Cape of Good Hope. Christian Philipp continued to write letters frequently from Hamburg and acted as an intermediary in passing letters between Leibniz and Hansen, his former general agent in Paris, who had moved to Oxford. Following several requests from Christian Philipp, Leibniz eventually passed on some information concerning the tumour that had killed the Duke (*A* I **3**, p 387). Among topics of discussion was the value of experiments; agreeing with Leibniz that a good experiment was worth a hundred discourses, Philipp expressed his admiration for the method of Bacon as seen in his *Sylva sylvarum* (*A* I **3**, p 354).

Although Leibniz had been confirmed in his office straight away and had been assured of the goodwill of the new Duke, he still dreamed of a post in Paris that would make him independent of the uncertainties surrounding such changes at the Court. This he confided to Huygens in a letter of 5 February, containing news of the Duke's death, his plans for the draining of the mines in the Harz and remarks on Fermat's demonstration of the law of refraction using the principle of least time (*GM* **2**, pp 36–8). Later in the year he expressed to Johann Lincker in Vienna an interest in the vacant post of Imperial Librarian, but he would only consider accepting an offer if the Emperor combined the Librarianship with the office of Privy Counsellor (*A* I **3**, pp 412–14). Lincker's efforts on his behalf were unsuccessful.

On the day of Johann Friedrich's funeral, Leibniz began a long correspondence with the Landgrave Ernst von Hessen-Rheinfels. The occasion was the Landgrave's request for the return from the Duke's Library of his privately printed and circulated book, *The sincere and discreet Catholic*. Leibniz explained that the book in question was not to be found in the Library, most likely having been in the private possession of the Duke, but that he had himself read the book with interest and approval when he was in Mainz with Boineburg (*A* I **3**, p 243). From the Landgrave, a convert to Catholicism keenly interested in religion and in close contact with the great theologian Antoine Arnauld, who had been so impressed with him in Paris, Leibniz could expect the princely support for his own plans relating to the reunion of the Churches which the death of Johann Friedrich had taken away from him.

Later in the month, he prepared a memorandum for the Duke repeating some earlier proposals and adding the suggestion of an Academy at

Göttingen for the nobility, after the pattern of the one in Turin, and recommending the Dutchman Jacques Ferguson as the teacher of mathematics. Then he asked for expenses to cover the cost of fodder for his horses and a servant on the journey that he would have to make to the Harz, where he expected to stay the whole summer, in order to set up the windmills, remarking that, if the project were established, he would not need to ask for such help (*A* I **3**, pp 61–2). As it turned out, Leibniz made only a brief visit to the Harz before making a journey to Walbeck near Helmstedt, in order to consult the miller who would build the windmills. From there he wrote on 17 June to Friedrich Casimir zu Eltz, the Inspector of Mines, suggesting ways of improving the operation of the ornamental fountains in Walbeck, which he had viewed with admiration, and announcing that he hoped to return to Osteroda in the Harz in about two weeks' time, when the miller would have completed his work in Walbeck and moved to Osteroda (*A* I **3**, p 63). At this point, however, believing perhaps that the arrangements for constructing the windmills were well in hand and could proceed, at least for a time, without his personal supervision, he took the opportunity to make a visit to Leipzig, planned the previous year, in order to attend to family affairs relating to inheritance.

Family affairs

During the visit to his family in Leipzig, Leibniz took the opportunity to meet again his friend Johann Daniel Crafft in Dresden. One outcome of this meeting was a plan for Leibniz to become known to the Emperor through Philipp von Hörnigk and the Margrave Hermann von Baden (*A* I **3**, pp 400–3). It was just at this time that he was writing to Johann Lincker concerning the vacant Imperial Librarianship in Vienna and the plan, if successful, would no doubt have enhanced his chances of attaining the status of Imperial Privy Counsellor to go with the office of Librarian, as he desired. To help Crafft in return for his good offices, Leibniz composed for him a memorandum on the advantage of a textile industry for submission to the Elector of Brandenburg (*A* I **3**, p 408).

Leibniz settled the affairs of inheritance with his brother Johann Friedrich on 10 July 1680. No sooner had he returned to Hanover than he received from his brother news of the death of Christian Freiesleben, whose widow, Clara Elizabeth, immediately claimed from Leibniz 476 taler in unpaid debts and in addition demanded that his father's library, which occupied a great deal of space, should be moved out of her home to another location or sold (*A* I **3**, pp 601–2). The library included not only the books of Leibniz's father but also those of his grandfather Wilhelm Schmuck and a mass of unbound volumes from the inheritance of his father's second wife, the daughter of the bookseller Bartholomaeus Voigt (*A* I **3**, p xlvii). For many years, Christian

Freiesleben, to whom Leibniz was related, had administered money for him, collecting the small rent from his mother's inheritance for example, and had also lent him money that was never paid back. Owing to the disorder of the papers on both sides, there ensued further disputes until Leibniz reached an agreement with Clara's brother Quintus Rivinus, who acted as her legal adviser. Though reluctant to dispose of the library, Leibniz eventually gave his brother power of attorney to arrange an auction (*A* I **4**, p 668), at the same time sending him a German poem, 'Jesus on the Cross', in honour of his piety (*A* I **4**, p 667). Having prepared an auction catalogue of 64 pages, Johann Friedrich Leibniz sold the books in September 1685 (*A* I **4**, pp 700–2). After expenses had been paid, 238 taler remained for Leibniz, which was not enough to cover the claims of Clara Freiesleben.

Earlier, the revenue department in Altenburg had paid out his share of the money owing to him there from his mother's inheritance and some other small claims, coming to 761 taler in all (*A* I **3**, pp xlvii, 614). This had covered his liability for the expenses of the construction of the windmills up to 1683. When he settled the debt to Clara Freiesleben in 1685, the Harz project again involved a heavy financial burden, so he attempted once more to exact an inheritance owed to him by the revenue department in Weimar. He commissioned his nephew Johann Friedrich Freiesleben, who was then studying in Jena, to pursue the claim. In this task, he was aided by Leibniz's cousin Aegidius Strauch, who had earlier asked Leibniz to assist him in obtaining a government office (*A* I **3**, pp 603–4). The young Freiesleben was the son of Leibniz's half-sister Anna Rosina, who had married Heinrich Freiesleben. He lacked experience in dealing with the bureaucratic officials in Weimar. Moreover, the claim was complicated by the political consequences of the territorial division of Saxony in 1672. When Johann Friedrich Freiesleben requested a loan from his uncle in 1687 in order to take his doctorate in law (*A* I **4**, pp 711–12) he was able to remind him of his dedicated efforts on his behalf, unfortunately resulting only in promises (*A* I **4**, p lv), and to claim that he had earned the help with his studies that had been promised to him.

Another member of the family is mentioned by Leibniz in correspondence at this time. Writing to his brother on 28 March 1687, Leibniz asked him to greet his sister's son, the sixteen-year-old Friedrich Simon Löffler (*A* I **4**, pp 714–15).

The Harz project

In a dry season, the supply of flowing water was insufficient to maintain the continuous working of the pumps in the Harz mines, so that the profits in a poor year might be as little as sixty per cent of those in a good year. The idea of using wind power to supplement the water power in fact came not from

Leibniz but from Peter Hartzingk, a Dutch engineer in the Mining Office at Clausthal, who died in 1680, but not before Duke Johann Friedrich had decided in favour of the scheme of Leibniz, who claimed that his windmills and pumps would be more efficient than those of Hartzingk (*A* I **3**, p xxxv). Hartzingk had proposed that, after working the pumps to drain the flood water, the flowing water should be collected in underground reservoirs and raised from there by the windmills to be used again in operating the waterwheels driving the draining pumps. The same water would thus be used over and over again, the storage reservoirs on the surface providing a reserve supply to make up the losses. In his memorandum of 9 December 1678, Leibniz had poured scorn on this idea, recommending instead the draining of the flood water by the direct operation of the windmills.

On returning to Hanover after settling his family affairs in Leipzig, Leibniz spent the month of August 1680 in Clausthal and Zellerfeld, the two neighbouring towns at the centre of the mining industry, during which he proposed a new plan (*A* I **3**, pp 66–80), in fact the scheme of Hartzingk which he had previously opposed. It was hardly a cause for wonder that the professional mining engineers in Clausthal should resent the interference of an outsider, who was in no sense a specialist but whose scheme had been imposed upon them in preference to that of their own expert. Leibniz's latest conduct in reviving the 'pump-storage' system of Hartzingk as if it were a new plan of his own must have seemed devious, to say the least, and they clearly believed that the financial reward he was to receive under the terms of the agreement ratified by the Duke was out of all proportion to the contribution he had made.

Although the Mining Office protested that the agreed scheme was for windmills working alternately with water power to operate the pumps, a commission led by the Privy Counsellor Otto Grote resolved that Leibniz should test both possibilities and this decision was confirmed by the Duke on 21 October 1680 (*MK,* p 62). The windmill whose construction had already been started at the Catharina mine should be used to operate the pumps directly, while the other two windmills should be built by the Zellbach stream to test the new plan. Again the cost was to be divided equally between the Duke, the Mining Office and Leibniz himself.

Between 1680 and 1686 Leibniz made over thirty visits to the Harz and spent there 165 weeks altogether. Besides the opposition of the Mining Office, which resulted in repeated appeals to the Duke, Leibniz also encountered considerable technical difficulties and unfavourable meteorological conditions.

An idea of the ill will towards him on the part of the mining community may be gleaned from letters he wrote in the spring of 1681 to Hieronymus von Witzendorff, who was to succeed Eltz as Inspector of Mines in 1683, and the Mining Counsellor Christian Berwardt, one of his most hostile opponents. To Witzendorff he complained about some of the unjust things he found were being said about him; for example, that he sought only his own interests and not those of the mines (*A* I **3**, pp 109–10). Leibniz was determined, however, that these people would not succeed in their plan to provoke him into unseemly disputes. Another complaint was that he had wasted time, in reply to which he reminded Berwardt that he had allowed neither the expense nor the rigours of the hard winter to prevent him from undertaking the necessary journeys to and from the Harz (*A* I **3**, pp 112–13).

Having returned to Hanover in the middle of September 1681 after a visit to the Harz lasting four or five weeks, Leibniz wrote a memorandum for the Duke reporting on the state of the work (*A* I **3**, pp 124–6). He also expressed concern to the Duke about the storage of the books of the Library pending the completion of the new building in a rear wing of the Palace. During his previous visit to the Harz earlier in the year, he had been kept informed of the progress on the building by his secretary Jobst Dietrich Brandshagen, who stayed behind in Hanover (*A* I **3**, pp 113–16). The books had been removed from the old rooms, but the new rooms still required masonry work and decoration. Meanwhile the books were not only inaccessible but inadequately protected from damage. It is probable that they were not placed on shelves again until 1683 or 1684. Also in his memorandum to the Duke, Leibniz recommended that an agreement should be sought with the representatives of the Emperor and Elector of Saxony, who would be attending a forthcoming political conference in Frankfurt, for an increase in the price of silver. This would be of benefit to the silver-producing states they represented and also, of course, to Hanover, especially if the output from the mines in the Harz could be significantly increased by the introduction of the new pumps and windmills. Leibniz had in fact received a commission from the Duke to accompany Otto Grote to the conference in Frankfurt or follow him soon after, as he remarked in confidence to Justus von Dransfeld, a teacher in Göttingen (*A* I **3**, pp 491–2). On 12 September 1681, Leibniz and Grote dined together at Osteroda in the Harz, where no doubt they discussed the impending conference as well as the progress in the building of the windmills (*A* I **3**, p 123). To his friend Christian Philipp, the Saxon ambassador in Hamburg, Leibniz confided that he had been reluctant to go to the political conference in Frankfurt, as it was known in advance that there was nothing to be done of a legal nature, and so he had delayed as long as possible, using the affairs of the Harz as a pretext (*A* I **3**, p 514). In the end, Grote found he had little need to call on Leibniz's assistance, so he was not pressed to go to Frankfurt. At the time of his memorandum to the Duke, Leibniz pleaded that he needed to spend several more weeks in the Harz in order to bring his project to such a stage that no doubt could remain as to its complete success. This he was anxious to do before making the journey to Frankfurt and he emphasised to the Duke that he sought only the honour of being useful (*A* I **3**, pp 123–4).

Leibniz was again in the Harz from the beginning of October 1681 until the middle of January 1682, when it seems that the work was suspended owing to the bad weather. A month later, he was anxious to return in order to put his men to work, as the weather was then improving. Before his departure, however, he requested a short audience with the Duke, remarking also that he would return to Hanover as soon as he could meet Casimir zu Eltz there, in order to settle the disputes, because he could not reason with the officials in the mines. To the Duke (*A* I **3**, pp 143–6) he complained that the work was frequently suspended for trivial reasons and the promised help was often refused; indeed the officials had actually obstructed the work by discouraging people from working for him with lies and threats. However, three windmills had been constructed and several times connected for tests, but only using the old pumps. He wanted the new pumps to be tested first with the old water power, in order to show that the pumps themselves were more efficient. Concerning the complaints of the mining officials respecting the escalating costs, Leibniz pointed out that the slight damage to the windmills was caused by lack of proper control, while most of the costs involved not the windmills, for which he alone was responsible, but the constructions for the draining and mining operations.

Despite many difficulties, Leibniz's enthusiasm remained undiminished, so that in March he could write a memorandum to the Duke (*A* I **3**, pp 149–66) on the general improvement of mining in the Harz. Having referred to the great wealth of the mines in the Harz, containing rich deposits of copper, iron, lead and silver, he gave a complete review, including proposals relating to topography, mining law, industrial management and chemistry.

In September 1682 Leibniz returned briefly to Hanover requesting von Platen to arrange an urgent meeting with the Duke and the mining officials for the resolution of yet further objections (*A* I **3**, pp 198–9). These did not relate to the force of the windmill itself, which was no longer in doubt, but to a balancing of the advantages and disadvantages of its use. He had inventions, Leibniz remarked, that would increase the advantages and decrease the disadvantages. For example, he could offer a windmill at half the usual cost that would work in all winds without the need of adjustment and he also had the means of transmitting the force to a great distance, notwithstanding obstacles and detours. To ensure smooth running he planned to introduce sails that would open and close according to the strength of the wind (*A* I **3**, p xxxviii) while the transmission of the force to the pumps would be effected by means of compressed air. In November 1682, he put the latter idea to Hieronymus von Witzendorff, who had accepted the office of Inspector of Mines left vacant by the death of Casimir zu Eltz. Leibniz flattered Witzendorff, saying that if Eltz had been of his mind, or if Witzendorff had come on the scene earlier, he would have achieved his project sooner and with less expense (*A* I **3**, pp 210–12). In reply, Witzendorff promised to assist Leibniz as much as he could (*A* I **3**, p 212) and, in February 1683, Leibniz wrote a poem for him (*A* I **3**, p 219). But the idea of using compressed air for the transmission of force had to be abandoned because pipes thick and strong enough could not be made.

By the middle of 1683, the costs had escalated to 2270 taler, whereas Leibniz had originally estimated 300 taler for the construction of a windmill. At last, the Duke had to agree with the mining officials that financial support for the project was no longer justified. This decision he communicated to Leibniz on 6 December 1683 (*A* I **3**, pp 237–8) whereupon Leibniz accepted the option to continue the work at his own expense until the end of 1684, in order to demonstrate the usefulness of his invention (*A* I **4**, pp 5–8).

Two series of tests of the windmills took place in 1684, the first in the spring and the second in the autumn. These were supervised by Leibniz's secretary Jobst Dietrich Brandshagen and another helper, Christoph Köhler. Having spent long periods in the Harz making the preparations, Leibniz had returned to Hanover when these were completed, where he was kept informed of progress by Brandshagen. Apart from interruptions caused by breakdowns, there were many occasions when the windmills could not be operated owing to lack of sufficient wind. Indeed, it was necessary to be ready at any time of the day or night to take advantage of any wind that might blow up. The first successful test in fact took place at night when the only official who could be located was a deputy at a neighbouring mine, who made a confused and indecisive report (*A* I **4**, pp 20–1).

The lack of wind had not been foreseen by Leibniz, for his enthusiasm seems to have caused him to overlook the wisdom of first investigating the prevalence, strength, direction and variation of the wind in the Harz. To overcome this difficulty he had proposed the construction of a new type of windmill that would work with little wind and on 31 January (*A* I **4**, p xliii) the Duke had granted him 200 taler for its construction. Leibniz's new invention was a horizontal windmill, which worked rather like a water wheel and would, he claimed, require no more attention than this traditional machine (*A* I **4**, p 43). One of the inventions he had devised to ensure the smooth running of his machines and reduce the danger of breakdowns was a speed regulator in the form of a 'feedback control' (Münzenmayer 1976, pp 113–19).

Although Leibniz believed the idea of a horizontal windmill to be original, a design for such a windmill had appeared in Fausto Veranzio's *Machinae novae*[3], published in Venice early in the seventeenth century. The idea may have been inspired by the guiding vanes (figure 5.1) often set up to direct the wind into the ventilation shaft of a mine. In the windmill designed by Leibniz (figure 5.2), guiding covers directed the wind, from whatever direction it

[3]Plate 13. There is a facsimile reprint of Veranzio's book (Munich 1965: Heinz Moos). For accounts of Leibniz's work on horizontal windmills, see Stiegler (1968) and Horst and Gottschalk (1973).

Figure 5.1 Guiding vanes for ventilation shaft. From G Agricola, *De rerum metallica* (Basel 1556). Courtesy of the British Library.

came, on to vanes attached to a wheel set up on a vertical axis. The advantages of the new windmill were that it was cheap, would work day and night in all winds with little attention and would weather storms without damage (*A* I **4**, pp 41–4). Only 20% of the wind energy, however, was converted into useful work, compared with 60% in the case of ordinary windmills. For this reason, the horizontal windmills were not really effective.

During the summer of 1684 the battle between Leibniz and the Mining Office continued unabated and when the second series of tests took place in the autumn, the weekly reports of the official observer contained nothing but complaints. For example, the wind became too weak to turn the machinery, or interruptions were caused by breakdowns needing lengthy repairs. Brandshagen's reports were slightly more encouraging, for on one occasion in November, the windmills were working continually for twenty-four hours (*A* I **4**, pp 129–30) and later in the month, for three days on end (*A* I **4**, pp 130–1). Then on 21 November (*A* I **4**, pp 133–4) Brandshagen reported that the horizontal windmill had been sufficiently advanced to test how it would turn without a load. Although there was scarcely enough wind to move the leaves on the trees, the windmill had worked well, and when

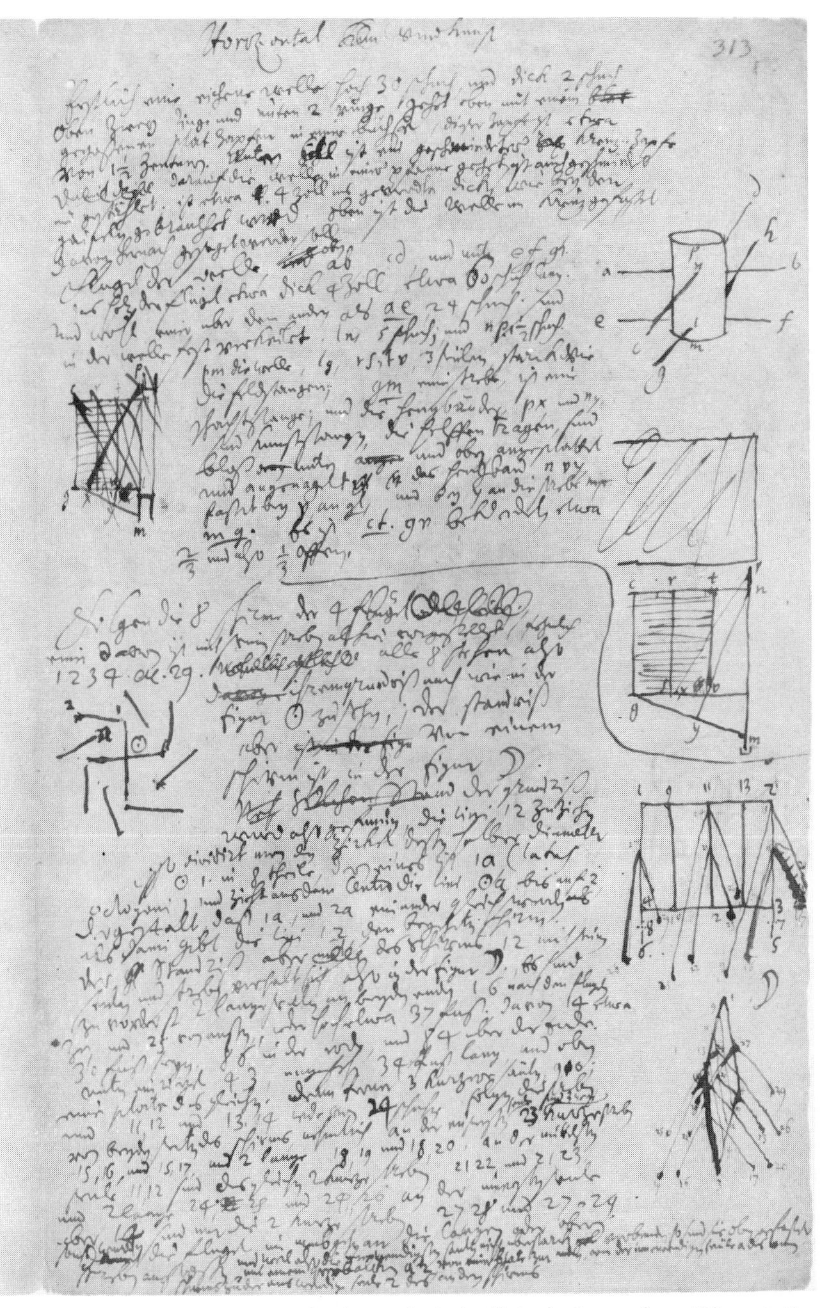

Figure 5.2 Leibniz's design for a horizontal windmill, including a plan of the rotating vanes and guiding covers. *LH* **XXXVIII**, Bl 313. Courtesy of the Niedersächsische Landesbibliothek, Hanover.

Brandshagen wished to stop it again, it took five strong workmen to bring it to rest.

On 1 December (*A* I **4**, pp 134–6) Leibniz sent a report to Otto Arthur von Ditfurdt, Witzendorff's deputy, asking him whether the official assessors agreed or not, and appealing to him to see that justice and reason prevailed rather than chicanery. Having arrived in the Harz himself shortly after, he informed the Minister Albrecht Philipp von dem Bussche in Hanover that there had been a slight delay with the horizontal windmill because the carpenter had constructed the guiding covers so that they hit against the rotating vanes, but this fault had been corrected since his arrival and he hoped that the new windmill would be raising water before the middle of December (*A* I **4**, pp 140–2).

Leibniz stayed in the Harz to supervise personally the third series of tests, which took place in January and February 1685, again with only partial success. The Mining Office demanded that he should demonstrate the capacity of his windmills to drain the mine more quickly or keep it dry longer than was possible with water power (*A* I **4**, p xl). In his last memorandum of justification, reviewing the history of the project from 1678, he rejected this obligation as not being part of the agreement (*A* I **4**, pp 176–84) so that the meeting of 20 February, as so often before, resulted in completely fruitless discussion. At this point, it was recognised in Hanover that the formerly expected results could not be obtained and that the project should be abandoned. Finally, on 14 April 1685, Otto Grote informed Leibniz of the Duke's decision that all further work on the construction of windmills should cease (*A* I **4**, p 189). Thus came to an end the project on which Leibniz had set such high hopes for providing the financial resources that would be needed for the realisation of another project, the universal characteristic, which he believed would have inestimable potential for the benefit of mankind. Though he had been beaten by nature and the obstinacy of men, his optimism remained undiminished.

A new attempt to formulate a logical calculus

Having abandoned his first attempt at a logical calculus based on arithmetic, Leibniz tried another approach in two manuscripts, *Specimen calculi universalis* and *Ad specimen calculi universalis addenda* (*GP* **7**, pp 218–27; Couturat 1903, pp 239–43), probably composed in 1680 while he was occupied with the beginnings of the Harz project. The results of these new endeavours he found to be much more coherent.

Terms are represented by letters, there is no equality sign and the theory is concerned only with universal affirmative propositions. Thus a is b means 'All a is b'. Leibniz poses two axioms (*propositiones per se verae*):

1 The principle of identity, *a* is *a* (animal is animal)
2 The principle of simplification, *ab* is *a* (rational animal is animal).

As a third axiom, he admits the principle of syllogism (*consequentia per se vera*): if *a* is *b* and *b* is *c*, then *a* is *c*. Logical identity of two terms is defined by the possibility of substituting one for the other without altering the truth value.

Among the theorems demonstrated by Leibniz is the following:

If *a* is *b*, then *ac* is *bc*.

Proof

ac is *a* by the principle of simplification
a is *b* by hypothesis
Therefore *ac* is *b* by the principle of syllogism
Also *ac* is *c* by the principle of simplification
Therefore *ac* is *bc* by composition of predicates of the same subject.

From this result Leibniz deduced another, which he called a remarkable theorem: if *a* is *b* and *c* is *d*, then *ac* is *bd*.

Proof

a is *b* by hypothesis
Therefore *ac* is *bc* by the previous theorem
Similarly *bc* is *bd*
Therefore *ac* is *bd* by the rule of syllogism.

This 'remarkable' theorem appears again in the *Principia Mathematica* of Russell and Whitehead.

First papers in the *Acta Eruditorum*

When the Leipzig professor of moral and political philosophy, Otto Mencke, visited Hanover in the spring of 1681, he discussed with Leibniz his plan for the publication of a new scholarly journal, the *Acta Eruditorum*, which would keep German scholars in touch with new publications. Referring to this conversation in a letter of 24 September 1681, Mencke asked Leibniz for a contribution. Publication of the journal commenced at the beginning of 1682 and Leibniz became a regular contributor, using the initials G.G.L. in place of his full name. His first paper in the journal, on the arithmetical quadrature of the circle, appeared in February 1682 (*AE* Feb 1682, pp 41–6). Then, in June, he contributed a paper on optics, 'Unicum opticae, catoptricae et dioptricae principium' (A common principle for optics, catoptrics and dioptrics) (*AE* June 1682, pp 185–90).

The unifying principle to which Leibniz refers in his paper on optics is that light travels along the path of least resistance. From this principle he deduces the laws of reflection and refraction. Consider, for example, the ray CEG (figure 5.3) which is refracted on moving from air into water. If m and n represent respectively the resistances of air and water, then CEG will be the path of least resistance provided $m\mathrm{CE}+n\mathrm{EG}$ is always less than $m\mathrm{CF}+n\mathrm{FG}$, where F is any point arbitrarily close to E. Taking $\mathrm{CH}=c$, $\mathrm{GL}=g$, $\mathrm{HL}=h$ and $\mathrm{HE}=y$, it follows that $\mathrm{CE}=\sqrt{(cc+yy)}$ and $\mathrm{EG}=\sqrt{(gg+yy-2hy+hh)}$. Since C and G are given, the problem is to determine the value of y for which $m\sqrt{(cc+yy)}+n\sqrt{(gg+yy-2hy+hh)}$ is a minimum. Using his still unpublished method of maxima and minima, which he claims shortens the calculation wonderfully in comparison with the methods known previously, we see almost without calculation, Leibniz remarks, that $my/\mathrm{CE}=n(h-y)/\mathrm{EG}$, or taking $\mathrm{CE}=\mathrm{EG}$, that $m/n=(h-y)/y=\mathrm{EL}/\mathrm{HE}$, so that the ratio of the sines of the angles of incidence and refraction is a constant depending on the optical resistances of the media.

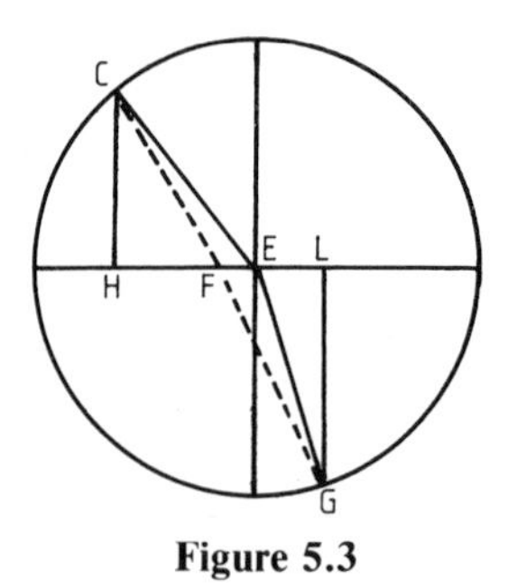

Figure 5.3

Leibniz claims that, in his paper, he has derived all the experimentally established laws of optics by mathematics, using a single principle which, when properly understood, comes from final causes. The ray itself does not consider the easiest path, but God has fashioned light so that this beautiful result naturally happens. Descartes and others who had rejected final causes, he believed, had made a mistake, for these could help us to find those properties of things whose inner nature might not be so clear to us. From all these considerations, Leibniz remarked, it was clear that the speculations of the ancients were not to be disdained. For it seemed probable to him that the great geometers Snell and Fermat, both well versed in ancient geometry, had arrived at their discoveries by applying to dioptrics the method that the ancients had used in catoptrics (Lohne 1966).

Towards the end of 1683, Leibniz contributed a paper on the discounting of bills—that is, the calculation of the present worth of a sum of money payable at a future date (*GM* **7**, pp 125–32). Then in 1684, at a time when the burden of the Harz project had been increased by the withdrawal of financial support, he contributed no less than five papers to the journal. The first of these, published in May, was in the form of a reply to Johann Christoph Sturm, who had asked for Leibniz's opinion concerning a paper on quadratures he had contributed two months earlier, in which he claimed to have used the method employed by Leibniz in his arithmetical quadrature of the circle. In his paper Leibniz pointed out to Sturm the insufficiency of Cartesian geometry, on which he had relied, as a basis for quadratures in general and then indicated his own method (*GM* **5**, pp 123–6), completing the exposition in a supplementary paper published in December (*GM* **5**, pp 126–7).

Of the remaining papers, two were concerned with mathematics. One of these, published in July, treated the strength of beams fixed to a wall and loaded at the free end (*GM* **6**, pp 106–12), a problem that had previously been investigated by Galileo, while the other, published in October (*GM* **5**, pp 220–6) with title 'Nova methodus pro maximis et minimis' (New method for maxima and minima), gave the rules for differentiating sums, products and quotients, including quantities involving fractional powers and irrationals, as well as the application to the determination of tangents and extreme values. At the end of the paper there is a solution of Debeaune's problem, in effect the solution of a differential equation to produce the logarithmic curve. Many years later, in a letter to the Countess of Kielmansegg Leibniz explained that it was the appearance of some papers of Tschirnhaus on the subject that had led him to reveal his method of differences after keeping it secret for nearly nine years (*D* **3**, p 458). To Arnauld, in a letter of 14 July 1686 (*GP* **2**, p 61), he explained that his method went further than those of Hudde and Sluse in permitting the convenient differentiation of quantities involving fractions or irrationals, besides having a more general application, allowing the determination of tangents to transcendental as well as algebraic curves. These remarks clearly indicate Leibniz's own perception of the innovations on which his claim to the invention of the differential calculus is based.

Just one month after his paper on the differential calculus, Leibniz contributed a paper on philosophy, setting out his definitive theory of knowledge, to which he frequently referred in later works. This carried the title 'Meditationes de cognitione, veritate et ideis' (Meditations on knowledge, truth and ideas) (*GP* **4**, pp 422–6) and was occasioned by the appearance in 1683 of Arnauld's *Des vraies et des fausses idées,* in which Arnauld developed an attack on Malebranche's theory of knowledge. Leibniz had not studied the controversy in detail, as we learn from a letter he wrote to Tschirnhaus (*A* II **1**, pp 541–2) in which he expressed surprise that Arnauld and Malebranche, who had been such good friends when he was in Paris, were attacking each other. To Tschirnhaus he confided his view that Malebranche had not penetrated deeply into analysis and the art of discovery in general. Arnauld, he thought, wrote with more judgment, but the real object of his paper was not to resolve the dispute between Arnauld and

Malebranche but to criticise and correct Descartes' theory of knowledge. Earlier sketches of the theory that Leibniz now described in detail for the public had been shown to Spinoza on the journey from Paris to Hanover and to Molanus and his monks in the debate with Eckhard.

The most perfect knowledge, Leibniz explains, is that which is both adequate and intuitive. It is adequate when every element of a distinct concept (that is, one which can be distinguished from all others) is known distinctly or when analysis is complete. It is intuitive when every element entering into the composition of a distinct concept can be thought of at the same time. Knowledge of a distinct primitive concept is always intuitive, whereas knowledge of composite concepts is for the most part what Leibniz calls blind or symbolic. As an example, he gives the case of the composite concept of a polygon with a thousand equal sides. In thinking of this, one does not always consider the nature of a side, of equality, or of a thousand, but uses these words as symbols of the ideas, remarking that one knows the meanings of the words but that their interpretation is not necessary for the present judgment. Because we are content with symbolic thinking, it often happens that we do not press the analysis of the concepts, believing that we have understood the symbols earlier, and consequently may overlook a contradiction in the composite concept. Leibniz therefore rejects Descartes' doctrine that what may be conceived clearly and distinctly in something is true and may be predicated of it. For what may seem clear and distinct in symbolic thinking is often obscure and confused. Instead, he adopts the position that an idea is true when the concept is possible and false when it implies a contradiction. In reply to the suggestion of Hobbes that truths are arbitrary because they depend on a free choice of definitions, he distinguished between nominal definitions and real definitions. The former contain only symbols for discerning one thing from others. A real definition, on the other hand, is one through which the possibility of the thing is ascertained, and such a definition is not open to free choice, since not all concepts can be combined with each other. A nominal definition does not therefore suffice for perfect knowledge unless it has been established by other reasons that the thing defined is possible. In particular, the argument for the existence of God proposed by the Scholastics and revived by Descartes is incomplete because it fails to establish that the concept of a most perfect Being is possible. But although Leibniz asserts that a most perfect Being is indeed possible and therefore necessary, he does not include his own demonstration in this essay.

In the following year, when there was much to be done in winding up the Harz project, the pattern of Leibniz's contributions to the *Acta Eruditorum* changed. He published only one paper—on the mechanics of inclined planes (*GM* **6**, pp 112–17)—again prompted by another publication on the same subject, but he also contributed no less than thirteen reviews. These included Philippe de la Hire's *Conic sections,* Ozanam's trigonometrical tables and the anonymous *Essays on physics, demonstrated by experiment and confirmed by Scripture* (Ravier 1937, pp 93–6).

In March 1686, Leibniz published a paper criticising the foundations of Cartesian physics, which will be discussed in a later section. One of the two reviews that he contributed in the same year concerned the *Algebra* of John Wallis (*AE* June 1686, pp 283–9). This work contained a method for calculating the curvature of curves, which inspired Leibniz to expound his own method in a paper that was printed immediately following the review (*GM* **7**, pp 326–9). Another paper, which he presented in June (*GM* **5**, pp 226–33) was also inspired by a book published in England. This was John Craig's book on quadratures, published in 1685, which Tschirnhaus had reviewed in the March issue of the *Acta Eruditorum*. Although Leibniz had reason to be pleased by Craig's employment of his own differential notation, he was surprised and no doubt concerned that Craig attributed to him a paper in the *Acta Eruditorum* of October 1683, which had actually been written by Tschirnhaus. In order to correct any false impressions concerning the principles of his own methods, he therefore sent to the *Acta Eruditorum* an exposition of his integral calculus, introducing the notation and showing quadratures to be special cases of the inverse method of tangents (*GM* **5**, pp 226–33).

Commerce, politics and the arts

Towards the end of 1680, when he had high hopes for the success of the Harz project, Leibniz wrote a memorandum for the Duke in which he suggested that there was considerable profit to be made out of an arrangement with the Dutch to refine the ore produced by the gold and silver mine which their East Indies Company had opened in Sumatra (*A* I **3**, pp 104–8). If, in addition, the refined ore could be bought in exchange for manufactured linen, which the East Indies Company needed in quantity, then the advantage would be greater. But speed was essential, for if the Dutch themselves discovered the secret of the refining process, the opportunity for the trade would be lost.

About the same time, Leibniz was trying to obtain from England an exemplar of the pressure cooker with safety valve that had been invented by Denis Papin in 1679. Papin's digestor, as it was called, enabled bones to be used for human nourishment and thus could provide cheap food for the poor. This invention prompted Leibniz to write a satirical piece, presumably for the Duke, in which the dogs quote Homer and the Scriptures in support of their right to the bones (*A* I **3**, pp 94–6). He mentioned the digestor in a letter to Christian Philipp, the Saxon ambassador in Hamburg (*A* I **3**, p 517) and in reply (*A* I **3**, pp 520–1) received the information that Papin was the same man who had invented the pump described in the works of Robert Hooke.

In connection with the deliberations of the Imperial Diet on the security of the Empire, Leibniz composed in the spring of 1681 a memorandum on ways of improving the organisation and morale of the army. Besides suggestions for such improvements as the provision of new weapons and inspiring leadership, he stressed the need for promoting the physical and mental welfare of the soldiers. Apart from the provision of adequate food, clothing and medicine, attention should be given to the problem of pay and above all to the provision of opportunities for free-time activities in sport and the organisation of constructive work, such as the draining of marshes, canal projects and the building of fortifications, to relieve the boredom of peace-time military training (*A* IV **2**, pp 577–602).

The occupation of Strasbourg by the French in the autumn of 1681 drew from Leibniz the bitter comment, 'the king needed it for the security of his kingdom; that is to say, to better maintain what he had stolen from the Empire, he had to steal more' (*FC* **3**, p 88). At this time there was another external threat to which Leibniz gave his attention. This was the danger from the spread of the plague which seemed to be advancing from the extremities of Europe towards the frontiers of Germany in the Harz. In a memorandum presumably written for the Duke (*A* I **3**, pp 131–6) he suggested that, as the physicians had not found a remedy, preventive measures of a political nature were needed. The greatest danger, he believed, came not from pollution of the air and the water in rivers and lakes, but from men and animals carrying the disease. Consequently, the greatest vigilance was needed to prevent entry into the country of infected persons and to ensure the detection and isolation of any who slipped through the net. If an outbreak should occur, then it was essential to restrict movement in and out of the infected area and even confine sufferers to their houses. Yet the preventive measures taken to minimise the spread of infection should be in accordance with the principles of natural justice and compassion. For example, healthy relations should be allowed to leave the house of a victim and the sufferers themselves should not be deprived of the consolation of friends, medical treatment or the necessities of life. Leibniz made detailed recommendations concerning arrangements for the supply of these needs with the least danger to those who would have to carry them out.

Through Christian Philipp, the Saxon ambassador in Hamburg, Leibniz became acquainted with Polycarp Marci. Soon after a meeting with Leibniz in March 1681 (*A* I **3**, p 465) Marci took up an appointment as Saxon ambassador in Stockholm where he remained until the end of 1683. During this time he kept Leibniz informed of events in Sweden. On 27 August 1681, however, he wrote to Leibniz on a personal matter of some interest, for in his reply, Leibniz expressed his views concerning the theatre and opera in particular. Previously we have seen him advocating the social value of rustic music and dancing, so the expectation is that he would also approve of opera performed in a large city like Hamburg. Pastor Anton Reiser had written a piece, *Theatromania,* condemning opera as the work of the devil (*A* I **3**, pp 496–7). Marci felt aggrieved because, as he reminded Leibniz, he had himself written the opera *Vespasian,* and he would welcome an impartial judgment from his distinguished friend. On receiving Leibniz's reply, Marci expressed his pleasure in finding that learned people agreed with him concerning the opera (*A* I **3**, pp 517–19). Leibniz described the opera as a powerful instrument to move the human spirit, by combining vigorous fancies, elegant expression, good poetry, grand music, beautiful pictures and artful movement to delight the inner as well as the two chief outer senses. Just as rhetoric could be used for bad as well as good purposes—and no-one would condemn rhetoric as such on this account—so could this newly devised means of expression. He therefore approved of opera in general, which he pointed out seemed to have had its origin in Church music, as a powerful instrument that could be used to govern the passions and incite feelings of honour, virtue and natural piety in the common man (*A* I **3**, pp 513–14).

At the invitation of the Landgrave Ernst von Hessen-Rheinfels, Leibniz in 1683 composed a political satire on Louis XIV and French imperialism. The Landgrave undertook to have the pamphlet printed anonymously, promising to withhold the identity of the author from the Duke if Leibniz so wished (*A* I **3**, p 275). In this writing, entitled *Mars Christianissimus* (*Most Christian War-God*) (*A* IV **2**, pp 446–502), Leibniz notes that 'the most powerful person in the world, excepting always the devil, is without doubt his Most Christian Majesty'. Taking note of the resolution made in France 'to recognise no longer any judge but the sword', he observed that while treaties and moral scruples bound ordinary men, there was a certain law superior to all others, conforming nevertheless to sovereign justice, which released the king from the obligations to observe them. With great irony Leibniz demonstrated that Louis, both by prophecy and miracles which he had performed, was entitled to absolute control over the general affairs of Christendom. In the concluding paragraph he offered a malicious reply to the question why Louis did not begin his beautiful designs by routing the Turks rather than afflicting the poor Christians. The reason was that his conscience constrained him to follow the rules of the New Testament, which commands that one begins with the Jews and then turns to the Gentiles. In imitation of this, Leibniz explains, the king will create for himself, by the reduction of the Christians, a sure passage to go one day to the infidels. Of Leibniz's many other writings against the French, none was as powerful and malevolent as this one.

At the request of Otto Grote, Leibniz worked between the winter of 1684 and the autumn of 1685 on another important memorandum for the Duke (*A* I **4**, pp 221–37). This concerned the establishment of a new Protestant Electorate. First, Leibniz presented the arguments for increasing the number of Electorates to nine. The Protestants were at a double disadvantage. On the one hand, they had only three Electorates—namely Saxony, Brandenburg and the Palatinate—whereas the Catholics had five, including three Prince Bishoprics, and on the other hand, all the Protestant Electorates could be lost by a change of religion on the part of the Elector or his successors, whereas the Catholic continuity of the Ecclesiastical Electorates was assured. Soon after the memorandum was written, the problem was accentuated by the death of the Elector of the Palatinate without male heir and the succession of a Catholic duke, thus reducing the Protestant Electorates to two.

Leibniz suggested that, as a first step, the idea should be put to the Emperor by means of discreet back-stage diplomacy. Moreover, the first approach should aim to dispose the Emperor and the Electoral College to the idea of the co-option of a Protestant prince in general, without mentioning any names. He expected opposition from the Ecclesiastical Electors but cooperation from most of the others. Even if the Emperor could act on his own without the consent of the Electors, their agreement was desirable in the interests of harmony. Leibniz favoured a direct approach to the Emperor and the Catholics rather than what might seem the proper procedure of first taking the proposal to the assembly of Ministers of State in Regensburg, where affairs touching the Protestants in general were decided. For if this path were followed, there would be unnecessary delay in waiting for the decision of this assembly, surely a mere formality, before the proposal could be put before the Catholics and the assemblies of the Empire.

Having suggested the best strategy, Leibniz next made out a good case for Brunswick-Lüneburg as the most suitable choice for a new Protestant Electorate. First, according to the best genealogies available, the princes of Brunswick or Este were descended from Charlemagne in the male line and at that time were universally acknowledged to have the highest rank after the Electors. Secondly, by virtue of political strength and geographical situation, Brunswick-Lüneburg would be best able to counterbalance the influence of France in the Electoral College. Of the four Electorates situated near the Rhine, three were Ecclesiastical and Leibniz presumed more flexible to the will of France, as the Prince Bishops had only their own persons and families to consider, whereas a secular Prince had the honour and inheritance of his House to maintain, which in Leibniz's view was the same as the conservation of the country.

Leibniz expressed the view that Brunswick-Lüneburg could probably count on the support of the secular Electors. In particular, the House had good relations with Saxony, the leading Protestant Electorate, and the Palatinate, while the recent marriage of the Duke's daughter, Sophie Charlotte, to the Elector of Brandenburg (on the occasion of which Leibniz had written a poem and designed a commemorative medal) (*A* I **4**, pp 120–5) had strengthened the ties between the two Houses.

The fight for the ninth Electorate divided the two branches of the House of Brunswick-Lüneburg in Hanover and Wolfenbüttel respectively. This became a matter requiring some delicacy on the part of Leibniz, for while he owed loyalty to Duke Ernst August in Hanover, he also became friendly with Duke Anton Ulrich and his brother Duke Rudolf August, who ruled jointly in Wolfenbüttel[4] and had common interests with Leibniz in matters of theology and Church reunion. He probably met Anton Ulrich for the first time in March 1683 during a short visit to Brunswick. On 23 August 1685, during a visit to Wolfenbüttel, he conversed with the Duke on the right of succession of the first-born prince (*A* I **4**, pp 206–8) and in the same year also started a correspondence with Rudolf August (*A* I **4**, pp 541–3).

Electoral status was eventually granted to Hanover in 1692, after Leibniz had made a long journey to the south of Germany and Italy in search of historical records that would enhance the Duke's claim.

Religion and Church reunion

On returning from his visit to Brunswick in March 1683, Leibniz had another meeting with Cristobal de Rojas y Spinola, the Bishop of Thina, who was visiting Hanover for reunion discussions with the Protestant theologians. Leibniz mentions this meeting in a letter to the Landgrave Ernst von Hessen-Rheinfels (*A* I **3**, pp 276–80) who had earlier sent him a writing on reunion matters with a request that he should also show it to Duke Anton Ulrich (*A* I **3**, pp 275–6). Although he planned to visit Wolfenbüttel in a few days, Leibniz went to Zellerfeld instead, from where he wrote to Duke Anton Ulrich on 7 May, sending him a number of papers on reunion matters he had received from the Landgrave Ernst.

In the autumn of 1683, the Landgrave Ernst attempted to convert Leibniz to Catholicism, setting out his arguments in the form of a letter to his 'esteemed and talented' friend (*A* I **3**, pp 324–7) which he invited him to show in confidence to Duke Anton Ulrich (*A* I **3**, p 328), who was also a Lutheran. Writing from the Harz on 11 January 1684, Leibniz explained his reasons for remaining a Lutheran (*A* I **4**, pp 319–22). Distinguishing between the interior communion of the Catholic Church (the Church as it should be) and the exterior communion (the visible Church), he accepted that the Church was infallible in all matters of faith necessary for salvation but objected that the visible Church also required its members to accept some errors in matters of science and philosophy, even when the contrary could be demonstrated. In these cases, the beliefs condemned by the Church —Copernicanism, for example—were not opposed to Scripture, tradition or the declarations of any Council. If he had been born into the Catholic

[4]The two Dukes ruled jointly from 1685 until the death of Rudolf August in 1704. From 1666 until 1685, Duke Rudolf August had ruled alone.

Church, Leibniz added, he would have remained a Catholic, unless he had been excommunicated, but as he had been born and brought up outside the Catholic Church, it would not be honest for him to join an institution which opposed propositions of science and philosophy which he held to be true and important.

Two months later Leibniz took the Landgrave Ernst to task for offending Philipp Jakob Spener,[5] the founder of Pietism, by comparing his sentiments towards the Emperor to those of a Christian of Turkey for an Ottoman Emperor (*A* I **4**, pp 323–6); for such a Christian could not pray for the taking of Vienna, whereas Spener could pray for divine help to enable the Emperor, though a Papist, to advance up to the Bosphorus.[6] Personal recriminations, Leibniz added, were in any case to be avoided, since they only invited replies in kind. For example, when Catholics claimed that God would hardly use a man like Luther for his purposes, the Lutherans could reply that it did not appear reasonable that some popes should have been vicars of Christ either. Assuring the Landgrave that his philosophical disagreements had nothing to do with Catholic theological dogma, Leibniz alluded to his project of the *Catholic demonstrations,* telling his friend of his wish to produce some day a writing on points of controversy between Catholics and Protestants. For the arguments to be received without prejudice, it would be necessary to conceal the fact that the author was not a Catholic. It was probably in the years immediately following that Leibniz produced his *Systema theologicum,* which considered the problems of reunion from the standpoint of a Catholic. This work was first published in 1845 (Ravier 1937, p 369).

Towards the end of 1683, the Landgrave again sought to convert Leibniz, remarking that his conversion would set a great example (*A* I **4**, pp 337–8). In the view of the Landgrave, the philosophical difficulties Leibniz had given as a pretext for remaining a Lutheran did not constitute an insuperable obstacle to his conversion, for he himself, though a good Catholic, also rejected certain condemnations of the Inquisition. However, Leibniz had already explained that, had he, like the Landgrave, already been a Catholic, he would not have left the Church on account of the condemnation of philosophical propositions having nothing to do with articles of faith. But he was not a Catholic, and that made his position different from that of the Landgrave.

On a lighter note, Leibniz wrote to the Landgrave Ernst on 20 January 1686 (*A* I **4**, pp 397–8) concerning an incident which he had found pleasing and curious. To celebrate the Nativity, an Italian oratorio had been sung in

[5]The movement founded by Spener, a Lutheran pastor, placed emphasis on the personal inner religious life and rapidly gained support in Germany.

[6]Taking advantage of an uprising in Hungary, the Turks in 1683 had invaded that country, crossed the Danube and besieged Vienna for sixty days. The siege was raised and the Turks defeated by the Polish army with the help of the Habsburg, Saxon and Bavarian contingents.

the Ducal Church in Wolfenbüttel. The Italian text included a word of praise for the reigning Pope Innocent XI, alluding to his efforts to unite Christians against the Turks. Although the sense of the words had escaped most of those present, Leibniz had no doubt that Duke Anton Ulrich had understood them. This was the first time, he believed, that a song of praise for a pope had been sung in a Lutheran Church.

Another strange event was reported by Leibniz to the Landgrave at the beginning of 1687 (*A* I **4**, pp 419–20), namely the appearance of a book entitled *Tuba pacis* by a Lutheran minister of Prussia (Praetorius), who wished that controversies should be settled by appeal to Scripture as interpreted by the Catholic Church and claimed that this idea was shared by some of the principal Lutheran theologians in Königsberg. However, the Landgrave was able to inform Leibniz (*A* I **4**, pp 423–5) that the author had already resolved to leave his religion and ministry when he wrote the book, so that it could not be taken as a serious contribution to the problem of reunion.

The invention of determinants

From the time of his meeting in Paris with Jacques Ozanam, Leibniz had concerned himself with problems of algebra, especially the solutions of cubic, biquadratic and higher degree equations. At an early stage, he recognised that the general solution of higher degree equations, which he and Tschirnhaus had sought in Paris, could not be achieved by elimination of intermediate terms and he reproached Tschirnhaus when in 1683 he published in the *Acta Eruditorum* a supposed solution based on this idea (*GM* **7**, p 5). In the same year Tschirnhaus published another paper in the *Acta Eruditorum* concerning the quadrature of algebraic curves, which led to a dispute with Leibniz and the suspension of the correspondence for several years (*GM* **7**, p 375). Tschirnhaus, according to Leibniz, had a habit of publishing the ideas of others as if they were his own (*GM* **2**, pp 51, 130) and claiming generality for results he had only demonstrated in a particular case (*GM* **2**, p 233).

Leibniz regarded algebra as a part of the *ars combinatoria* (combinatorial art). This was evident in the dialogue already mentioned in which he used the ideas of arithmetic and algebra to demonstrate Plato's 'recollection' theory of learning. The character Charinus (Leibniz himself) leads the boy through the multiplication of algebraic sums to the combinatorial formula

$$\binom{n}{k} + \binom{n}{k-1} = \binom{n+1}{k}$$

which serves to construct Pascal's triangle and thus classify algebra as part of the combinatorial art. Having raised the question of the relation between multiplication and dimensions, Charinus teaches the calculation of areas and volumes, then constructs four- and five-dimensional physical concepts before finally giving a monologue on rational algebraic forms approached through the combinatorial art.

From the time of his arrival in Hanover, Leibniz worked on the solutions of systems of linear equations and the elimination of unknowns to arrive at conditions of consistency for both linear and higher degree equations. It was soon after the disagreement with his friend Tschirnhaus and the decision of the Duke to suspend further financial support for the Harz project that Leibniz reached the decisive stage in the development of his own original work in algebra, namely the invention of determinants and the discovery of their properties, which could serve for the solution of these problems. The manuscript (Knobloch 1972) in which, after many trials (Knobloch 1980) an acceptable theory of determinants is achieved, is dated 22 January 1684.

The practice of using letters to represent coefficients in equations failed to show their order and relations. To remedy this defect, Leibniz introduced symbolic or 'fictitious' numbers to represent the coefficients. For example, a pair of linear equations with general coefficients would be written $10+11x+12y=0$, $20+21x+22y=0$. This notation, which he used from 1678, corresponds to the modern notation $a_{10}+a_{11}x+a_{12}y=0$, $a_{20}+a_{21}x+a_{22}y=0$, where the function of the 'coefficient symbol' a is simply to distinguish the symbolic numbers from the arithmetical numbers.

In the key manuscript of 1684, Leibniz considers in succession systems of linear equations from one equation with one unknown up to five equations with four unknowns, arriving at the general form of solution commonly known as Cramer's rule. Although he had discovered the rule for the formation of the terms as early as 1678, it was only now, as he himself recognised, that he had found the correct rule for the signs.

For the case of two equations with one unknown, $a_{10}+a_{11}x=0$, $a_{20}+a_{21}x=0$, the result of eliminating x is

$$\begin{vmatrix} a_{10} & a_{11} \\ a_{20} & a_{21} \end{vmatrix} = 0.$$

Leibniz writes the determinant, to which he does not give a special name, as $\begin{matrix}+10.21\\-20.11\end{matrix}$. Then he shows how the notation can be successively simplified, first to $\begin{matrix}+{\scriptstyle 1}0.{\scriptstyle 2}1\\-{\scriptstyle 2}0.{\scriptstyle 1}1\end{matrix}$, next to $\begin{matrix}+0.1\\-1.0\end{matrix}$ (the small numbers being understood) and finally to $\overline{0.1}$. It is easy to show that $\overline{0.1}=-\overline{1.0}$. With this notation, the solution of two linear equations in two unknowns becomes $\overline{0.2}+\overline{1.2}x=0$, $-\overline{0.1}+\overline{1.2}y=0$. The result of eliminating two unknowns from three linear equations is written $\overline{0.1}{}^{3}2-\overline{0.2}{}^{3}1+\overline{1.2}{}^{3}0=0$ and also in the alternative forms $\overline{0.1}{}^{2}2-\overline{0.2}{}^{2}1+\overline{1.2}{}^{2}0=0$ or $\overline{0.1}{}^{1}2-\overline{0.2}{}^{1}1+\overline{1.2}{}^{1}0=0$. Each of these is equivalent, in modern notation, to

$$\begin{vmatrix} a_{10} & a_{11} & a_{12} \\ a_{20} & a_{21} & a_{22} \\ a_{30} & a_{31} & a_{32} \end{vmatrix} = 0.$$

Evidently they represent expansions along the third, second and first row respectively. It should of course be noted that the notation $\overline{0.1}$, for example, can represent

$$\begin{vmatrix} a_{10} & a_{11} \\ a_{20} & a_{21} \end{vmatrix} \quad \text{or} \quad \begin{vmatrix} a_{10} & a_{11} \\ a_{30} & a_{31} \end{vmatrix} \quad \text{or} \quad \begin{vmatrix} a_{20} & a_{21} \\ a_{30} & a_{31} \end{vmatrix}$$

and that, in Leibniz's notation, these can be distinguished only by the context. In place of the above forms for the third order determinant, he simply writes $\overline{0.1.2}$. In general, $\overline{1.2. \ldots .n}$ represents the determinant

$$\begin{vmatrix} a_{11} & \cdots & a_{1n} \\ \cdot & & \cdot \\ \cdot & & \cdot \\ a_{n1} & \cdots & a_{nn} \end{vmatrix}.$$

Leibniz easily deduced the principal properties of determinants involving interchanges of rows and columns. He also achieved in general the elimination of the unknown from two higher degree equations by reducing this problem to one concerning a system of linear equations (Knobloch 1974, p 38).

Although he composed a very considerable number of manuscripts on these problems, Leibniz published nothing on the theory of determinants proper. In 1700 and again in 1710, however, he published his method of using fictitious numbers to represent coefficients in equations (Knobloch 1974, p 37). Also in later years he discussed this method with a number of correspondents, including Jakob Bernoulli, the Marquis de L'Hospital and Jakob Hermann. Besides the system described above, he employed many others, born out of his wish to develop the art of invention (Knobloch 1982).

A remarkable error of Descartes

Criticism of Descartes had been a common feature of many of the early papers Leibniz contributed to the *Acta Eruditorum*. For example, his own methods of tangents and quadratures had brought to light the limitations of Cartesian geometry, while the demonstration of the laws of reflection and refraction using the principle of least resistance indicated that Descartes was mistaken in rejecting final causes in physics. Again, the incompleteness of the ontological argument for the existence of God proposed by the Scholastics and revived by Descartes was exposed by Leibniz in the paper setting out his own definitive theory of knowledge in opposition to that of Descartes.

Thus far Leibniz had directed his criticism against important but particular aspects of Descartes' philosophy. However, in his 'Brevis demonstratio erroris memorabilis Cartesii' (Short demonstration of a remarkable error of Descartes), published in the *Acta Eruditorum* in March 1686 (*GM* **6**, pp 117–23), he struck at the very foundation of Cartesian physics; namely, the Cartesian principle of the conservation of motion. Leibniz first remarks that a number of mathematicians had estimated the force of bodies in motion by the product of mass and velocity, having observed that, in the five common machines[7], the mass and velocity compensated for each other. Since it was reasonable to believe that the sum of the motive forces should be conserved in nature, Descartes, who held the motive force and quantity of motion to be equivalent, was led to assert that God conserved the same quantity of motion in the world. Concerning the conservation of force there was no disagreement. For Leibniz acknowledged that we never see force lost by one body without being transferred to another, while the arguments of Pardies on the impossibility of perpetual motion and Huygens' defence of his principle that the centre of gravity of a system of bodies could not rise of its own accord, which had been the subject of a controversy with Catelan between 1681 and 1684, confirmed his belief that the total quantity of force in the universe could not increase of its own accord. Leibniz located Descartes' error in the assumption that motive force and quantity of motion are equivalent. In order to show the great difference there is between these concepts and so demonstrate the fundamental error in Cartesian physics, he selected a case in which, although the forces were generally admitted to be equal, analysis nevertheless revealed the quantities of motion to be quite different.

Citing the example of a pendulum, Leibniz supposes that, in general, a body falling from a certain height possesses the force necessary to lift it back to the original height in the absence of air resistance or other external interference. He also supposes that the same force is needed to raise the body A (figure 5.4) of 1 pound to the height CD of 4 feet as to raise the body B of 4 pounds to the height EF of 1 foot. Cartesians as well as other philosophers and mathematicians, he remarks, admit these assumptions. It follows that the body A, after falling from C to D, will have the same force as the body B after falling from E to F. To calculate the velocities he uses the law established by Galileo, that the velocity is proportional to the square root of the distance fallen. This implies that, at the end of the respective falls, the velocity of A is twice the velocity of B, so that the quantity of motion of A

[7]These were the lever, windlass, pulley, wedge and screw.

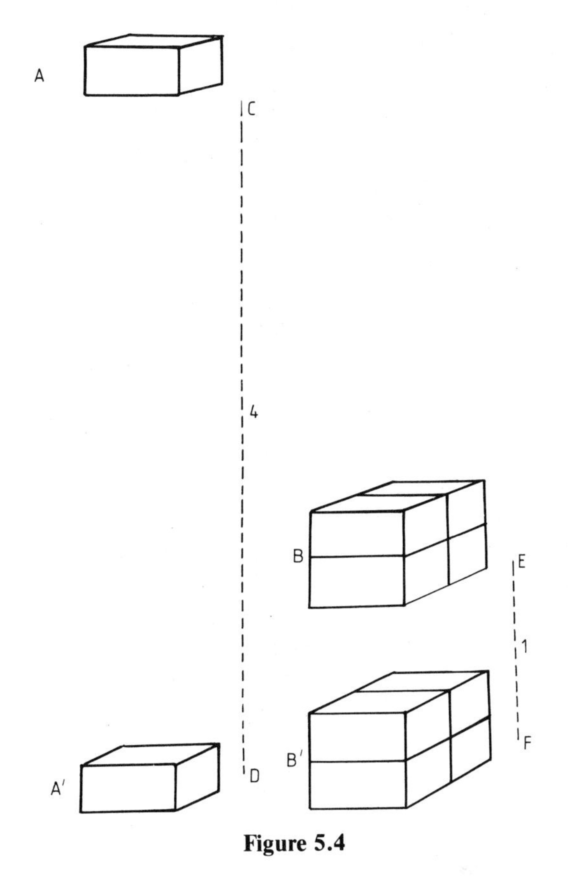

Figure 5.4

(which has a mass of 1 pound) is half the quantity of motion of B (which has a mass of 4 pounds). To Leibniz it therefore seemed that force is to be estimated from the quantity of the effect it can produce; for example, from the height to which it can raise a body, but not from the velocity which it can impress on the body. The force of a body, he states, is proportional to the cause or effect of its velocity; that is, to the height producing the velocity or capable of being produced by it or to the square of the velocity. It is merely accidental, he adds, that in the five common machines, where there exists an equilibrium (since the difference of mass of the two bodies is compensated by the difference of velocities), the force can be estimated by the quantity of motion. From the vantage point of the present day, we can see that Leibniz was using the term 'force' to connote different concepts which, at that time, he had not distinguished.

In September 1686, six months after the original publication in the *Acta Eruditorum,* there appeared a French translation of the 'Brevis demonstratio', to which was appended a criticism by the Abbé Catelan (*GP* **3**, pp 40–2) in the Dutch journal *Nouvelles de la république des lettres.* Catelan declared that it was not Descartes (as Leibniz had claimed) but Leibniz himself who had been led astray by too great faith in his own genius. In Catelan's view, the quantities of motion in the two falls described by Leibniz were unequal because the times of fall were different.

Leibniz was pleased with the opportunity given him by Catelan's criticism to correspond with the editor of the *Nouvelles de la république des lettres,* Pierre Bayle (*GP* **3**, pp 39–40), to whom he sent a reply for publication in the journal. This appeared in February 1687 (*GP* **3**, pp 42–9). Catelan's view, as interpreted by Leibniz, was that the Cartesian principle of the equivalence of motive force and quantity of motion was restricted to cases in which the forces were acquired in equal times, such as the five common machines. He could not imagine that Catelan had ever come across any Cartesians, at least among those who passed for geometers, who approved of this restriction. If Bayle were to consult some able Cartesians among his friends, Leibniz had no doubt they would confirm that they always estimated force by quantity of motion and held the total force to be conserved. It was not his wish, Leibniz added, to profit from the feebleness of Catelan's defence of the Cartesian position.

Among Cartesians, Catelan was the only one, so far as Leibniz knew, who supposed that the motive force of a body, which they estimated by its quantity of motion, depended on the time taken to acquire it. The force obviously depended only on the present state of the body. If two equal bodies had acquired the same velocity, one by a sudden impact and the other by a descent occupying some interval of time, it would be absurd to say that these bodies had not acquired the same motive force. In any case, the times of descent in the example he had described could be made equal by allowing the two bodies to move along planes inclined at appropriate angles.

At this point, Leibniz presented further clarification of his objection to the Cartesian principle. If a body of 4 pounds moving with 1 degree of speed along a horizontal plane were to transfer its motion to a body of 1 pound, the speed of this body would be four degrees of speed, which would enable it, on engaging a string causing it to move like a pendulum, to rise to a height of 16 feet. But a body of 4 pounds needs the same force to rise 1 foot as a body of 1 pound needs to rise 4 feet. The transfer of motion in accordance with the Cartesian principle would therefore result in a three-fold increase in the force out of nothing, which is absurd. Following this new refutation of the Cartesian principle, Leibniz introduced his own; that there is always a perfect equivalence between the full cause and the whole effect. Since the whole effect of the motive force is to raise a body to a certain height, this effect becomes a measure of the force. Thus, by implication, Leibniz takes the measure of force to be mv^2.

From his new principle of the equivalence of cause and effect, Leibniz also drew a metaphysical consequence; namely, that the force or power (*puissance*), though measured by the future effect, was something real in the present, so that it was necessary to admit in bodies something besides magnitude and velocity. For further information concerning his criticism of the Cartesian notion of body and his own theory of knowledge, he referred the reader to his 'Meditations' in the *Acta Eruditorum* of May 1684. Huygens, it may be noted, in 1668 had taken the sum of mv^2 to be conserved in elastic collision.

In June 1687 Catelan contributed another feeble attempt to defend the Cartesian principle while restricting its application to cases in which the velocities were acquired in equal times. Leibniz replied in a paper published in the September issue of the *Nouvelles de la république des lettres* (*GP* **3**, pp 49–51), where he challenged Catelan to find the line of uniform descent. Leibniz proposed this nice problem, inspired by Catelan's wish to measure force in terms of time, so that, as he remarked, the controversy should give some occasion for the advancement of science. Predictably, Catelan did not respond, but a solution by Huygens, without demonstration, was published in the October issue. Leibniz himself published a solution with demonstration of its correctness (but without the derivation of the line) using the infinitesimal calculus in April 1689 in the *Acta Eruditorum* (*GM* **5**, pp 234–43).[8]

Correspondence with Arnauld concerning the *Discourse on metaphysics*

Writing to the Landgrave Ernst von Hessen-Rheinfels on 11 February 1686 (*A* I **4**, p 399), Leibniz remarked that, being at a point where for some days he had nothing to do, he had composed a small *Discourse on metaphysics* (*GP* **4**, pp 427–63), on which he would like to have Arnauld's opinion. He enclosed a summary (*GP* **2**, pp 12–14) for the Landgrave to send to the French philosopher and theologian, who, since 1679, (following the death of the Duchess of Longueville, the most influential patron of Jansenism at the Court) had lived in voluntary exile in the Spanish Netherlands. In his reply, Arnauld (*GP* **2**, pp 15–16) questioned the utility of a writing that would be universally rejected; in particular, he had been shocked by Leibniz's concept of an individual substance, which seemed to him to deny God's freedom. For Arnauld argued that, if Leibniz's supposition were true, that the concept of Adam includes all that will happen to him and his posterity, it would deprive God of freedom of action now with regard to the human species. To the

[8]The *vis viva* controversy is usually regarded as having been resolved in 1743 by D'Alembert, who described it as a dispute about words (Hankins 1965). However, D'Alembert had little to do with the termination of the controversy, either theoretically, practically or historically, and it was only in the middle of the nineteenth century that the simultaneous use of momentum and kinetic energy for the solution of problems in elastic collision began to appear in textbooks (Iltis 1970).

Landgrave (*A* I **4**, pp 400–2) Leibniz remarked that he received this reply with a mixture of amusement and sadness that the good Arnauld appeared to have lost part of his wits. He now understood why Malebranche and other friends had lost their patience with him.

Leibniz explained to Ernst that Arnauld reasoned like the Socinians, who conceived God in a human manner, imagining that God made choices as each situation arose—this would indeed limit his freedom, for his present choices would be restricted by those made earlier—whereas God's choices were interrelated and made together. Moreover, the concept of Adam included his free actions. All this, Leibniz pointed out, he had expressly stated in the summary. For having said that the concept of an individual contained everything that would happen to him, so that the concept provided the *a priori* proofs or reasons for the truth of each event, he added that these truths, though certain, were nevertheless contingent, being based on the free will of God and of creatures. It was true that their choice always had its reasons, but these 'inclined without necessitating'.

On 13 May 1686, Arnauld (*GP* **2**, pp 25–34) apologised to Leibniz for his tactlessness and sought further clarification of the notion of an individual concept. Having responded to the request in a letter of 14 July 1686, Leibniz added an exposition of some of the other important doctrines in his *Discourse* (*GP* **2**, pp 37–59). First he refers to the principle of logic which facilitates the *a priori* demonstrations from its complete concept of everything that happens to an individual substance. This is the principle that, in every true affirmative proposition, whether necessary or contingent, the concept of the predicate is contained in that of the subject. From this principle Leibniz infers the principle of sufficient reason, or, as he calls it here, the common axiom that there is a reason for everything that happens. To Arnauld (*GP* **2**, p 62) he expressed the view that it was not always within our power to carry out the analysis needed for an *a priori* demonstration. In a manuscript dating from this time, however (Couturat 1903, pp 376, 388–9), Leibniz made a distinction between necessary and contingent truths which has an important bearing on his metaphysics. There he explains that an *a priori* proof of a proposition demands an analysis which shows how the concept of the predicate is contained in that of the subject. In the case of necessary truths or truths of reason, this analysis involved a finite number of operations and was therefore within the capacity of men. But in the case of contingent truths or truths of fact, an infinite analysis was needed, which only God could supply. A real distinction between necessary and contingent truths, such as the one indicated here, is a prerequisite for Leibniz's conception of freedom. A truth is necessary if the opposite would imply a contradiction. The opposite of a contingent truth would not imply a contradiction, but there must be some reason why it is true. The principle of sufficient reason supplies the *a priori* reason, but this is in the nature of a cause which involves the will without compulsion. Consequently, contingent propositions have *a priori* proofs of their own truth but do not have demonstrations of necessity.

Another consequence of Leibniz's doctrine of truth, as embodied in the principle that, in every true proposition, the concept of the predicate is contained in that of the subject, is that every individual substance expresses the whole universe after its own fashion and is like a world apart, independent of everything except God. This independence however, as Leibniz explains to Arnauld, does not prevent commerce between substances, for as all created substances are produced by God in accordance with the same plan and are an expression of the same universe, they harmonise exactly among themselves. Interaction between substances should therefore be understood as a reflection of this harmony rather than a real physical influence. Thus Leibniz interprets the statement that one acts on another to mean that one is a more distinct expression than the other of the cause or reason for the changes in both, in the same way that we attribute motion to a vessel rather than the whole sea. Many philosophers, he adds, have been obliged to agree that, when it is a question of the union of mind and body or the active and passive relationship of one mind with another, direct interaction is inconceivable. Leibniz dismissed the doctrine of occasional causes, which had been introduced to resolve this difficulty, because it seemed to him to require a continual miracle, whereby God constantly changed the laws of bodies on the occasion of the thoughts of the mind. There remained only his hypothesis of the concomitance or harmony between substances. What happens to the mind is born to it in its own depths, without its having to adapt itself subsequently to the body, any more than the body to the mind. Each obeying its own laws, the mind acting freely and the body without choice, they agree one with another in the same phenomena.

Finally Leibniz explains to Arnauld that if the body is a substance and not a simple phenomenon like the rainbow, it cannot consist simply of extension but must possess something that one calls substantial form, which is akin to soul. Nevertheless, he was as corpuscular as anyone could be in the explanation of particular phenomena. Nature, he remarked, should always be explained along mathematical and mechanical lines, provided one knows that the principles of mechanics do not depend on mathematical extension alone but upon certain metaphysical reasons.

In September 1686 Arnauld (*GP* **2**, pp 63–8) expressed his satisfaction with Leibniz's explanation concerning the complete concept of an individual, which he had at first found so shocking, but asked for more light to be thrown on the hypothesis of concomitance or harmony between substances and the substantial forms of bodies. In this letter Arnauld also referred to Leibniz's 'Brevis demonstratio', mentioning that, according to his recollection, though he had not studied these matters for twenty years, Descartes in his statements about machines did not take velocity into account. Finally, he enquired of Leibniz the latest news concerning the calculating machine and the watch on which he had been working during his years in Paris.

In his reply of 8 December 1686 (*GP* **2**, pp 73–81), Leibniz suggested to Arnauld that he had himself clarified the hypothesis of harmony. On the question of substantial forms, however, Leibniz elaborated his ideas in a series of detailed answers to Arnauld's questions. The first point raised by Arnauld was that, since the body and mind are two distinct substances, it seems that one is not the substantial form of the other. To which Leibniz replied that, in his view, the body in itself (that is, the cadaver) is not a substance, for like a heap of stones, it lacks substantial unity. Moreover, he was able to point out that his position was consistent with the declaration of the Fifth Lateran Council (1512–17), that the soul is the substantial form of the body. Another difficulty raised by Arnauld was that, if the substantial form were to give unity to a body, it would have to be unextended, indivisible and hence indestructible, like the soul. Leibniz agreed that substantial forms had the properties stated by Arnauld, and he was inclined to believe that all births of animals lacking reason and not meriting a new creation were merely transformations of another animal which was alive but imperceptible; as an example he mentioned the changes observed in the silkworm. Thus all the crude souls would have been created at the beginning of the world, according to the fecundity of seed mentioned in Genesis, but rational souls were totally different from those already mentioned, being capable of reflection and imitating in miniature the nature of God. These needed a special creation.

Concerning bodies in general, Leibniz indicated to Arnauld that he could not say with absolute certainty whether these—for example, the sun, earth, moon, trees or even animals—were animate, or at least substances or even simple aggregates of many substances. But if there were no bodily substances (that is, bodies united by substantial forms), it would follow that bodies were only true phenomena like the rainbow, for matter being endlessly divisible, it would not be possible to arrive at anything that could be described as an entity, unless this were an animate machine whose substantial form created substantial unity independent of the external union of contiguity. Moreover, he added, if there were no bodily substances, then it would follow that, apart from man, there was apparently nothing substantial in the visible world.

As the correspondence continued through 1687, Leibniz answered further queries from Arnauld, occasionally introducing new ideas but usually reiterating points already made. In his letter of 9 October 1687 (*GP* **2**, pp 111–29) he makes clear to Arnauld that matter considered as the mass in itself[9] is only a pure phenomenon or well-founded appearance, as are also space and time, while the real entity is the animate substance to which the

[9]That is, as he remarks in his summary at the end of the letter, the mass when we consider only what is divisible in it.

matter belongs. Again he explains to Arnauld that he does not hold every substance to be a mind; the worm does not think. As we have not experienced the functions of other forms, he explains, we can have no clear notion of them. On the basis of anatomical analysis, however, Malpighi was inclined to believe that plants can be included in the same category as animals, while the experiments of Leeuwenhoek had shown an almost infinite number of little animals in the smallest drop of water, and Swammerdam's work seemed to confirm the idea of preformation and transformation of animals that his theory implied.

On several occasions Leibniz thought of publishing the correspondence with Arnauld and revised it for this purpose but nothing came of the idea. There is a clear reference to the correspondence and the *Discourse* at the beginning of the first published account of his metaphysics, which he contributed to the *Journal des Sçavans* in 1695 under the title 'Système nouveau de la nature et de la communication des substances', for there he remarks that he conceived the system some years before and discussed it with one of the greatest theologians and philosphers of the time. That Leibniz himself regarded the *Discourse on metaphysics* as marking a decisive stage in the development of his philosophy is shown by his remark to Thomas Burnet in a letter of May 1697, that he was only satisfied with the philosophical ideas he had held from about 1685(*GP* **3**, p 205). Although many of the ideas had been germinating from his university days and are to be found in earlier works—for example, the subject–predicate logic in *De arte combinatoria,* the idea of harmony in the *Confession of nature against atheists,* the principle of sufficient reason in the letter to Wedderkopf, substantial form in the letter to Thomasius and the idea of the best in a note written after his meeting with Spinoza—the *Discourse on metaphysics* is the first of Leibniz's writings to bind them together into a coherent system.

There are indications that Leibniz intended the *Discourse on metaphysics* as an introduction to his projected *Catholic demonstrations*; for, at the same time as he was replying to Arnauld's queries, he was also making requests to the Landgrave Ernst to obtain from Arnauld a declaration that the opinions he expressed (even if Arnauld believed them to be false) contained nothing contrary to the Catholic faith. Elsewhere, Leibniz reminded the Landgrave, Arnauld had written that the Church should not make difficulties concerning philosophical opinions that had nothing to do with the faith (*A* I **4**, pp 404–6). Arnauld and the Landgrave, however, had more interest in Leibniz's conversion than in his plea for religious tolerance with respect to philosophical views. Thus Arnauld remarked to the Landgrave (*A* I **4**, pp 443–4) that Leibniz had very curious views about physics, which seemed to him scarcely defensible, and that it would be preferable if he gave up such speculation and applied himself to the choice of the true religion. The Landgrave in turn warned Leibniz (*A* I **4**, p 444) that he could not set any hopes on his salvation if he did not become a Catholic.

The aftermath of the Harz project

Despite his disappointment when the Duke decided that the project for draining the mines with windmills should be abandoned, Leibniz could not immediately detach himself from the Harz (*A* I **4**, p xliv). Soon after the decision, he expressed his opinion to the Duke (*A* I **4**, pp 197–8) that the mining officials would one day recognise the utility of his project, especially the advantage of the horizontal windmill, which he believed he had clearly demonstrated. Thereafter he made renewed efforts to find ways and means of improving mining technology. For example, he proposed a device with containers rising and falling on an endless chain, so that the water power could be applied entirely for the purpose of raising the ore (*A* I **4**, pp 210–11). His remark to Otto Grote in a letter of March 1686 (*A* I **4**, pp 259–60) that the mining officials agreed with him in conversation but suffered loss of memory when they had to make decisions in the Mining Office, indicates that he did not receive any more cooperation with his new invention than he had received in the case of the windmills.

If the device could be made to work, the increase in efficiency would be considerable, for as Leibniz pointed out, the weight of the single chain employed in the usual technology could, in a deep mine, be greater than that of the ore being raised, whereas his device entirely eliminated the waste of power in raising the chain (*A* I **4**, p 261). A successful test was evidently carried out, for in a letter to Grote of 1 April 1686 (*A* I **4**, p 264), Leibniz declared that he was well pleased. The mining officials, he added, were slow to accept new ideas, but the great advantage of his invention having been clearly demonstrated, he expected they would begin to see reason. Finally, he remarked to Grote that, when he had been reimbursed for the money he had spent, he would say adieu to the mines, for his only desire had been to obtain the satisfaction of demonstrating something of utility and consequence, which they could use if they wished. The accounts for the whole Harz project were finally settled in December 1686, when Leibniz received a balance of 300 taler from the Mining Office in Clausthal (*A* I **4**, p 305).

The many visits Leibniz made to the Harz mountains in connection with his projects in mining technology provided him with opportunities to pursue his interest in geology. The most fruitful visit in this respect, however, took place in the autumn of 1685 when he was no longer preoccupied with the problems of pumps and windmills. On this occasion he spent about a week sightseeing and seeking geological specimens in company with his secretary Brandshagen and another helper Christian Essken, travelling in a robust farm cart. They visited caves containing bones and teeth of prehistoric animals. Leibniz obtained a few specimens, which he described in his *Protogaea,* written in 1691 (*D* **2**, 2, p 181) where he supplements his geological findings in the Harz with the observations he made in Southern Germany, Dalmatia and Italy in the course of his long journey in search of historical records.

A new commission

When Duke Ernst August succeeded his brother in January 1680, Leibniz had written to Franz Ernst von Platen, suggesting the composition of a short but accurate history of the House of Brunswick-Lüneburg that would place the emphasis on the more recent period (*A* I **3**, p 20). It was at this time that he first gave serious attention to the problem, his interest having been aroused by the history of the Guelfs published in 1677 by Philipp Jakob Spener (*MK,* p 59). His attention was again turned to the problem of the Duke's ancestors, when, in April 1685, he received from the Court poet Bartolomeo Ortensio Mauro, who was then with the Duke in Venice, a request for an opinion concerning an extract from a work on genealogy by the Abbot Teodoro Damaideno, who claimed to trace the House of Este back to the Romans (*A* I **4**, pp 495–6). In his report to the Duke (*A* I **4**, pp 191–6) Leibniz pointed out that history and genealogy had been reduced to a science, particularly by Dutch and French historians—he was himself influenced especially by Jean Mabillon—so that accurate documentation based on original sources and contemporary authors was demanded.[10] He did not think the House of Este could be traced back to the Romans and he would not himself contemplate going back over two thousand years, for the scholars of their time would find the idea of such an enterprise as amusing as that of a certain theologian of Venice who wished to trace the Habsburgs back to Noah's Ark. If he could fill in two gaps, Leibniz informed the Duke, he could trace the House back to about AD 600. But for this purpose he would need to undertake journeys in search of the sources, which were usually to be found in monasteries.

Having received from Otto Grote the Duke's decision concerning the end of the Harz project on 14 April 1685, Leibniz seems to have turned to historical research with growing enthusiasm. On 22 May 1685 (*A* I **4**, pp 197–8) he reminded the Duke that, in conformity with his wishes, he had occupied himself for some time with the history of the House of Brunswick-Lüneburg. To proceed further, however, he would need to undertake journeys in search of source materials and he requested from the Duke the financial security that would enable him to work on the history without having to ask continually for expenses. Leibniz was supported in his attempt to secure financial independence by Grote, who informed the Duke that Leibniz would be willing to undertake the writing of the history of the House of Brunswick-Lüneburg if his salary were to be converted into a pension for life (*MK,* p 75). The Duke agreed to this suggestion in a formal document of 10 August 1685 (*A* I **4**, pp 205–6). In consideration of his acceptance of the new commission, he was also relieved of the usual duties of the Chancellery, granted travelling expenses and a permanent secretary, besides being given the rank and title of Privy Counsellor for life.

[10]On Leibniz as a prime founder of modern historiography, see Meinecke (1959).

From this time on, the composition of the history of the House of Brunswick-Lüneburg became Leibniz's principal official duty. It gave him the opportunity to travel and meet influential statesmen and scholars, while it did not prevent him from continuing to make his own contributions to science and philosophy. In the autumn of 1685 Leibniz received further information concerning the genealogies of Damaideno from Gerhard Corfey, the Secretary of War at the Court in Hanover, to whom he expressed the view that, although Damaideno's proofs might be considered good in Italy, they would not be accepted elsewhere (*A* I **4**, pp 213–16). In December he sent a detailed account of his criticisms to Damaideno himself (*A* I **4**, pp 534–9). Then early in January 1686 (*A* I **4**, pp 544–5) he wrote to Daniel Papebroch, a Jesuit in Antwerp, from whom he hoped to receive advice concerning medieval sources.

By the following year, Leibniz had already started on his search in the vicinity of Hanover. Thus on 30 April 1687, he wrote to the Landgrave Ernst from Göttingen, remarking that he had been travelling continually for fifteen days, taking in Administrative Districts and Abbeys, in search of historical documents (*A* I **4**, pp 431–2). A week or two later, he told the Landgrave that he planned a long journey that would last all the summer (*A* I **4**, pp 432–9) and that he hoped he would be able to pay him a visit in Rheinfels at that time. In fact he spent the summer visiting libraries and archives in Wolfenbüttel, Lüneburg and Celle, besides making short stays in Brunswick. Then in the autumn he made preparations for a journey to southern Germany and Vienna. As a result of his finds in Augsburg, however, this journey in search of the origins of the House of Brunswick-Lüneburg had to be extended to Modena in Italy.

6
Grand Tour to Southern Germany, Austria and Italy
(1687–1690)

When Leibniz set out from Hanover towards the end of October 1687, he carried with him recommendations from Bartolomeo Ortensio Mauro, the Duchess Sophie's secretary and Court poet, to the famous musician and diplomat Agostino Steffani and his brother, the Bavarian Undersecretary of State, Ventura Terzago, both of whom he hoped to meet in Munich (*A* I **4**, p 657). While the primary aim of his journey was the search for historical documents relating to the origin of the House of Brunswick-Lüneburg, Leibniz lost no opportunity for meetings with scholars and visits to museums and other places of interest on the way, for the purposes of which he made many detours and extended visits. Among his many interests, including geology, mining technology, natural history, reform of the coinage and Chinese culture, the reunion of the Churches occupied a prominent place in his thoughts at this time, especially after his conversations with Cristobal de Rojas y Spinola, since 1685 Bishop of Wiener-Neustadt, about forty miles south of Vienna. Besides composing numerous memoranda, including plans for an Imperial College of History and a General Reference Library in Vienna, and successfully completing an important diplomatic and political commission for the Duchess Sophie in Modena, Leibniz also found time to contribute to the *Acta Eruditorum* papers in which he applied his infinitesimal calculus to the problems of resisting media and planetary motion and also to draft the first part of a treatise on dynamics. If anyone believed he had wasted his time, Leibniz wrote to Otto Grote (*A* I **5**, pp 325–6) following an illness that had kept him indoors for three weeks, they did him an injustice, for he had often spent the whole night reading manuscripts, adding that it was not necessary for a man to live but only to work and fulfil his obligations. Although this last remark was written in a particular context, it could be said to describe Leibniz's general attitude of complete dedication to his role as a scholar in the service of mankind.

On 1 November, having arrived in Hildesheim, Leibniz made two visits; first to the Capuchin Dionysius Werlensis, who reported on his conversations with Bishop Rojas y Spinola concerning the problems of Church reunion, and secondly to the museum of natural curiosities of the late Dr Friedrich Lachmund, where he saw fossils of plants and animals which he later evaluated in his *Protogaea*. The next day, he viewed another natural history collection in the library at Kassel,[1] then spent four days in Frankenberg with the Hessen director of mines, Johann Christian Orschall. With his host he conversed on Henry More and Johann Baptista van Helmont and found diversion in reading a work of the Protestant mystic Valentin Weigel, *Der goldene Griff: das ist, alle Dinge ohne Irrthum zu erkennen*. In Marburg, his next stop, Leibniz visited the Church of St Elizabeth and conversed with Dr Johann Jakob Waldschmidt, the personal physician to the Landgrave Karl von Hessen-Kassel, on experiments such as the killing of animals by injecting air into their veins and the observations of plants placed in a vacuum. Just before the middle of November, he reached the first highlight of his journey south, when he arrived at the small fortress Rheinfels, to pay his promised visit, somewhat later than originally planned, to his friend the Landgrave Ernst.

Guest of Landgrave Ernst von Hessen-Rheinfels

Every day during over two weeks as the guest of the Landgrave Ernst von Hessen-Rheinfels, Leibniz conversed with his host on historical and religious themes, especially the problems concerning Church reunion. At the end of the visit Leibniz presented to his host a memorandum (*A* I **5**, pp 10–21) containing his own thoughts on the best way to achieve the reconciliation they both desired. Leibniz saw in Church reunion a political problem, whose solution was necessary if Europe was to regain unity, especially in the face of French aggression (*A* I **5**, p xxxv). Following the revocation of the Edict of Nantes by Louis XIV in 1685, the persecution of the Protestants in France had been intensified, so that they had been deprived of all civil rights. A number of German states had formed a defensive alliance in 1686, known as the League of Augsburg, but clearly the political isolation of France and the political stability of England would be greatly promoted by a general reconciliation and reunion of the Catholic and Protestant Churches.

Of all the methods of reunion that had been proposed, Leibniz declared the one negotiated with the approval of the Emperor between Rojas y Spinola, Bishop of Wiener-Neustadt, and some Protestant theologians, to

[1]The exhibits in these two collections are catalogued by Leibniz in *RJ*, ff 25–9.

be the most reasonable, although he held the view that, without the support and enthusiasm of a great personage, such as the Pope, Emperor or leading Prince, either Catholic or Protestant, the practical difficulties would be insurmountable.

Tolerance, though a necessary beginning, was not sufficient by itself. Conferences and discussions, however, had usually proved fruitless, as had controversies in writing, because the proponents had sought to impress their own side rather than reach a solution. The great merit of the method proposed by the Catholic Bishop Rojas, after consultation with Protestant theologians, was that it could accommodate the principles of both Catholics and Protestants. As far as he had been able to understand, Leibniz explains, the great principle of the Catholics is that a Christian belongs to the internal communion of the Church and is neither heretic nor schismatic when he is in the mind of submission, quick to believe and desiring to apprehend what God reveals, not only in the Scriptures but also in the interpretations, supposedly revealed through divine guidance, laid down in a legitimate ecumenical Council. From which it follows that one who is in this mind of submission when he believes some heresy in ignorance will not on that account be a formal heretic, nor, when he is excommunicated, a schismatic. Although the Council of Trent was tacitly recognised by the Catholics at that time, there were Catholics, Leibniz pointed out, who rejected other Councils and had not on that account been declared to be heretics. The principle of the Protestants was to be found in the Confession of Augsburg. Now all who have adhered to this Confession are bound, by the declaration in the introduction, to accept the judgment of the Church determined by a General Council convened and conducted in due form. Their position is that such a Council has not yet taken place, for they reject the Council of Trent as illegitimate. As a first step towards reconciliation, Rojas had asked in the Courts of several Protestant Electors and Princes for a positive declaration of their willingness to submit to the judgment of a General Council and for their views concerning the form such a Council should take in order to avoid any disagreement as to its legitimacy. On the supposition that a favourable response were forthcoming, he had also raised the possibility of finding the means for a preliminary but true reunion, so that the Protestants of Germany and Hungary especially could be reconciled with the Catholics pending the future Council, despite their rejection of the Council of Trent.

If, in rejecting the Council of Trent, the Protestants were in error through ignorance, then this was not heresy. As for the particular doctrines defined in the Council of Trent, it was clear that the holding of particular errors, provided they were not contrary to the general principle of the Catholics, did not make a heretic, otherwise several saints who were not aware of these errors would have been heretics. There was a precedent, Leibniz remarked, for the Catholics to accept that the Protestants were not ready to be persuaded to approve the Council of Trent and yet to accept them as fellow

Christians. This was set by St Paul who, in circumcising Timothy, made a concession to the Jews of that time whom he recognised were not ready to be persuaded of the validity of the Council of Jerusalem. The Catholics, he added, should have no difficulty in according to the Protestants such things as the marriage of priests, the orders of ministers and the communion of the laity in the two kinds, but in return the Protestants would not have to condemn fellow Christians who practised other rites and believed other dogmas decided at the Council of Trent.

During his stay in Rheinfels, the Landgrave communicated to Leibniz a suggestion of Karl Paul von Zimmermann, Privy Counsellor of the Archbishop of Cologne, that he might consider becoming Chancellor of the diocese of Hildesheim. Of course his present duties and religion would not permit him to accept, but in order to have a permanent record for his own satisfaction, he asked the Landgrave to repeat in writing what he had told him concerning Zimmermann (*A* I **5**, p 4). The Landgrave sent Leibniz a note to this effect, excusing himself for neglecting his guest on that morning owing to a slight indisposition (*A* I **5**, pp 7–8). When he said goodbye to the Landgrave on 2 December, Leibniz took with him a recommendation to the Elector Philipp Wilhelm of the Palatinate (*A* I **5**, pp 23–5).

The road to Munich

On leaving Rheinfels Leibniz made his way to Frankfurt am Main, where he arrived about the middle of December. During his stay of three days he inspected the natural history collection and books on caterpillars of Maria Sibylla Merian[2] (*MK*, p 84) and renewed his acquaintance with the orientalist Hiob Ludolf (*A* I **5**, p 25), to whom he mentioned the plan proposed by Franz Christian Paullini and W E Tentzel for the setting up of an Imperial College of History. Also he discussed with Ludolf the recently published Confucius edition of P Couplet and other Jesuits, which the Frankfurt publisher Johann David Zunner had just received from Paris. He looked forward to the day when the original Chinese and the means of understanding it would be accessible in Europe. By good fortune Leibniz was able to meet again another old acquaintance, the Swedish diplomat Esaias von Pufendorf, with whom he had discussed the philosophy of Descartes in

[2]Maria Sibylla Merian, *Der Raupen wunderbare Verwandelung und sonderbare Blumen-Nahrung*, 2 vols (1679–83 Frankfurt am Main). This work contains a wonderful collection of coloured engravings. At the beginning of 1695, Crafft mentioned to Leibniz in a letter from The Hague that he had seen them in Amsterdam, where the artist was then living. He also reported that they were incomparably better than the lifeless products of the colour-printing process (needing retouching with water colours) that had been invented by Johannes Teyler (Grieser 1969, p 226).

Hamburg and who was passing through Frankfurt on his way to Paris. It seems that, while he was in Frankfurt, some of Leibniz's friends hinted to him the possibility of marriage with a rich young spinster if he could gain the favour of her uncle, the Court Chaplain Hermann Barckhaus (Bodemann 1895, p 224).

Before leaving Frankfurt in the company of a young scholar, Friedrich Heyn, who had agreed to assist him on the way to Vienna by making extracts from manuscripts and rare books, Leibniz reported on his meetings and discussions there to his friend the Landgrave Ernst (*A* I **5**, pp 25–8), remarking that the next stage of his journey would take him to Aschaffenburg. In his reply, the Landgrave told Leibniz that there had been continuous wind and rain since he left Rheinfels (*A* I **5**, pp 33–5). He was evidently surprised that Leibniz had decided not to visit the Elector of the Palatinate in Heidelberg after all.

The attraction for Leibniz in Aschaffenburg was the collection of historical manuscripts belonging to his late Jesuit acquaintance Johann Gamans, who had planned to write a history of Mainz, including the dioceses of the suffragan bishops in Halberstadt, Hildesheim and Eichsfeld. Leibniz remarked to Otto Grote (*A* I **5**, pp 305–7) that there was no-one in Aschaffenburg qualified to study this valuable collection of historical documents and recommended that any opportunity to acquire it for Hanover should be taken.

On 25 December, four days after leaving Aschaffenburg, Leibniz arrived in Würzburg. Here he visited the monastery of the Benedictines, which possessed two libraries, one for new literature and one for old books, and also the Jesuit libraries. Apart from viewing interesting books, he collected information concerning the bishops of Würzburg[3] and Bamberg, and heard that, in Schweinfurt, there was a German chronicle going back to 1009 (*RJ*, f 3r).

From Würzburg, Leibniz travelled through Fürth, where he noted in his diary that there were living five hundred Jews with their own school, arriving in Nuremberg on 31 December. Besides visiting old friends from his student days, he spent his time seeing the sights of the town, which included an unusual collection of weapons and the Town Hall, in which were to be found Dürer's 'Adam and Eve' together with his impressive masterpiece 'The Four Evangelists'.[4]

Leaving Nuremberg on 7 January 1688, he arrived in Sulzbach two days later, where he was the guest of the well-known Cabbalist and alchemist Christian Knorr von Rosenroth, who, in 1680, had published a German translation of the *Hypothesis physica nova* without Leibniz's permission.

[3]These were members of the Schönborn family.
[4]'Adam and Eve' is now in the Prado, Madrid, and 'The Four Evangelists' in the Alte Pinakothek, Munich.

One of the topics of conversation was Knorr's work *Messias puer,* which never actually appeared in print, consisting of the life of Jesus as a boy, based on evidence from the old Cabbalists. Leibniz was impressed by Knorr and his work, as his laudatory comments to the Landgrave Ernst, Molanus and Ludolf testify (*A* I **5**, pp 43, 109, 235). During his stay in Sulzbach, Leibniz also met Elias Wolfgang Talientschger de Glänegg, a collector of fossils, with whom he discussed mineralogy and visited the lead mines nearby (*RJ*, f 6).

From Sulzbach Leibniz made a diversion through part of northern Bohemia in order to visit his friend Johann Daniel Crafft. On 20 January he wrote to the Landgrave Ernst (*A* I **5**, pp 39–44) from Chodenschloss on the frontier, where his expectation of finding his elusive friend had not been fulfilled, enclosing letters for the Landgrave to send to Arnauld and Huygens. To Huygens he remarked that he had not expected to see the problem of the line of uniform descent, which he had proposed to Catelan, honoured by a solution from him (*GM* **2**, pp 39–40). Towards the end of January he at last tracked down Crafft at Graupen (*A* I **5**, p 55), where the two friends discussed the extraction of ore, gold-washing, dye-manufacture, technical innovations and coinage reform, laying plans for projects they hoped to present to the Emperor during a combined stay in Vienna later in the year (*MK*, p 86). Leibniz immediately drafted a memorandum on the reform of the coinage (*A* I **5**, pp 47–51) and in April communicated his thoughts to the Minister Albrecht Philipp von dem Bussche (*A* I **5**, pp 98–105) and the Duke himself (*A* I **5**, pp 114–29). Although the coins of the Empire and those of Brunswick had the same nominal value, the Brunswick coins contained more refined silver, so that Brunswick lost money each year by minting coins from the finer silver mined in the Harz. As all the silver came from the mines of Austria, Saxony and Brunswick, it would be to the advantage of these powers, Leibniz suggested, if they introduced equivalent weights of fine silver in coins of the same nominal value and agreed on the price of silver.

The return from Graupen in the first half of February took him to Freiberg, Marienberg, Annaberg and Ehrenfriedersdorf, where he saw a new pumping installation designed by the English engineer Kirkby, then through Geyer, Aue and Schneeberg to Karlsbad. This route was chosen in order to visit a number of mining towns, where he made contact with the mining officials and collected information concerning the quality, kind and quantity of minerals in each place, besides endeavouring to acquire a general picture of the mining techniques employed. From Karlsbad he made his way back to Sulzbach and then on to Amberg where he arrived on 21 February and stayed for about a week, during which he saw a theatrical performance and visited a round chapel called 'Maria Hülfe' on the nearby mountains, where there were twelve crutches hanging behind the altar, left by the lame who had been cured (*RJ*, f 8).

From Amberg Leibniz journeyed on to Regensburg, arriving on 12 March. Here he stayed a fortnight, taking in the atmosphere of the town where the permanent Imperial Diet held its deliberations, and renewing his acquaintance with the Imperial Secretary (later Archivist of the Prince Bishop of Passau) Philipp Wilhelm von Hörnigk, whom he had met in Hanover during the visit of Rojas y Spinola in 1679. On 25 March, Leibniz informed the Landgrave Ernst (*A* I **5**, pp 63–79) that he would now proceed to Munich and Augsburg, remarking that he would be able to receive letters through Hörnigk. The next day he left Regensburg, having deposited some of his luggage with Hörnigk, and arrived in Munich at about noon on 30 March.

Munich and Augsburg

Having settled into the 'White Swan' in the Weinstrasse, Leibniz called on the Kapellmeister Agostino Steffani, who obtained from the Elector permission for him to visit the library. He was introduced to the librarians on Monday 5 April (*A* I **5**, pp 80–1) by Steffani. On that day and the next he worked in the library; meanwhile, on the instructions of the librarians, he addressed a formal request to the Elector for a more extended use of the library (*A* I **5**, pp 93–4). To his amazement, the Elector, acting on the advice of certain Counsellors who suspected Leibniz's motives, denied him any further access to the manuscripts. The embarrassed Steffani promised to resolve the problem but his representations were evidently unsuccessful (*A* I **5**, p 95). He had no difficulty, however, in arranging for Leibniz to see the Duke's residence and art gallery. The visit to these attractions cost him fourteen gulden[5]. Leibniz filled the rest of the week with sight-seeing in the town, noting especially the paintings in the churches, frescos on the walls of houses and four factories which employed many small boys and about forty girls all dressed in blue (*RJ*, ff 10v–11r).

In the course of his two visits to the library, Leibniz read the German manuscript which formed the basis of the work on Bavarian history commissioned by the Elector Maximilian and published in Latin by the famous historical writer Johann Turmair, also known as Aventin. The German version contained references to sources that were lacking in the printed work and Leibniz saw that these related to an old manuscript in a monastery at Augsburg.

On Sunday 11 April he left Munich and arrived in Augsburg at noon on the following day. With the assistance of the town trustee, Dr Daniel Mayr, he located the codex 'Historia de Guelfis principibus', though not without some difficulty, in the Benedictine monastery. Study of this manuscript

[5]The gulden was equivalent to two-thirds of a taler.

produced what was probably the most significant historical discovery of his tour, for it enabled him to demonstrate beyond all doubt the common origin of the Guelfs and the Counts of Este. Previously this had been suspected but lacked proof. He found that the Counts were designated *Estensem* in old but clear writing, so that the designation *Astensem* to be found in Aventin, which had thrown doubt on the connection of Brunswick-Lüneburg with the House of Este, was shown to be a corruption. Immediately after this discovery of the common origin of the Guelfs and the House of Este, Leibniz wrote to Francesco de Floramonti, the ambassador of Brunswick-Lüneburg in Venice (*A* I **5**, pp 129–30), hoping to obtain an introduction to the Court in Modena that would enable him to pursue further researches on the origin of the House of Este.

After some sight-seeing in Augsburg, which included visits to the arsenal, waterworks and cathedral, Leibniz returned to Munich, where he arrived at noon on Easter Monday, 19 April. A stay of ten days gave him the opportunity of communicating with friends and Court officials in Hanover before going on to Vienna. First, he replied to a letter from the Duchess Sophie, which was waiting for him on his return to Munich. She had been overjoyed on seeing his letter to her secretary Mauro relating how Steffani and Terzago had assisted him, for since his departure, no one had heard from him, and it had been rumoured in Hanover that he had gone to the next world in order to search there for the origin of the House of Brunswick (*K* **7**, pp 13–14). To Sophie Leibniz remarked (*K* **7**, pp 10–13) that he was pleased to have nearly finished his researches in Bavaria, for some of the Counsellors of that country—evidently those who had effected his exclusion from the library—were a little uncivil. He also described two curious events he had witnessed on his travels. On Good Friday, he had seen a procession crossing a bridge in a village near Munich. In the middle of the procession, four men surrounded and beat a man representing the Saviour. One of the four, on passing below a crucifix on the bridge, instead of beating the actor, struck the crucifix, which seemed very strange. Then on Easter·Monday, he found that it was the custom for the preacher to read a story, called an 'Easter fairy tale' (*Oster-mährle*). The one he heard was taken from a German book of humour but changed a little.[6]

To his friend Molanus, Leibniz (*A* I **5**, pp 107–9) reported that he had seen manuscript annotations on the whole Bible, consisting of relevant quotations from secular authors, written by the lawyer Georg Remus. Leibniz described his historical discoveries and the difficulties he had encountered in Munich in letters to the Minister Albrecht Philipp von dem Bussche (*A* I **5**, pp 98–105) and the Duke Ernst August himself (*A* I **5**, pp 114–29), to whom he explained that the journey to Munich was

[6]The book was a well known work by Hans Jacob Christoffel von Grimmelshausen, *Der abenteuerliche Simplicissimus*, first published in 1669.

undertaken expressly for the purpose of inspecting the manuscripts of Aventin, which he had known to be there. He also communicated his discoveries to Otto Grote (*A* I **5**, p 139), who was pleased to receive news from him after the months of silence. After congratulating him on his demonstration of the common origin of the Houses of Brunswick and Este, Grote pressed him to return to Hanover as the Duke intended to make an opera of the history of Henry the Lion, for which his presence and advice would be essential (*A* I **5**, pp 138–40).

Leibniz left Munich at the end of April, travelling to Passau, where he boarded a ship to take him down the Danube. He passed Linz on 5 May and arrived in Vienna three days later.

Vienna

On arriving in Vienna Leibniz paid a visit to the Privy Counsellor and ambassador of Hanover and Celle, Christoph von Weselow, to whom he offered his legal assistance with the establishment of the claims of Brunswick-Lüneburg—in opposition to those of Brandenburg—on the Dukedom of East Friesland. Leibniz was able to provide the documentary proof of the basis of the claim—that, in the past, ancestors of the House had ruled in Friesland in the feminine line—for which Weselow had already applied without success to Hanover (*A* I **5**, pp 142–3). To Sophie he reported on 30 May that the Princess of East Friesland, Christine Charlotte von Württemberg, had had an audience with the Empress, and that Weselow now hoped for a resolution in her favour (*K* **7**, p 18).

In a letter to Otto Grote (*A* I **5**, pp 143–5), Leibniz reported that on his way to Vienna he had seen some monasteries founded by the ancient Guelfs close to the river Inn and also ancient documents and statues of the ancestors of the House of Bavaria. Since Weselow had been unable to help him owing to arthritis which had kept him in bed, he would attempt to make the acquaintance of the Emperor's Librarian, Daniel Nessel, in order to gain access to the manuscripts of the Imperial Library.

After viewing the treasury in the Castle, Leibniz made his first visit to the Library, where he saw a German Bible written on parchment by order of King Wenceslas of Bohemia, a Lutheran Latin Bible with annotations in Latin and German, and two volumes of mathematics in Chinese with many figures printed on silk (*RJ*, ff 17r–19r). Towards the end of May, Prince Karl Philipp, who had fought against the Turks as an officer in the service of the Emperor, arrived in Vienna with a letter for Leibniz from his mother, the Duchess Sophie (*K* **7**, pp 14–16). In his reply (*K* **7**, pp 16–19) Leibniz commented on the political situation and the disposition of the armies against the Turks. On a more personal note he expressed the hope that the baby due to be born to her daughter Sophie Charlotte in July would give

Brandenburg a new Electoral Prince. Also he confided to the Duchess his intention to visit for two or three days the Bishop of Wiener-Neustadt, Cristobal Rojas y Spinola, who would not, he thought, be sorry to see him.

Leibniz set out for Wiener-Neustadt and his meeting with Rojas on the Tuesday after Whitsuntide. For the first time he was able to see the extensive correspondence the Bishop had conducted on the problems of Church reunion, including the authentic pieces which made clear that the Pope, Cardinals, the General of the Jesuits and the Master of the Palace (traditionally a Dominican concerned with the censure of books and doctrines) had approved his plans and negotiations (*A* I **5**, pp 676–7). Reporting to Sophie on his visit (*K* **7**, pp 37–40) to Wiener-Neustadt, Leibniz sent a copy of the letter the General of the Jesuits had written to the Emperor's confessor on the subject in 1684 (*K* **7**, pp 40–1) and he defended Rojas against the criticism he had heard, that having secured temporal advantages, he had abandoned his pious designs. On the contrary, Leibniz told the Duchess, Rojas was anxious to take up the threads of the negotiations again, as soon as there was some prospect of success. The time indeed seemed opportune, for Bossuet had informed Rojas that the King of France would not oppose his plans and both Pope and Emperor were well disposed to the idea, while the theologians of Hanover had already shown enthusiasm. The Duchess could make a great contribution by using her influence to gain the support of Berlin, which had previously voiced its opposition through the late Count of Rabenac. A concerted effort was made by Leibniz and the Bishop to enlist Sophie's assistance. Rojas himself wrote to Sophie and both Rojas and Leibniz enclosed copies for Sophie in the letters they wrote to Molanus. Also Leibniz reported to the Landgrave Ernst (*A* I **5**, pp 174–86) what he had learnt on his visit to Wiener-Neustadt.

In a letter written from Berlin on 24 August 1688 (*K* **7**, pp 44–5), where she had gone for the christening of her grandson, the Electoral Prince born to her daughter Sophie Charlotte in July, Sophie conveyed to Leibniz that the present mood in Berlin was not conducive to the serious consideration of the question of the reconciliation of the religions. An expensive funeral had recently taken place for the former Elector and the town was full of Huguenot refugees from France who cried 'anathema' to any mention of Rome. Also in the formulary of prayer said in Church, she reminded Leibniz, God was thanked for deliverance from Papist blindness. However, she had spoken to the Elector of Brandenburg, her son-in-law, to the Prince of Anhalt (who did not pass for a great theologian) and to two ministers about the matter and they had not raised any objections. Evidently they did not show much enthusiasm. Sophie remarked to Leibniz that Molanus had suggested to her that there should first be a reconciliation of the Catholics with the Lutherans before undertaking the more difficult task of a reunion with the Protestants generally. In his reply to this letter (*K* **7**, pp 46–7), Leibniz congratulated Sophie (he also wrote to Sophie Charlotte (*K* **7**, p 48)) on the birth of the Prince, and remarked that he hoped to be in Hanover before the end of the year.

In September 1688 Louis XIV devastated the Palatinate for the second time and declared war on the Empire under the pretext that the Emperor intended making peace with the Turks in order to attack France. This was the beginning of the war of the League of Augsburg, a defensive alliance against France. Writing to Leibniz on 16 September (*K* **7**, pp 49–50), Sophie declared her belief that the Prince of Orange would soon sail with a formidable fleet to protect the Reformed religion in England. Meanwhile the Dukes of Celle and Wolfenbüttel and the Elector of Brandenburg were sending armies to Holland to counter the danger of attack by France. It was hoped in Holland, she added, that peace would be made with the Turks so that the Emperor would have his hands free to deal with France. These political developments did not, however, diminish the efforts being made for the reunion of the Churches. Sophie informed Leibniz that she had replied to Rojas, who would no doubt show her letter to him.

Throughout the winter, Leibniz reported to Sophie on the political situation in Vienna. He believed (*K* **7**, pp 50–2) the Emperor would continue to push back the Turks if possible, though a Turkish emissary was on the way to Vienna. As to the negotiations with Rojas, he believed that the Protestant princes would be blameworthy if they failed to take advantage of the disposition of Pope and Emperor to grant what they should themselves be seeking with enthusiasm. For they had all to gain and nothing to lose, while the Catholic princes had little reason to concern themselves with the affair. Sophie, on her part, kept Leibniz informed of political developments in the north. In her letter of 4 November (*K* **7**, pp 56–9) she told him that William of Orange had set sail for England at the end of October with fifty ships. Some time previously, she had sent to England what had been agreed in Hanover concerning the reunion question, but not surprisingly, in view of the religious strife there, it had not been well received. With greater sense of realism than Leibniz had on this subject, she advised him that the time was not opportune for pursuing the plan of Rojas.

It was through Rojas that Leibniz gained introductions to the officers of the Imperial Court, in particular to the Chancellor Theodor Althet Heinrich von Strattmann, who in 1669 had been responsible for the printing of his pamphlet on the election of the Polish King (*A* I **5**, p xxxviii), and Count Gottlieb Amadeus von Windischgrätz, who later became Vice-Chancellor. On his return to Vienna from Wiener-Neustadt in the middle of June 1688, he obtained permission to use the Imperial Library. At one time he had borrowed more than thirty manuscripts which, among other things, confirmed what he had discovered in Augsburg concerning the origins of the House of Brunswick-Lüneburg (*A* I **5**, pp 232–4).

Meanwhile he received from the Vice-Archivist Georg Michael Backmeister in Hanover news of the removal of the Ducal library and his own personal possessions to a house in the town, so that the building could be demolished to make way for a new Opera House. The transfer, he was assured, had been supervised by his former servant Balthasar Reimers, who had taken special care of his personal belongings.

Leibniz found in the Imperial Library also some manuscripts concerning Lorraine. These he described to Claude François de Canon, the Minister of Duke Charles of Lorraine, whose claims he wished to support with documents (*A* I **5**, pp 206–7). On being driven out of his lands by the French, the Duke had taken refuge with the Emperor, to whom Leibniz also mentioned the material he had found on Lorraine (*A* I **5**, pp 271–4).

In September Leibniz had an opportunity to meet with delegates of Hungarian mining towns, especially Schemnitz, who were visiting Vienna, and to hear from them details of the mining there. He planned a visit to Hungary to see the mines for himself (*A* I **5**, pp 233, 251), but this visit was prevented by illness (*A* I **7**, p 354). He learnt that the Hungarians desired to buy lead from the Harz, as the Poles, from whom they had previously obtained their requirements, had increased the price. He reported this to Otto Grote (*A* I **5**, pp 232–4) as a commercial possibility that should be of interest to Hanover and offered advice concerning economical transport by water.

In the second half of September (*MK*, p 91) Leibniz and Crafft, who had arrived in Vienna the previous month (*A* I **5**, pp 205, 208), visited Wiener-Neustadt to see their common friend Rojas, and no doubt took the opportunity to discuss the proposals Leibniz intended to submit to the Emperor, especially those concerning coinage and textiles. Immediately before this visit Leibniz composed a manuscript on motion in a resisting medium, which formed part of the basis of a paper published in the *Acta Eruditorum* early in the following year (*LH* **35**, IX, 5, ff 22–5). Towards the end of October, he fulfilled an ambition of twenty years' standing when his request for an audience with the Emperor was granted (*A* I **5**, p 270). He made a number of proposals which were then developed in several memoranda. These included the project of an Imperial College of History (*A* I **5**, pp 277–80), reform of the coinage, reorganisation of the economy, improvement of trade and textile manufacture, establishment of an insurance fund, the taxing of luxury clothes (already in operation in Hanover), the setting up of a central State Archive, the conclusion of a State Concordat and the formation of a general reference library (*A* I **5**, pp 339–43). A little later, he made a proposal for the lighting of the streets of Vienna with oil lamps (using rape-seed oil), which Crafft would put into operation (*A* I **5**, pp 391–2).

During the months of November and December Leibniz was ill, suffering from what he described as the most severe attack of catarrh he had had in his life. This was accompanied by headache and a cough, which became worse if he ventured into the cold air, and loss of appetite, which made him weak.

Thanks to the consideration of the Librarian Daniel von Nessel, however, he was able to use books and manuscripts and even the catalogue of the Imperial Library in his room at the 'Steyner-Hof' (*A* I **5**, pp 314–15, 348–50). Writing to Otto Grote on 30 December (*A* I **5**, pp 325–6) he remarked that, owing to his illness, he had to abandon his plan to visit the Hungarian mines and was thinking only of his return journey to Hanover. In a letter to the Duchess Sophie, confirming that her son Friedrich August had been promoted to Major-General in the Imperial Army, he remarked that, if it had not been for his illness, he would have been on his way home (*K* **7**, p 62).

At the beginning of January 1689, just before he planned to return to Hanover, Leibniz discussed with the Imperial Privy Counsellor Gottlieb Amadeus von Windischgrätz (*A* I **5**, p 344) the proposals for a new Concordat between Pope and Emperor relating to the election of ecclesiastical princes and other Imperial prelates, and he composed a memorandum for him on the subject (*A* I **5**, pp 380–5). Then, in the middle of the month, his plans were suddenly changed, as a result of the arrival of a reply from the Brunswick-Lüneburg ambassador in Venice, Francesco de Floramonti, to his request for an introduction to the Court of Modena, informing him that Duke Franz II would allow him to use the archives for historical and genealogical researches. Writing to Otto Grote on 20 January 1689, he explained the change of plan, remarking that he would travel to Italy as soon as possible, taking the direct route to Venice, from where he would set out for Modena (*A* I **5**, pp 360–2). When he was in Augsburg, he related to Grote, he had asked Floramonti for assistance and the ambassador had consulted Count Dragoni, who made a special journey to Modena on Leibniz's behalf. The lack of response from Dragoni after several months had been taken by Floramonti and Leibniz himself as a tacit refusal, so the letter Leibniz had just received was a pleasant surprise. He explained to Grote that, since it would take at least five weeks to receive a reply from Hanover, and as he was confident that the Duke would approve his plan to visit Modena, he had decided to leave for Italy as soon as possible, in order not to waste time in Vienna. Another week, he thought, would be needed before he was well enough to travel.

Nearly three weeks passed before he actually set out from Vienna on the journey south, having first written letters to the Duchess Sophie and to von dem Bussche. On 5 February, he informed the Duke's Minister (*A* I **5**, pp 392–7) that he would leave in three days' time, for he had to hurry across the Alps before the snow melted and the road became muddy. The chief purpose of his letter to Bussche, however, was to report his confidential political conversations with Count Windischgrätz concerning the rather mistrustful relations between Hanover and Vienna. Before his arrival in Vienna, Sophie had warned him that, on account of the alliance that had existed between Hanover and France for many years, he might not be received in the friendliest terms (*K* **7**, p 15). In fact, he quickly won the confidence of the ministers in Vienna and assured Sophie (*K* **7**, p 18) that the alliance was regarded by some as only a neutrality which tended to affirm the peace, while the Empire should thank the alliance if this could contribute to the restoration of Holstein from the Danes. In the new situation created by the French invasion of the Palatinate and declaration of war on the Empire, there was a need to improve the understanding between Hanover and Vienna. Windischgrätz was of the view, Leibniz told von dem Bussche, that suspicion had been increased by the secrecy of Hanover, whose intentions could only be guessed, whereas other princes communicated their intentions to the Emperor, enabling him to concert with them the operations of the campaign that was soon to begin.

With his letter to Sophie (*A* I **5**, pp 366–7) Leibniz enclosed a copy of a letter of Rojas on the reunion problem. He asked Sophie to speak to the Duke about it and attempt with him to pass on the idea to Brandenburg, as if the suggestion came from them and not Rojas. Although he did not agree with all Rojas had said, Leibniz thought his plan reasonable and useful. In her reply (*K* **7**, pp 68–9) she expressed her disappointment that instead of coming home to see the opera of Henry the Lion, he was going further to find the origin of the House of Brunswick, so that 'those who come after us have not to seek ours'. Sophie evidently did not take as serious a view of these matters as did Leibniz. Remarking that she had difficulty in remembering the names of the heroes of history, she told him she was inclined to agree with Solomon, that all is vanity. Again, the enterprise of Rojas was laudable and his reasoning admirable, but all the world, she perceived, did not take the trouble to reason or understand reason. There was need to hope for an extraordinary revelation on this subject, she believed, and with a fine sense of humour declared that, as Christianity had entered the world through a woman, she would be overjoyed if the reunion of the Churches could be effected by herself.

After a last minute delay of two days, Leibniz set out for Italy on 10 February 1689, travelling first to Wiener-Neustadt, where Rojas gave him a recommendation to Cardinal Decio Azzolini in Rome (*A* I **5**, p 682). He then continued to Graz and Trieste, visiting the famous mercury mine at Idria on the way. From Trieste he probably travelled by ship to Venice, where he arrived on 4 March. It was while he was on this journey that his paper on the motions of the planets, described by him as an elegant specimen of his infinitesimal calculus, was published in the *Acta Eruditorum*.

In Venice, where he spent a month waiting impatiently for a letter from Count Dragoni to establish his relations with Modena, Leibniz began to compile, mostly from memory, a catalogue of books for the proposed Imperial reference library in Vienna (*A* I **5**, pp 428–62), which he completed on his travels over the next few months and sent to Strattmann.

To Sophie (*K* **7**, pp 70–2) Leibniz reported that he had visited her son Prince Maximilian Wilhelm and that, while waiting for letters from the

Court of Modena, he intended to visit Rome, where he hoped to see Queen Christina of Sweden; for despite the rumours of her death, he was pleased to be able to inform the Duchess that she was out of danger. He also mentioned to Sophie that, when he left Vienna, it was not known that Baron von Platen was due to arrive. Indeed, it was said in Vienna that one never knew where one was with her husband the Duke, because he never communicated his intentions to the Emperor. However, Leibniz added, the arrival in Vienna of the Grand Marshall, who had the Duke's confidence and knowledge of his secret affairs, should result in a better understanding between the Duke and the Emperor.

Papers on resisting media and planetary motion

In his letter to Otto Mencke (*NC* **3**, pp 3–5) accompanying the papers he was sending for publication in the *Acta Eruditorum,* Leibniz explained the circumstances of their composition. Having been out of touch with new publications owing to his travels, he was delighted to receive from a friend some issues of the *Acta Eruditorum,* in one of which he came across the review of Newton's *Principia* (*AE* June 1688, pp 303–15). Three topics especially caught his attention and these were the subjects of his papers. First, there was the problem of optic lines, concerning which he had long been in possession of very elegant methods of his own. Second, there was the problem of the resistance of the medium, a subject on which he had contributed some propositions to the Académie Royale des Sciences in Paris twelve years before. Third, the review of Newton's work had stimulated him to allow a hasty extemporisation to appear concerning the causes of the planetary motions, as other occupations would prevent him from achieving the more careful comparison of theory and observations that he had in mind.

As an illustration of his methods concerning optic lines, Leibniz offered the solution of a problem in catoptrics, making use of the principle of Tschirnhaus reported in the *Acta Eruditorum* in 1682. The second paper, 'Schediasma de resistentia medii, & motu projectorum gravium in medio resistente' (Papers on the resistance of the medium and on the motion of heavy projectiles in a resisting medium), was more substantial and incorporated the propositions worked out in Paris. Some propositions on this theme are contained in a manuscript dated 'Hyeme 1675' (Hess 1978, pp 206–10).

In the introduction to the paper printed in the *Acta Eruditorum,* Leibniz remarked that, although Galileo and his followers had neglected the resistance of the medium, a theory, to be of use in the practice of ballistics, must take account of this resistance. Moreover, he suggested that the differential calculus provided an invaluable aid in the building of such a theory. Only the results are given in the published paper, though the detailed derivations employing the principles and notation of the calculus exist in manuscript (Aiton 1972a). Resistance, according to Leibniz, was of two types, absolute and respective. Absolute resistance arises from the rubbing of the fluid against the solid and is similar to the friction between solid surfaces. It consists of a force between the body and the individual particles of fluid, independent of the velocity. The dynamic effect of the absolute resistance—that is, the decrement of velocity caused by the rubbing of a stream of particles—is in proportion to the number of particles encountered and hence to the distance traversed. Taking equal elements of time, the effect of the absolute resistance is therefore in proportion to the velocity. Respective resistance arises from the impact of the fluid on the body. The force of impact on the body of individual particles is proportional to the density and velocity, but again, the effect of the impact of a stream of particles is also proportional to the number of particles encountered, that is, to the element of the distance traversed. Taking equal elements of time, the effect of respective resistance is therefore proportional to the square of the velocity.

Although he believed that the absolute and respective resistances acted together, he considered a joint treatment to be too complicated, so he only investigated the motion of bodies subject to either one or the other. In the case of absolute resistance—that is, resistance in proportion to the velocity—he succeeded in determining the equation of the trajectory, by first finding the horizontal and vertical motions and then combining these components. Using a similar argument, he attempted to determine the trajectory in the case of respective resistance—that is, resistance in proportion to the square of the velocity—but his result was erroneous, as Huygens pointed out to him, because he had wrongly supposed that, in finding the horizontal and vertical components of the motion, the resistances could be taken to be proportional to the squares of the horizontal and vertical components of velocity respectively. Probably this slip, which he immediately acknowledged when it was pointed out to him, was a result of the distractions arising from the circumstances of his life in Vienna at the time of writing.

In his paper, 'Tentamen de motuum coelestium causis' (Essay concerning the causes of the motions of the heavenly bodies) (*GM* **6**, pp 144–61), Leibniz explained the planetary motions in terms of the action of three forces. First, he supposed the planet to be carried by a harmonic vortex with the sun as centre; that is, a vortex in which the layers circulate with speeds inversely proportional to their distances from the centre. The name is derived from the property that, on taking the distances from the centre in arithmetic progression, the speeds of the layers are in harmonic progression. While circulating with the vortex, the planet was moved to and fro along the rotating radius vector by the combined action of two contrary forces, a centrifugal force arising from the circulation of the planet with the vortex and a gravitation towards the sun, which, like the magnetic attraction, was in Leibniz's view undoubtedly derived from the impulsions of fluids.

From the defining property of the harmonic vortex, he deduced that, whatever might be the law of attraction, the areas of the sectors swept out by the planet were necessarily proportional to the times, in accordance with Kepler's empirical law. In his calculation of the centrifugal force, Leibniz made a mistake, taking the effect of the force in an infinitesimal element of time to be a uniform motion from the curve to the tangent drawn from the preceding point instead of a uniformly accelerated motion. The error came to light in 1704 as a result of correspondence with Pierre Varignon, when Leibniz recognised that he had underestimated the centrifugal force by a factor of two.

For a body moving in the circular arc M_1P (figure 6.1), the effect of the centrifugal force in the element of time dt, as conceived by Leibniz, was a uniform motion VP. Hence the centrifugal force could be expressed by VP, or as it follows from the geometry of the figure, by PN or D_1T_1. But in the harmonic vortex, $r^2\,d\theta = h\,dt$, where h is a constant, so that arc $M_2T_1 = r\,d\theta = h\,dt/r$, where $r = \odot M_2$. Now $M_2D_1 =$ arc M_2T_1 approximately. Also $D_1T_1 = (M_2D_1)^2/2r$ approximately. Consequently, the measure of the centrifugal force D_1T_1 may be taken as $(h^2/2r^3)dt^2$. This means that, in a harmonic vortex, the centrifugal force engendered by the circulation is inversely proportional to the cube of the distance from the centre.

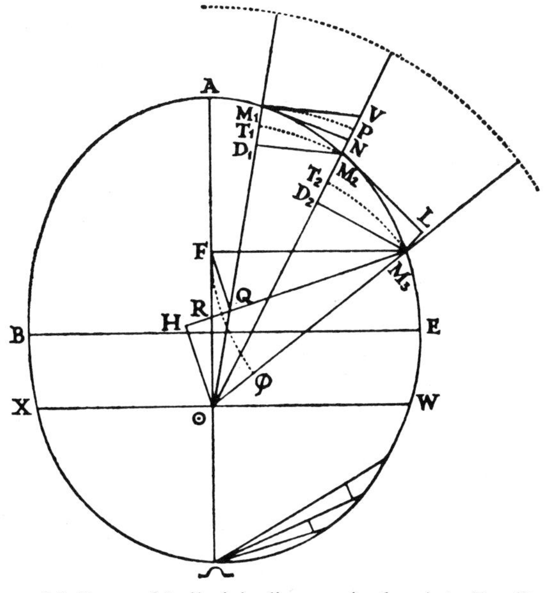

Figure 6.1 Copy of Leibniz's diagram in the *Acta Eruditorum*.

There were two ways of resolving the element of motion of the planet in its orbit into components. Thus the infinitesimal element M_2M_3 may be regarded as compounded of the inertial motion M_2L and the motion LM_3 resulting from the new impression of gravity, or, alternatively, as the motion of the circulation D_2M_3 (strictly T_2M_3) and the radial motion M_2D_2 (strictly M_2T_2) already acquired. For Leibniz, the first resolution was purely hypothetical, for it represented what would happen in the absence of the harmonic vortex, when the solicitation of gravity (as he called it) would be the sole cause of the deflection from the instantaneous inertial path. Consequently this resolution provided the means for the calculation of the effect of gravity. It should be noted that the motion LM_3 resulting from the effect of gravity was a motion along a fixed line instantaneously coinciding with the rotating radius vector. No rotation, and hence no centrifugal force, was involved. Reality was represented by the second resolution, for Leibniz supposed the planet to be moved transversely by the harmonic vortex and simultaneously along the rotating radius vector by the combined action of the centrifugal force (engendered by the circulation) and the solicitation of gravity.

In manuscript notes written in preparation for the composition of the essay, the 'tangent' M_2L is described as equal and parallel to M_1M_2 (*LH* **35**, X, 1, f 17r). In other words, the 'tangent' was taken to be a prolongation of the chord M_1M_2. It seems indeed that Leibniz represented the planetary orbit by a polygon with infinitesimal sides, supposing, for example, that the planet was deflected from the chord M_1M_2 into the chord M_2M_3 by an instantaneous impulse acting at the point M_2. This interpretation is supported by a manuscript diagram (*LH* **35**, X, 1, f 16r) and moreover resolves a number of puzzles in the essay, which would otherwise appear as seemingly trivial but inexplicable errors. Figure 6.2 serves to clarify Leibniz's intention. The motion in the chord M_2M_3 (representing the motion in the arc M_2M_3) is compounded of a uniform motion in the 'tangent' M_2L (conceived as a prolongation of the chord M_1M_2) and a uniform motion in LM_3. Although this representation of motion in a curve permits the calculation of the centripetal force needed, it is open to a logical objection which Leibniz never recognised. For the difference between the chord and the arc in the middle of the infinitesimal element of time is of the same order of magnitude as the line representing the force. Leibniz's principle that all motions are compounded of uniform rectilinear motions (*GM* **6**, p 502) is true, but in this case the curve must be represented by a polygon whose sides are infinitesimals of the second order.

To calculate the acceleration of the planet along the rotating radius vector, Leibniz proceeds as follows. From M_1 and M_3 (figure 6.2) construct M_1N and M_3D_2 perpendicular to $\odot M_2$ and M_3G parallel to LM_2. The letter G, he later explained to Huygens, was omitted from his diagram (figure 6.1) by the printer, but its position was defined in the text. From Kepler's law of areas, it

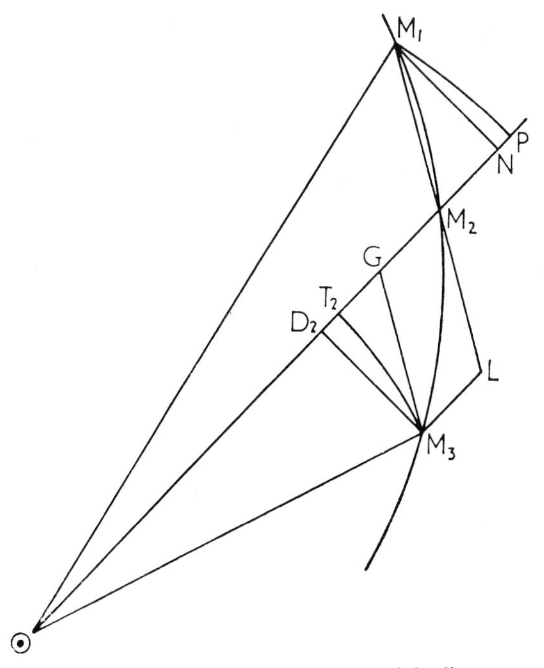

Figure 6.2 An interpretation of Leibniz's diagram.

follows that $M_1N = M_3D_2$, so that the triangles M_1NM_2 and M_3D_2G (figure 6.2) are congruent. Now the difference of the radii $\odot M_1$ and $\odot M_2$ is $PM_2 = NM_2 + NP = GD_2 + NP$, and the difference of the radii $\odot M_2$ and $\odot M_3$ is $T_2M_2 = M_2G + GD_2 - D_2T_2$. Consequently, the difference of the differences $PM_2 - T_2M_2 = \mathrm{dd}r = NP + D_2T_2 - M_2G = 2D_2T_2 - M_2G$, since, on neglecting third differences, $NP = D_2T_2$. Since D_2T_2 represents the centrifugal force and M_2G ($= LM_3$) the solicitation of gravity, the result may be written $\mathrm{dd}r = 2$ times centrifugal force − gravity. On substituting the value of the centrifugal force already determined, this becomes

$$\mathrm{dd}r = (h^2/r^3)\mathrm{d}t^2 - \text{gravity}.$$

Supposing now that the planetary orbit is an ellipse with the sun in a focus, he deduces from the geometry of the ellipse, using the rules for differentiation given in the paper of 1684, that

$$\mathrm{dd}r = [(h^2/r^3) - (2h^2)/(ar^2)]\,\mathrm{d}t^2$$

where a is the latus rectum of the ellipse. Comparing this with the previous result, it follows that, in the case of an elliptical orbit, the effect of gravity, M_2G, equals $[(2h^2)/(ar^2)]\,\mathrm{d}t^2$, so that the attraction is inversely proportional to the square of the distance. The calculation is correct. Leibniz errs only in supposing that $(h^2/r^3)dt^2$ represents the effect of twice the centrifugal force engendered by the harmonic circulation instead of the centrifugal force itself.

There remained the problem of the motion of the different planets in accordance with Kepler's third law and the detailed explanation of gravity as an effect of the impulsions of fluids. These matters Leibniz promised to treat in a separate publication (see Aiton 1972b, 1984).

Rome

From Venice Leibniz travelled along the coast as the only passenger in a small boat. Later he told an interesting story about this trip (*MK*, p 95). When a storm blew up, the sailors, believing that he did not understand their language, agreed between themselves to throw him overboard and divide his possessions between them. Without allowing them to see that he had understood their intentions, he brought out a rosary he had with him and pretended to pray. Whereupon one of the sailors, perceiving that their passenger was not a heretic, persuaded the others not to kill him. Having arrived in Ferrara, he travelled with the post-coach through Bologna and Loreto to Rome, where he arrived on 14 April 1689. A few days later, Queen Christina of Sweden died unexpectedly, and the almost simultaneous death of Cardinal Azzolini, to whom Rojas had given him a letter of introduction, prevented him from gaining access to her collection of manuscripts.

On 24 April Leibniz addressed a letter to Melchisedech Thevenot (*A* I **5**, pp 680–1), the Royal Librarian in Paris, regretting that the changed political circumstances over the last few months had dashed his hopes of being able to visit France and assuring his friend that he had not neglected mathematics and science. Indeed, he had sometimes profited from the solitude of inns by achieving certain thoughts that he had sketched out earlier. He mentioned two in particular. First, he claimed to have discovered a general method of solving equations up to the fifth degree (the basis of which he had explained to Huygens) though the calculations were involved and tedious. The other discovery to which he referred was that of the cause of the rules of the motions of the planets, supposing these were carried by the fluid heavens, as ocean currents carry vessels.

Concerning his immediate plans, Leibniz remarked to Thevenot that he would make a short visit to Naples before the summer heat and then return to Rome where he would stay for several weeks before going on to Florence and Modena. Before he set out for Naples, he had conversations in Rome with the astronomer Adrien Auzout, who had been a founder member of the Paris Academy of Sciences, and he also drafted for Descartes' biographer Adrien Baillet some notes on the life and work of the French philosopher (*GP* **4**, pp 310–15).

Leibniz spent about a week in Naples, ascending Mount Vesuvius, meeting lawyers and the historian Lorenzo Crasso, through whom he gained entrance to the Royal Archives—where he saw manuscripts on the history of the fourteenth century Duke Otto of Brunswick, who married Queen Johanna of Naples—before returning to Rome in the second half of May.

During the summer Leibniz met the Jesuit missionary Claudio Filippo Grimaldi, who was on a visit to Rome before returning to Peking as President of the Chinese Bureau of Mathematics in succession to Ferdinand Verbiest. With Grimaldi Leibniz was able to discuss questions of cultural exchanges between Europe and China, and he put together a list of enquiries in the hope of receiving information concerning language, ethnic groups and the state of technology in China. He expressed particular interest in the attempts of the Jesuits to reach China overland. Later, in the preface of his *Novissima Sinica,* he related what Grimaldi told him in Rome. The Chinese Emperor, princes, relations and high officials were instructed in the Palace by Verbiest daily for three or four hours on mathematical instruments and books. The Emperor understood Euclid and with the help of trigonometry could calculate the motions of the heavenly appearances.

Several times during his stay in Rome Leibniz visited the Vatican library, where he saw many diplomatic papers and chronicles, including information concerning Queen Johanna of Naples (*A* I **5**, p 665). Cardinal Casanata even invited him to become custodian of the library (an office usually held by a cardinal) but as he declined the expected change of religion, the idea came to nothing. From his contacts with the Curia, Leibniz formed the opinion that they showed a liberal receptiveness to the new science. For on certain days, at the suggestion of Cardinal Barbarigo, discussions of scholars, to which he often contributed, took place in the Collegium propagandae fidei on all problems of research in natural science (*A* I **7**, pp 495–9).

Pope Innocent XI, who had supported the plan of Rojas for the reunion of the Churches, died on 12 August. Two months earlier, Leibniz had shown his concern by composing for him a short prayer-poem. When the Conclave met to choose a successor, Leibniz came in touch with some of the visiting cardinals, especially the French, and after the election on 16 October, no doubt hoping for a continuation of the policy on reunion, greeted the new Pope Alexander VIII with a long poem (*A* I **5**, pp 477–83).

While in Rome Leibniz became a member of the Accademia fisico-matematica. Besides the astronomer Adrien Auzout mentioned already, he met the founder Giovanni Giusti Ciampini and other members, including Francesco Bianchini and Giovanni Battista del Palagio. With Ciampini he shared an interest in chronology. Through Palagio he met the Jansenist Amable de Tourreil, who was introduced to him under the pseudonym Antonio Alberti. This Jansenist reported to the Landgrave Ernst that Leibniz's philosophy and intelligence held him in a religious indifference, while some of the things he had seen in Rome would not lead him to the true

Church (*A* I **5**, pp 686–7). Perhaps he had in mind experiences such as Leibniz's visit to the catacombs when the Pope's Secretary, Rafael Fabretti, showed him phials containing the blood of martyrs. Another Jansenist confidant of the Landgrave Ernst, Louis-Paul de Vaucel, who lived in Italy under the pseudonym Dubois, also observed Leibniz from a distance and reported what he saw to Rheinfels.

Leibniz's meetings with the members of the Accademia occasioned his dialogue 'Phoranomus seu de potentia et legibus naturae', which includes a preface in defence of the Copernican theory. It was probably soon after the composition of this dialogue that Leibniz read Newton's *Principia* for the first time. The annotations[7] in his own copy, which concern questions relating to centripetal and centrifugal forces, planetary orbits and resisting media, the principal subjects of the papers he wrote after reading the review, suggest that they were written in Rome at the same time as he began to compose his own treatise on the foundations of mechanics and the laws of motion (*GP* **4**, p 412).

Shortly before his departure from Rome in the second half of November, which was delayed by a slight illness, Leibniz at last obtained access to the papers of Queen Christina of Sweden. Formerly a student of Descartes, she had lived in Rome for thirty years, leaving a rich legacy of manuscripts, coins, paintings and statues. When he left for Florence, Leibniz took with him recommendations to Cosimo della Rena and Vincenzo Viviani.

A treatise on dynamics

In his 'Phoranomus' (phoronomy or kinematics)[8] (Gerhardt 1888), Leibniz explains how his reflections on motion led him to the formulation of a science of force or action, which he called dynamics. This new science he developed in detail in his *Dynamica de potentia et legibus naturae corporeae* (*Dynamics. On force and the laws of natural bodies*) (*GM* **6**, pp 281–514).

When he had believed extension and impenetrability to be the only essential qualities of bodies, Leibniz explains in the 'Phoranomus', he had concluded a natural inertia in bodies to be impossible, so that in a vacuum or on a level surface without resistance, a body at rest would have to acquire the speed of another body striking it, however small that body might be. For the difference at every instant between a stationary body and a moving body was that the moving body possessed a certain *conatus* or tendency to start its

[7]The example shown in figure 6.3 concerns an elliptical orbit with force directed towards a focus. It seems that Leibniz was attempting to demonstrate that QT is inversely proportional to SP, in accordance with the definition of his harmonic circulation (Fellmann 1973, p 79).

[8]The preface, omitted by Gerhardt, has been published in Couturat (1903), pp 590–3.

[50]

SECT. III.

De motu Corporum in Conicis Sectionibus excentricis.

Prop. XI. Prob. VI.

Revolvatur corpus in Ellipſi: Requiritur lex vis centripetæ tendentis ad umbilicum Ellipſeos.

Eſto Ellipſeos ſuperioris umbilicus S. Agatur *SP* ſecans Ellipſeos tum diametrum *DK* in *E*, tum ordinatim applicatam *Qv* in *x*, & compleatur parallelogrammum *QxPR*. Patet EP æqualem eſſe ſemiaxi majori *AC*, eo quod acta ab altero Ellipſeos umbilico H linea *HI* ipſi *EC* parallela, (ob æquales *CS*, *CH*) æquentur ES, *EI*, adeo ut EP ſemiſumma ſit ipſarum PS, P*I*, id eſt (ob parallelas *HI*, P*R* & angulos æquales *I*P*R*, *HPZ*) ipſorum PS, PH, quæ conjunctim axem totum 2*AC* adæquant.

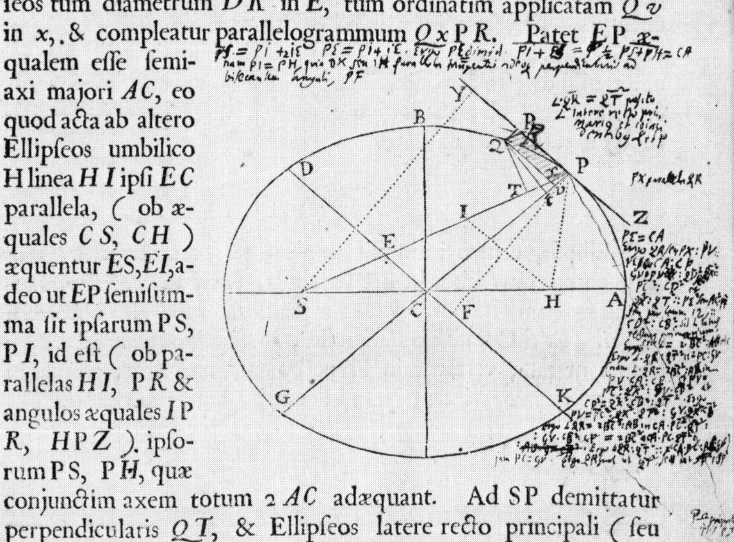

Ad SP demittatur perpendicularis *QT*, & Ellipſeos latere recto principali (ſeu $\frac{2\,BC\ quad.}{AC}$) dicto *L*, erit L x *QR* ad *L* x P*v* ut *QR* ad P*v*; id eſt ut P*E* (ſeu *AC*) ad *PC*: & L x P*v* ad *Gv*P ut L ad *Gv*;

Figure 6.3 An example of Leibniz's annotations in his copy of Newton's *Principia*. Courtesy of Dr E A Fellmann, Basel.

course (employing, as he remarked, the expression of Erhard Weigel) and there was nothing in the concept of body that could give rise to a diminution of this *conatus* or prevent its reception by the stationary body. In the case of a collision of two moving bodies, of course, there would be a partial balancing of contrary *conatuses* and a consequent diminution of motion. A kinematics of this kind, taking no account of the size or mass of bodies, could not account for the inertial resistance of bodies clearly revealed by experience. Moreover, the laws of impact could not be deduced, because the concept of bodies, defined in terms of extension and impenetrability, could not contain their future actions, nor in consequence the laws of their own movements. For these and other reasons, he had concluded that the nature of matter was not yet sufficiently known to us, and that we could not account for the force of bodies if we did not put in them something other than extension and impenetrability. He had therefore introduced force as a primitive concept and found in the evaluation of forces, subject to the metaphysical principle that the total effect is always equal to the entire cause, a thread of Ariadne with which to escape from the labyrinth.

The long treatise *Dynamica* is divided into two parts, abstract dynamics and concrete dynamics (a division similar to that employed in his earlier *Hypothesis physica nova*) and is set out formally in the manner of Euclid, like Newton's *Principia*. In the introduction, Leibniz rejects the Cartesian principle of the conservation of motion as an error based on a confusion of the concepts of motion and force and he restates the conclusion already published in his 'Brevis demonstratio', that the active force of a body is proportional to the square of the velocity. This result is demonstrated in four different ways. The first proof uses Galileo's law of fall as a demonstrated lemma in conjunction with the axiom that the force needed to lift 4 pounds 1 foot is the same as that needed to lift 1 pound 4 feet. The axiom of the impossibility of perpetual motion and the axiom that the centre of gravity of a system of bodies cannot rise of itself provide the bases for two more proofs. Finally, the result is demonstrated *a priori* from the abstract concept of action (*GM* **6**, p 346). This 'action' is measured not only by the effect of the active force of bodies but also by the vigour with which the effect is produced (*GM* **6**, p 354). For example, if bodies move on a horizontal plane without friction, the effect of their active force is the distance that they move (in a given time) and the vigour is the speed with which this distance is traversed. Suppose L is the action of a body A, which moves a distance 1 foot in 1 second, M the action of a body B, which moves a distance of 2 feet in 2 seconds, and N is the action of a body C, which moves a distance of 2 feet in 1 second. The bodies are assumed to have equal mass. Then $N=2M$, since the effect (distance) is the same in each case but is accomplished in C twice as fast as in B. Also $M=2L$, since, although the speeds are the same, the effect (distance) in B is twice that in A. Consequently, $N=4L$, so that action is proportional to the square of velocity. Since action is proportional to force, it follows that the force of a body is proportional to the square of the velocity and hence to the height to which the body can rise. In symbols, for a body moving with velocity v a distance s along a horizontal plane in time t, action $= msv = mv^2t$. Leibniz was especially pleased with this *a priori* demonstration of the true measure of the active force of bodies, which he had previously deduced *a posteriori* from phenomena.

In the second part of the *Dynamica,* Leibniz establishes his principles of the conservation of the total active force in the universe and the equality of the total cause and complete effect, besides introducing some new dynamical principles concerning the internal interactions of a system of bodies and the motion of the system as a whole. First, he distinguishes between absolute, progressive and relative force, each of which, he supposes, is conserved in the same degree in the universe, or in each isolated system. Relative force is that by which the members of the system of bodies can interact among themselves, and the progressive force is that by which the system as a whole can act externally. From the conservation of the relative force, it follows that the centre of gravity of the system is not affected by the internal interactions, while the conservation of the progressive force entails, in modern terms, the conservation of momentum. Taken together, the progressive and relative forces constitute the absolute force of the system: that is, mv^2 (*GM* **6**, p 495).

Florence and Modena

For several weeks after his arrival in Florence at the beginning of December 1689, Leibniz was assisted almost daily by the distinguished librarian Antonio Magliabechi, whom he thanked with a poem (*A* I **5**, pp 485–7). Through Cosimo della Rena, whose historical studies centred on the princes of Tuscany (including among them a few ancestors of the dukes of Brunswick), Leibniz learnt that a monk from Pisa, Teofilo Marchetti, had reported the existence of old documents relating to the history of Este and also tombs of former dukes in the Carmelite monastery at Vangadizza. This proved to be very useful information, as nothing was known of it in Modena (*A* I **5**, pp 665–6).

With Vincenzo Viviani, Galileo's last student, Leibniz discussed mathematical problems. He found Viviani to be full of admiration for the new infinitesimal method. Rudolf Christian von Bodenhausen, tutor to the sons of the Grand Duke Cosimo III, was so impressed that he offered to prepare a fair copy and take care of the printing of the first part of the *Dynamica,* which Leibniz had drafted in Rome, but this did not in fact happen. When he left Florence for Bologna on 22 December, Leibniz carried with him a recommendation from Magliabechi to the physicist Domenico Guglielmini, who provided the opportunity for him to conduct long conversations with the anatomist Marcello Malpighi whose intellect and culture impressed him (*A* I **7**, p 353).

On 30 December, after many diversions, Leibniz at last arrived in Modena. On the same day he wrote to Johann Friedrich Linsingen, who had formerly been Chamberlain to Duke Johann Friedrich and was then an Imperial Privy Counsellor in Vienna, pressing his own claim for an appointment as Imperial Privy Counsellor at some time in the future (*A* I **5**, pp 495–6). He stayed in Modena for five weeks, interrupted by a short return visit of two or three days to Bologna in the second week of January. At an audience, Duke Franz II promised him full support through the officers of the Court for his historical researches. Towards the end of his stay, he worked day and night on acts and documents (*A* I **7**, p 59) which included almost illegible manuscripts. The eye-strain he suffered was not in vain. For as he reported to Duke Ernst August (*A* I **5**, p 666), he had completely determined the true connection between the Houses of Brunswick and Este.

The return to Venice

On 2 February 1690 Leibniz left Modena, arriving the next day in Parma, where he met Benedetto Bacchini, the editor of the *Giornale de' Letterati*. After a stay of three days with Count Dragoni in Brescello, he travelled in a large barge down the river Po to Ferrara, from where he made an excursion to the Carmelite monastery at Vangadizza, called La Badia, to inspect the documents and tombs he had learnt about in Florence. Besides the tombs of the old Margraves of Este, he also found that of Countess Kunigunde, the first wife of the Margrave Albert Azzo II, whose son Guelf IV (1070-1101) was Duke of Bavaria. Countess Kunigunde was thus ancestress of the new Guelf family. Leibniz copied her epitaph from the parchment codex, 'Regula monasteri Abbatiae Vangadiciensis' (*A* I **6**, p 343). A few days after this highly rewarding visit to Vangadizza. he arrived again in Venice, where he stayed two months, holding discussions with a number of scholars, including the architect Pietro Andreini, the coin-collector Girolamo Conaro and the astronomer Michel Angelo Fardella[9], whom he later attempted to entice to Wolfenbüttel. From Prince Maximilian Wilhelm's physician he heard that there was an Englishman living in Modena, Nathan Lacy, who could obtain fresh water from sea water (*A* I **5**, p 519). Another interesting encounter was his meeting with the famous geographer Vincenzo Coronelli, at that time working on his 'Atlante Veneto'.

Early in March, Leibniz interrupted his stay in Venice for about a week, in order to visit the heartland of the House of Este, including the small towns of Monselice and Este. In the ruins of the Church of the Franciscans in Este, he found the monument of the Margraves Taddeo and Bertoldo of Este, which had escaped the recent destruction by fire.

On his return to Venice, Leibniz found a letter from the Duchess Sophie, written on 3 February (*K* **7**, p 80), asking him to undertake a political commission designed to secure the marriage of one of the daughters of Duke Johann Friedrich into the House of Modena. For some time the idea of such a marriage had been promoted on behalf of Hanover by Count Dragoni but,

[9]On the relations of Leibniz and Fardella, see Femiano (1982).

in Sophie's view, the negotiations had been conducted badly, owing, she thought, to the discomfort suffered by the Count from an illness. Leibniz, she believed, would be more successful. While he was still in Modena, Leibniz reported to Sophie that he had it in mind to make some hint that might turn the thoughts of the Duke towards the idea of marriage with one of the princesses (*K* **7**, p 77). Having now received her explicit request to use his powers as an advocate, he presented a well reasoned case for the alliance to the Chancellor Camillo Marchesini (*K* **7**, p 82). Besides mentioning the virtues of the princesses and the fact that they were not entirely without fortune, he emphasised that such an alliance would revive the old connection between the two Houses. The plan came to fruition in 1695 when Princess Charlotte Felicitas, the elder daughter of Johann Friedrich, married Duke Rinaldo of Modena, who had succeeded Franz II in 1694. Later Leibniz negotiated with his former student Philipp Wilhelm von Boineburg, who became an Imperial Privy Counsellor in Vienna, for the marriage of the Roman king Joseph (afterwards Emperor Joseph I) with the younger daughter of Duke Johann Friedrich, Princess Wilhelmine Amalia (Guhrauer 1846 **2**, p 106). In a postscript to her letter, Sophie mentioned to Leibniz that his library had been converted into an opera house, Rojas had written to say he wished to pay another visit to Hanover to conduct negotiations for Church reunion, and that the French had not yet burned their lands, a pointed allusion to the devastation of the Palatinate, which was still fresh in the memory.

On 23 March, the day before he left Venice, Leibniz wrote his last known letter to Antoine Arnauld. In this letter, besides telling Arnauld about his travels and discoveries, Leibniz developed some important additions to his own philosophy.

The last known letter to Arnauld

It will be recalled that, before setting out on his journey, Leibniz had sought the views of Arnauld on the ideas he had developed in his *Discourse on metaphysics,* and that, after some initial misunderstandings, Arnauld had expressed his general agreement, while at the same time remarking to the Landgrave Ernst von Hessen-Rheinfels that Leibniz had some curious opinions about physics. As a result of his discussions and reflections during the journey, he had been able to introduce some clarifications, on which he now invited Arnauld's comments (*GP* **2**, pp 134–8).

Since his earlier letters to Arnauld, Leibniz had made two important advances in the formulation of his philosophical system; first, he partly resolved the ambiguity concerning the nature of corporeal beings that had remained in his previous account, and, second, he interpreted the individual concept as a law governing a continuous series. A body, he now believes, is not, properly speaking, one substance but an aggregate of substances. Consequently, there will be found throughout a body indivisible substances incapable of generation and corruption, similar to souls. All these substances, he adds, always have been, and always will be, united to organic bodies capable of diverse transformations. Moreover, each substance contains in its nature a law of the continuation of the series of its own operations and everything that has happened and will happen to it. In the letter, these clarifications are incorporated by Leibniz in a summary of the principles of his philosophy. These include the pre-established harmony, which explains not only the union of body and soul but also the operation of one substance on another, besides the principle that every substance expresses the whole universe from its point of view. Intelligences or souls capable of knowledge of eternal truths and of God, Leibniz adds, possess many privileges which free them from liability to the upheavals of bodies, so that for them moral laws must be added to physical laws. Together, these souls form the commonwealth of the universe, of which God is the monarch.

Turning to physics, Leibniz summarised for Arnauld the chief conclusions he had reached in opposition to Descartes. These were first, that force is different from motion and must be measured by the quantity of its effect, and second, that quantity of motion is not conserved, otherwise perpetual motion would ensue, but that the total quantity of force is conserved in the universe. He then added some new statements concerning absolute, relative and progressive force, taken from his *Dynamica.* In the absence of further explanation, these new statements must have seemed rather enigmatic to Arnauld. Leibniz clarified the meaning of these ideas in his 'Specimen dynamicum', published in the *Acta Eruditorum* in 1695 (*GM* **6** , pp 234–54).

To conclude his account of recent discoveries, Leibniz explained to Arnauld what he believed he had demonstrated concerning the motion of the planets. First, he postulated that every motion of a body in a curve or with varying speed was engendered by the surrounding fluid, from which he inferred that the planets were carried by fluid spheres, adding in his own copy of the letter that, following the ancients and Descartes, these could be called vortices. From the hypothesis of the combined action of a centripetal force (with an arbitrary law) and a harmonic vortex, he had then deduced Kepler's law of areas, and supposing finally, in accordance with the observations, that the planetary orbit was an ellipse, he found that the radial motion needed (in combination with the harmonic circulation) to produce this orbit required an inverse-square law of attraction.

In a copy of the letter revised with a view to publication (*GP* **2**, pp 134–8), Leibniz included some speculations concerning the cause of gravity. Gravity, he supposed, was similar to the magnetic attraction and was caused by material rays trying to move away from the centre and consequently pushing bodies lacking this tendency towards the centre. Comparing the rays of attraction with those of light, he inferred that, in the same way that bodies are illuminated, they will likewise be attracted in proportion to the inverse square of the distance.

The letter was sent to Arnauld in the usual way through the Landgrave Ernst, who wrote to Leibniz on 11 June 1690 (*A* I **5**, pp 588–9), telling him that Arnauld had been forced to leave Brussels and take refuge in Holland, on account of his quarrels with the Jesuits and the danger from this quarter if he remained in the Spanish Netherlands. It seems that Arnauld did not reply to this letter and that no further correspondence ensued in the interval up to the death of the great theologian in 1694.

The journey home

Leaving Venice on 24 March 1690, Leibniz travelled to Mestre, then over the Brenner Pass to Innsbruck, from where he posted a last letter of farewell and thanks to Antonio Magliabechi in Florence (*A* I **5**, pp 563–4), who had been very helpful to him. When he arrived in Augsburg, it seems that he received from the Duke Ernst August some commissions to be executed in Vienna on his return journey. After a short stay in Regensburg, he therefore turned east, visiting Philipp Wilhelm von Hörnigk in Passau, where he had become Archivist of the Prince Bishop (*A* I **5**, p 571) and then going on to Vienna, where he arrived towards the end of April (Guhrauer 1846 **2**, pp 108–9).

It is curious that, in his letter to Sophie of 11 May, there is no mention of the Duke's instruction to return through Vienna. He tells Sophie (*K* **7**, pp 84–5) that he decided to pass through Vienna because he had left all his clothes there and also papers which he did not wish to risk having conveyed unaccompanied.

During his stay in Vienna, Leibniz had further conversations with Rojas, the Bishop of Wiener-Neustadt, concerning the latest details of his plan for the reunion of the Churches. These included the idea of a joint diplomatic approach by Brunswick-Lüneburg and Saxony to the Emperor for a more sympathetic treatment by him of the Protestants in Hungary. In the context of the moves towards reunion, Rojas believed that such a plea for moderation might be more successful than previous intercessions by the ambassadors of Saxony and Brandenburg. As Duke Ernst August was about to meet the Elector of Saxony in Leipzig and would therefore have an opportunity to pursue this idea, Leibniz decided that a confidential report on his conversations with Rojas should be communicated to him without delay. At first, he planned to meet the Duke personally in Karlsbad (*A* I **5**, p 573), but when he realised that he could reach neither Karlsbad nor Leipzig in time, he reported to the Duke in a letter instead (*A* I **5**, pp 577–9). In support of the proposed diplomatic move, he reminded the Duke that a letter from the former Elector of Saxony to the Emperor on behalf of the Hungarian

Protestants had been the occasion of the first negotiations of Rojas on matters of religion.

Having been told by the Chancellor Strattmann on his first visit to Vienna that the Emperor was willing to take him into his service as Imperial Court historiographer, Leibniz used the opportunity of his return visit to address a letter on the subject to the Emperor, seeking further information concerning his intentions and the conditions of such an appointment (*A* I **5**, pp 574–5). Count Windischgrätz encouraged him to believe that there was a real possibility of an appointment at the Court in Vienna on acceptable terms (*A* I **5**, pp 576–7).

During his short stay in Vienna, Leibniz also wrote a paper, 'De causa gravitatis, et defensio sententiae auctoris de veris naturae legibus contra Cartesianos' (*GM* **6**, pp 193–203), for the *Acta Eruditorum*. In the covering letter to the editor, Otto Mencke (*A* I **5**, pp 572–3), he explains that, on his return from Italy, he had seen a copy of the *Acta Eruditorum* for April 1689, containing a criticism by Denis Papin of his demonstration of the error made by Descartes in supposing that the quantity of motion was conserved in collisions. Papin, he added, had not understood his demonstration and was in error concerning the explanation of gravity. However, he had replied with indulgence and moderation. Leibniz's paper was published in May 1690. In the same issue of the *Acta Eruditorum* Jakob Bernoulli used the differential notation and introduced the term 'integral' to describe the transition from the differential conditions to the solution. Leibniz himself decided only slowly and reluctantly to adopt Bernoulli's term, which appeared to him to obscure the sense of the operation (*MK*, p 103).

In his letter to Sophie of 11 May, Leibniz explained his change of plan for the journey home. Having abandoned hope of meeting the Duke in Karlsbad, he had decided to leave Vienna in the next week, taking the route through Prague and Dresden (*K* **7**, pp 84–6). From there he passed through Leipzig without stopping to visit his relations residing there, arriving in Hanover in the first half of June after an absence of two years and seven and a half months.

7
Hanover under Duke Ernst August
(1690–1698)

On his return to Hanover, Leibniz seems to have wished to make it known that he had not been idle during his long absence. For example, he gave an account of his historical researches and the important results he had achieved in letters to the Prime Minister of Celle, Andreas Gottlieb von Bernstorff (*A* I **5**, pp 601–2) and the Archivist and Counsellor of Justice, Chilian Schrader (*A* I **5**, pp 602–3). Again, in his letter to Otto Grote (*A* I **5**, pp 599–600) claiming the balance of his expenses, he enclosed an extract from a letter of Ramazzini to Magliabechi as evidence of his diligence and the esteem with which he was held in Italy (*A* I **5**, p 685). Altogether he had spent 2300 taler, or about 2½ taler per day, which included the expenses of a secretary and servant. Before setting out, he had received 300 taler and further payments had been made from time to time, so he was confident that the payment of the balance would present no difficulty to the Chancellery.

Within eight or ten days of his return, Leibniz replied to the latest letter of the Landgrave Ernst von Hessen-Rheinfels (*A* I **5**, pp 590–2). After expressing his concern that Arnauld was treated so badly by those who should honour him, namely the Jesuits in France, he explained to his noble friend that he had found the Jesuits in Rome to be reasonable, especially Father Tolemei, the Procurator General, and Father Grimaldi, the missionary who was returning to China. Grimaldi, he added, had been refused permission to travel overland through Russia, as he had wished, and would have to go by sea from Portugal. Concerning affairs nearer home, he told the Landgrave that Duke Ernst August was ready to take a small force to the Netherlands if it appeared that something of consequence could be gained. The Augsburg alliance, he thought, would take some action on the Rhine, though he doubted they would retake Philipsburg.

In his next letter to the Landgrave, written on 14 July 1690, he defended

Arnauld for his stand against the Jesuits on the question of relaxed morals. Evidently with the intention of contrasting Arnauld's standing as a scholar with that of his opponents, Leibniz remarked to the Landgrave that the number of real scholars among the Jesuits was very small. He had not met one in Bavaria or Vienna who had a profound knowledge of history. Since the beginning of the century their scholarly prestige had fallen and the misunderstandings between the Jesuits of Rome and France would further diminish their reputation. Nevertheless, he confided, he had an affection for the religious orders and deplored the dispute between the Jesuits and the Jansenists.

Towards the end of August Leibniz was in attendance on Duke Ernst August at Brunswick, where he saw performances of the Italian operas *Orfée* and *Hermione* and the German opera *Julia*. In nearby Wolfenbüttel, he was shown the great collection of books from the time of the Reformation by Duke Rudolf August. It was on the occasion of this visit that the two Dukes, Rudolf August and Anton Ulrich, invited him to become Director of their library in Wolfenbüttel. From Brunswick he followed the Court to Celle, where Duke Georg Wilhelm, at the suggestion of his friend the Duchess Sophie, granted him an annual pension of 200 taler in support of his work on the history of the House of Brunswick. Sophie also supported his request for permission to become Director of the library in Wolfenbüttel. The Duke Ernst August, to whom Leibniz expressed his intention to visit the library in Wolfenbüttel from time to time in any case, because it contained many books that he needed for his history of the House of Brunswick, gave his consent in a conversation on 10 October 1690 (*A* I **6**, p 3). By this time Leibniz had already submitted a formal report to the Duke on his historical finds (*A* I **5**, pp 662–8) and prepared various drafts of a 'Brevis synopsis historiae Guelficae' (*P*, pp 227–55).

Before the end of the year he had found permanent quarters for his visits to Wolfenbüttel with the Chamberlain Johann Urban Müller (*A* I **6**, p 11), which he occupied until March 1692 (*A* I **7**, p 55), when he moved to the house of Pastor Justus Lüders, dining throughout in the local school for the nobility. During the following years he made numerous visits to Wolfenbüttel, staying for weeks at a time, sometimes travelling to Hanover for a conference and returning immediately to Wolfenbüttel. It became his custom to spend Christmas in Wolfenbüttel.

On his return from Wolfenbüttel about the middle of January 1691, Leibniz presented his outline of the planned history of the House of Brunswick to Duke Ernst August, remarking rather too optimistically that he believed the work could be completed in two years if he received support (*A* I **6**, pp 22–31). From the beginning of 1691 he was in fact paid an extra allowance of 150 taler (in addition to his pension of 600 taler as Privy Counsellor) for the employment of a secretary to help him with the historical work. On the recommendation of his brother Johann Friedrich, he employed in this capacity, until the summer of 1693, Gottfried Christian Otto, who had to abandon his law studies in Strasbourg owing to the loss of his property following the French invasion of the Rhineland (*A* I **6**, p xxiii).

Soon after his return from Italy, Leibniz received another commission from the Duchess Sophie. As he mentioned to the Landgrave Ernst in a letter of 13 October 1690 (*K* **7**, pp 86–7), Sophie had asked him to enter into communication with Paul Pellisson-Fontanier concerning his *Réflexions sur les différends de la religion avec les preuves de la tradition ecclésiastique*, which had been published in Paris in 1686 and became influential in several editions. In Leibniz's view, Pellisson, a converted Huguenot at that time historiographer of Louis XIV, had not dealt adequately with the situation of those who had been excommunicated unjustly. The correspondence with Pellisson was conducted through the Duchess Sophie, her sister Louise Hollandine, Abbess of Maubuisson, and the secretary of the Abbess, Marie de Brinon. From September 1691 the French Church leader Jacques-Bénigne Bossuet, Bishop of Meaux, also joined in the discussions, which concerned questions of reunion and the compatibility of the contemporary science with Christian doctrine.

Also in the autumn of 1690, Leibniz took the opportunity afforded by the appointment of Baron Ludwig Justus Sinold von Schütz as Brunswick ambassador in London to send a letter to Henri Justel, who was at that time librarian to King William, and so renew his contact with England. He asked Justel (*A* I **6**, pp 263–7) for news, as he had been out of touch with England for several years. For example, the latest number of the *Philosophical Transactions* he had in his possession was dated 1678. From the republic of letters he reported to Justel on the recent publication of Huygens' *Traité de la lumière* with its extensive supplement, *Discours sur la cause de la pesanteur*. These works of Huygens, together with Newton's *Principia*, were in his view the most important of their kind since Descartes, for the poor Cartesians only copied and paraphrased the thoughts of their master in these matters. From the works of Huygens, he could draw support for his own retention of the vortices, of which Leucippus, Bruno and Kepler had spoken long before Descartes, for Huygens had employed vortices to explain terrestrial gravity. The vortices, he remarked to Justel, were not so objectionable as Newton thought, and he had given a demonstration in the *Acta Eruditorum* serving to reconcile the vortices with Newton's mathematics. Without a common deferent vortex he could conceive of no reason why the planets should rotate in the same direction and very nearly in the same plane, phenomena that were observed in the satellites of Jupiter and Saturn as well as in the planets. No doubt Leibniz expected that Justel would discuss his letter with members of the Royal Society and bring to their notice his paper on celestial motions in the *Acta Eruditorum*. However, Justel had been very ill and unable to attend meetings of the Royal Society for a long time (*A* I **6**, pp 300–2).

Two years later, through the good offices of Justel, Leibniz came into contact with the mathematician and astronomer Edmond Halley, who had become Secretary of the Royal Society. This was about six months after Fatio de Duillier, in a letter to Huygens (*HO* **10**, p 214) sowed the seeds of the notorious priority dispute by expressing surprise that Leibniz, in his papers on the calculus in the *Acta Eruditorum*, did not say a word of acknowledgment to Newton. For according to Fatio, Leibniz derived his calculus from what Newton had written to him on the theme. In March 1693 Leibniz addressed his first direct letter to Newton—the earlier communications had passed through intermediaries at the Royal Society—and Newton replied rather belatedly in October 1693, making the excuse that Leibniz's letter had been mislaid. There was no hint of animosity in this exchange.

Towards the end of 1692, Leibniz tried to help his old friend Johann Daniel Crafft, who had lost his patron, by securing for him an appointment at the Courts in Hanover or Wolfenbüttel. A modest pension, he explained to the Duke Ernst August (*A* I **8**, pp 108–9) would suffice to engage the services of this able man, then on a visit to Hanover, who had extensive experience of the organisation and practice of all kinds of manufacturing industry, including textiles and ceramics. Although Leibniz had the support of the Brunswick-Lüneburg ambassador in Dresden, who had known Crafft from the time of his service in Saxony, his pleas to the Dukes of Hanover (*A* I **8**, pp 99–100) and Wolfenbüttel (*A* I **9**, pp 45–7) on behalf of his friend received no response. The two friends met again in Hamburg at the end of September 1693, where they discussed plans for the setting up of a brandy distillery in the Netherlands. In November of the following year, Leibniz visited Amsterdam for further discussions of the plan, on which several memoranda were presented to King William III of England. On the return journey, he made an excursion to the Castle Arnstein, near Eichenberg, where the von Bodenhausen family had lived since 1430. There he met Wilke von Bodenhausen, a brother of the Abbot in Florence, who agreed to help Crafft with his project (*A* I **10**, p 646). This seems to have been the last meeting of Leibniz with his friend Crafft, who died in Amsterdam in April 1697.

Leibniz maintained contact with Italy mainly through Magliabechi and von Bodenhausen. Brosseau remained his general link with Paris, though the correspondence with Pellisson, through the intermediaries of the Duchess Sophie and Madame de Brinon had opened up another route. He continued to correspond with Christiaan Huygens in The Hague on problems of mathematics, gravity and planetary motions until Huygens' death in 1695. A letter to Jakob Bernoulli, which had been waiting for him in Hanover since the end of 1687, opened up a correspondence with the Bernoulli brothers Jakob and Johann in Basel, who had already made great strides in the application of his differential calculus. Towards the end of 1691 Johann Bernoulli visited Paris, where he instructed Guillaume François Antoine de

L'Hospital, Marquis de Sainte-Mesme and du Montellier, in the differential calculus, first in Paris and then for about three months as a guest of the Marquis on his estate at Oucques (Spiess 1955, p 137). L'Hospital was not a novice, having read and understood Leibniz's original paper in 1688, but Bernoulli's instruction no doubt helped him in the writing of the text-book he had been contemplating (Spiess 1955, p 133). At the end of 1692 he entered into correspondence with Leibniz. Two years later, having heard that Leibniz was preparing a writing *De scientia infiniti*, he advised Leibniz of his own plan, remarking that he had begun to write out detailed explanations and proofs of the rules at the time he first came across the paper in the *Acta Eruditorum* (*GM* **2**, p 250). L'Hospital's excellent work—the first text-book on the differential calculus—entitled *Analyse des infiniment petits*, was published in Paris in 1696, with a promise of a work on the integral calculus to follow. In the preface, L'Hospital quoted Leibniz's own acknowledgment, in the *Journal des Sçavans* of 30 August 1694, that Newton had discovered something similar to the differential calculus, but he added that Leibniz's notation rendered the applications easier and more speedy (L'Hospital 1696, p xxx).

A constant stream of articles and reviews in the *Acta Eruditorum* and the *Journal des Sçavans* continued to flow from Leibniz's pen. In 1692, through Magliabechi, he contributed an article to the *Giornale de' Letterati* in Modena. At the Brunswick fair in 1692, he met Wilhelm Ernst Tentzel, editor of the *Monatliche Unterredungen*, published in Hamburg, with whom he discussed the philosophy of Malebranche. Besides sending a few contributions to this journal, he corresponded with Tentzel on a variety of themes up to 1706. Also in 1692, he met Henri Basnage de Beauval, who succeeded Pierre Bayle as the editor of the Dutch journal, *Nouvelles de la république des lettres,* changing the title to *Histoire des ouvrages des savans*. Basnage visited Hanover in June of that year. Soon afterwards Leibniz sent him a copy of the review he had written for the *Acta Eruditorum* of the fourth part of Pellisson's *Reflexions,* where Pellisson, without his permission, had included extracts from his letters. Leibniz was concerned that his comments on the delicate matter of religion, which had not been intended for publication, should not be misinterpreted to his disadvantage, and he therefore asked Basnage to make his position clear if he should review Pellisson's book in his influential journal (*GP* **3**, pp 83–7). Basnage had already published an article of Leibniz on an historical controversy before their meeting in Hanover (*A* I **6**, pp 381–4). Despite the burden of his work on the history of the House of Brunswick and his various political duties, Leibniz found time not only for his voluminous correspondence and the development of his mathematics and philosophy but also for the pursuit of other interests as diverse as China, linguistics, public health and technical innovations. At last in 1694, with the help of a competent artisan, a workable prototype of the calculating machine had been produced, capable of multiplying figures of 12 digits, and the artisan was engaged to produce further machines (*D* **6**, 1, p 59). Also in this year, despite his earlier unhappy experiences, Leibniz sought yet again to introduce new pumps and techniques of utilising water power into the Harz mines (*MK*, pp 121–2). Among other technical problems to which he gave his attention was the improvement of carriages (Gerland 1906, pp 236–41).

In September 1695, he confided to Vincent Placcius, from whom he sought help with the history of Brunswick, that it was impossible to describe the extent of his distractions. While his time was taken up with the search for historical documents, he had so many new thoughts on mathematics and philosophy, and knowledge of other literary novelties that he didn't wish to lose sight of, that he often didn't know what to do first. Two years later, in a letter to Andreas Morell, the famous numismatist (Bodemann 1895, p 190), he complained that, owing to his numerous duties and the great quantity of his papers, many letters vanished in the mass before they were answered.

Although he had made discoveries of fundamental importance relating to the origin of the House of Brunswick and worked hard in the final stages of the negotiations leading to the granting of Electoral status for Hanover in 1692, Leibniz had to wait until 1696 for the promotion that he believed he deserved. In this year he was appointed Privy Counsellor of Justice, a judicial office ranking just below that of Vice-Chancellor. He then received annually 1000 taler from Hanover, 200 taler from Celle and 400 taler from Wolfenbüttel (*MK,* p 140).

Salary and status, however, were not everything. A few months before his promotion, he had written of his life in Hanover to the Scottish nobleman Thomas Burnet of Kemney (*GP* **3**, pp 174–9), who had visited Hanover in 1695, where the possibility of Hanoverian succession to the English throne through the female line was already in mind. Owing to his sedentary life, Leibniz explained, he was often indisposed, and as a remedy he thought of taking exercise from time to time and making some small journeys. The great disadvantage he found of living in Hanover, rather than in a large town like Paris or London, was the difficulty of finding anyone to talk to. There was no-one at the Court, he confided to Burnet, with whom he could talk on learned matters, and without Sophie, who was not at all formal with him, he would have even less opportunity of intellectual conversation. Reacting to a rumour in England of his death, Leibniz added that, if death would give him enough time to complete the projects he had already started, he would promise in return not to start any more, and even if he worked diligently, he would nevertheless gain by the contract an extension of life. But he doubted if death would take account of his wishes or the progress of science.

History and politics

Having completed the most original and interesting stage of his historical research, beginning with his discovery in the Benedictine monastery in Augsburg of the manuscript that put beyond all doubt the common origin of the Guelfs and the Counts of Este, a confirmation that led to further researches in the archives of Modena and finally to the location of the tomb of Countess Kunigunde, a key figure in the history of the Guelfs, in the Carmelite monastery at Vangadizza, Leibniz still had before him the arduous task of assembling all the less interesting details that would have to be included in a definitive history of the House of Brunswick. For the purpose of collecting such material he made many visits to the archives in Brunswick, Celle and especially Wolfenbüttel, besides corresponding with scholars who might help, seeking personal meetings with these where possible. For example, in April 1691 he sent to Huldreich von Eyben, a judge of the supreme court in Wetzlar and an expert on medieval history, a transcript of Kunigunde's epitaph (*A* I **6**, pp 436–43). A month later, he spent about four weeks in Wolfenbüttel, with visits to Brunswick and Celle, where he sought out the archivist Chilian Schrader, to whom he had communicated his findings immediately after his return from Italy. Later in the year he was interested in the question whether the seals on old documents could decide whether the animals on the Brunswick coat-of-arms were lions or leopards. In the presence of the archivists of Celle, Wolfenbüttel and Hanover, the three branches of the House of Brunswick, he identified the animals depicted on the seals of the old documents to be lions (*MK,* p 112). Shortly afterwards, in answer to a query, he wrote a memorandum on the genesis of the Brunswick coat-of-arms for the Vice-Chancellor in Hanover, Ludolf Hugo (*A* I **7**, pp 15–16).

Leibniz became Director of the Wolfenbüttel Library at the beginning of 1691, an office he held for the rest of his life. His duties were not concerned with the daily routine of the Library but with long-term organisation and planning. At Brunswick in the summer he met his eventual successor as Director, Lorenz Hertel, at the time a diplomat in Brunswick, who served under him as Librarian from 1705. By the end of 1691 he had initiated as a first objective the preparation of a general alphabetical catalogue of authors. This was completed in 1699.

The first fruit of Leibniz's researches among the manuscripts in the Library at Wolfenbüttel was his *Codex juris gentium diplomaticus,* published in Hanover in 1693. An advance announcement of this important work had been given in December 1692 by Tentzel in his *Monatliche Unterredungen* (Ravier 1937, pp 127–8) and another, with a copy of the title page, appeared in the *Acta Eruditorum* in March 1693, two months before the actual publication. The work consists of a collection of mostly unpublished constitutional documents from the twelfth to the fifteenth century, with a preface expounding Leibniz's theory of a natural law based on Christianity that should govern relations within and between states. In a letter to Basnage de Beauval of 26 October 1692 (*GP* **3**, pp 89–93), he explained how the work came to be written. When the Imperial Librarian in Vienna, Daniel von Nessel, had in 1690 published a catalogue of international treaties, he had asked Leibniz for information concerning omissions. As he began to collect worthwhile pieces, he conceived the idea of bringing out his own collection of documents. A supplementary volume, with title *Mantissa codicis juris gentium diplomatici,* was published in 1700.

Progress on his major project, the writing of the history of the House of Brunswick, had been much slower than Leibniz evidently anticipated when he remarked to the Duke at the beginning of 1691 that he hoped to complete the work in two years. For at the end of this time he was still only at the stage of seeking the Duke's agreement to realise the history of the Guelfs for the period 768 to 1235 in the form of annals (*MK*, p 125). The history of the House of Brunswick was in fact never completed, and although three volumes eventually appeared, this task continued to bind him like the stone of Sisyphus, as he put it in a letter to the Jesuit mathematician Adam Kochanski (*MK*, p 121), to the end of his life.

On almost every political question that arose in Hanover, Leibniz's advice was sought. For example, on 23 December 1690 Otto Grote called him to Hanover from Wolfenbüttel to discuss the genealogy of the Dukes of Anhalt with the Lüneburg Secretary Johannes Walther, a question that arose out of the rival claims of Brunswick-Lüneburg and Anhalt to the succession in Sachsen-Lauenburg. This was a rather trivial matter compared with the great political events that were to take place in Hanover over the next few years. The first of these was the conspiracy of Prince Maximilian Wilhelm with the Master of the Hunt Otto Friedrich von Moltke in December 1691 against the order of primogeniture proclaimed by his father. The Prince's intention was to claim Hanover for himself, leaving only Celle to the Electoral Prince, his brother Georg Ludwig. Prince Maximilian was at that time next in line after Georg Ludwig, his elder brother Friedrich August (as well as a younger brother Carl Philipp) having been killed in battle in 1690. The plot was discovered by the Hanoverian ambassador in Vienna, Johann Christoph Limbach (*A* I **7**, p xxvii), who was rewarded by promotion to Counsellor of Legation, and the Duke also received a warning from his daughter Sophie Charlotte, the Electress of Brandenburg (Guhrauer 1846 **2**, p 145).

The chief conspirator, Otto Friedrich von Moltke, was arrested and imprisoned on the evening of 5 December 1691. During the night he tried to escape by climbing a rope. This broke, however, and he was seized by a sentry and led back to his prison. There was mounting tension among those surrounding the Court as the gates of the Palace remained mysteriously closed until the following midday. Describing the affair to the Landgrave Ernst in a letter of 13 January 1692, Leibniz mentioned that the conspirators had even started to negotiate with Vienna, Berlin and Copenhagen, adding that he didn't know whether the Duke would listen to pleas for clemency (*A* I **7**, pp 242–3). In a further letter of 30 January (*A* I **7**, pp 256–62), he reported that the Prince himself had been taken under arrest to Hamelin, where he would remain pending his rehabilitation. After several years, having renounced all claims, he received his freedom and lived in voluntary exile in Vienna. The chief conspirator, Otto Friedrich von Moltke, was not so fortunate, for the Duke did not in fact grant clemency. On 15 July 1692 he was executed and his body given to his widow for burial (Guhrauer 1846 **2**, p 146).

The ninth Electorate finally came into being on 23 March 1692 when the Emperor conferred Electoral status on the lands of the Duke of Calenberg (Hanover) and the Duke of Celle. Leibniz had been concerned with the final stages of the negotiations (*A* I **7**, pp 84–6) and thus participated in the completion of the scheme he had initiated when, in 1684, Otto Grote had asked him to prepare a memorandum setting out the case for a new Protestant Electorate. To commemorate the agreement between the Houses of Celle and Hanover concerning the Electorate, he designed a medal (*A* I **8**, pp 104–5). When Otto Grote, in December 1692, received the Electoral coronet on behalf of Duke Ernst August from the Emperor in Vienna, he made a speech which included an historical summary written by Leibniz (*MK*, p 120).

The tragic events which took place in Hanover in 1694 involved the Courts of Hanover, Celle and Wolfenbüttel. These were the mysterious murder of the Commandant of the Guard in Hanover, Count Philipp Christoph von Königsmarck, and the imprisonment of Princess Sophie Dorothea, wife of Prince Georg Ludwig, heir to the Electorate of Hanover, in the Castle at Ahlden.[1] The story goes back to 1681 when Duchess Sophie of Hanover, through her influence on Duke Georg Wilhelm of Celle, brought about the engagement of his daughter Sophie Dorothea and her son Georg Ludwig. The Princess had been engaged to the heir of Wolfenbüttel, the son of Duke Anton Ulrich, before her father, without the knowledge of her mother, disposed of her hand in a political alliance that would ensure the reversion of the Celle lands to Hanover. The Duchess of Celle was deeply distressed by the wedding, which took place secretly on 28 November 1682. Inevitably this marriage between the beautiful, witty and affectionate Sophie Dorothea and the cold, stern and formal Georg Ludwig was far from happy. After the Elector's mistress, Countess von Platen, accused the Princess of having a lover in Count von Königsmarck, and her husband in consequence doubled his harshness towards her, she begged her parents for a place of refuge, only to be reproached by her father. In these circumstances she resolved to escape

[1]The story is told in an anonymous writing, *Histoire secrette de la Duchesse d'Hanover* (1732 London).

to a convent in France and turned to the Count for help. On the night chosen for the flight, 2 July 1694, the Count was seized by four masked men as he stepped into the Gallery in the Palace, whereupon he was murdered and his body buried in the vaults. The next morning the Princess was told by her confidant, Fräulein von dem Knesebeck, that the Count could not be found and two weeks later she was told by Count von Platen that he was dead. Her letters found in the possession of the murdered Count complained of the unjust treatment she had received at the hands of her husband and also ridiculed the Elector, who had exchanged his Sophie for an unworthy mistress.[2] On 17 July, the Princess was taken under guard in a closed coach to the castle of Ahlden, near Celle, where she remained for the rest of her life, completely separated from her children. Although, for political reasons, the Elector wished for a reconciliation, she refused to return to her husband and pressed for a divorce. A special divorce court was set up in Hanover under the presidency of von dem Bussche. As director of the Church Council, Molanus took a leading part in the hearings which were concluded on 28 December 1694 with the granting of the divorce (Guhrauer 1846 **2**, pp 148–52). Although Leibniz was present in Hanover at the time, there is no evidence that he played any role in the divorce proceedings. The official story was that the Princess had left Hanover in order to live apart from her husband. Concerning the mysterious disappearance of Königsmarck the court professed to have no knowledge.[3]

In the following year the heir of Wolfenbüttel, Prince August Wilhelm, married the Princess Sophia Amalia von Holstein-Gottorp. Reporting to the Electress Sophie on the reception in Brunswick (*K* **8**, pp 6–8), Leibniz said he had been surprised to hear the Princess described as heir of Denmark and Norway at a sermon in the Cathedral. But he had been assured by the Prince himself that she had been declared heir to the crown of Denmark when she was only three years old, apparently to exclude her brothers, the princes of Gottorp, both born and unborn. She was the daughter of the elder sister of the king.

A month later, Leibniz was present at the wedding in Hanover of Princess Charlotte Felicitas of Brunswick-Lüneburg, daughter of Duke Johann Friedrich, to Duke Rinaldo d'Este of Modena. The ceremony took place on 28 November 1695, five years after Leibniz, on behalf of the Duchess Sophie, had promoted the idea of such an alliance at the Court in Modena. The religious ceremony was conducted according to the Roman rite but in

[2]The correspondence has been published by Schnath (1952).

[3]In the spring of 1695, presumably at the request of Sophie, Leibniz made some comments on a report of the affair, which the Duchess of Orleans had brought to the notice of her aunt. Described by the Duchess as false, this report probably originated with the Danish ambassador to Wolfenbüttel. Leibniz supported the official Hanoverian position, which laid the blame for the separation and divorce on the Princess alone (*A* I **11**, pp 50–5).

such a way that the same words (except perhaps the words of consecration) could have been pronounced by a Protestant ecclesiastic. Describing the event in a letter to Thomas Burnet (*GP* **3**, pp 164–71), Leibniz remarked that he had designed a medal to celebrate this new union of the Houses of Brunswick and Modena after a separation of more than 650 years. On the occasion of the wedding, the Court in Hanover published a *Lettre sur la connexion des Maisons de Brunsvic et d'Este*, in which Leibniz summarised the chief results of his historical researches in Augsburg, Modena and Vangadizza. He also prepared an Italian translation, which was published at the same time (Ravier 1937, pp 23–4). This drew from Magliabechi an expression of surprise that Leibniz could write Italian with such elegance and insight into the spirit of the language (Bodemann 1895, p 162).

As in previous years, Leibniz visited Brunswick for the fair in the summer of 1695, taking the opportunity, as he remarked in a letter to Sophie (*K* **8**, pp 1–3), to work also in the Library at Wolfenbüttel. Besides visits to the opera and the theatre to see the actor Christian Bressand, whom he described to Sophie as a second Molière, he also had interesting conversations with the numismatist Count Anton Günther von Arnstadt concerning three coins of Henry the Lion. One of these, which served to explain the other two, had a picture of the Duke and a bishop on the front and a picture of St Peter carrying a key in one hand and a fish in the other on the reverse. To Sophie he explained that the ecclesiastic depicted on the front was the Archbishop of Bremen, whose Cathedral Church was dedicated to St Peter. These coins, he added, would please Molanus and serve as illustrations for the history of the House of Brunswick. Also at the fair he met Andreas Morell, a Swiss who had been imprisoned in the Bastille in an attempt to make him change his religion. After his release he had been given permission to make impressions of the old coins in the Royal Collection in Paris. He had altogether 20 000 impressions, including almost all the Greek and Latin coins still surviving from the ancient world, which he intended to engrave and publish in a volume.

Sophie, who had been staying at the hunting lodge in Linsburg, was very pleased to receive from Leibniz a letter so full of interest. She compared his activities among the rather sterile troop in Brunswick to that of the bees who suck honey everywhere they find it (*K* **8**, pp 3–4). However, she warned him that Count von Arnstadt was suspected of being dishonest in his dealings with coins. Having returned to Hanover, Leibniz wrote to Sophie on 11 September 1695, looking forward to seeing her in good health when she returned from Linsburg, and also her daughter Sophie Charlotte, who would be visiting Hanover with her husband.

Besides pursuing his own historical researches, Leibniz was always ready to encourage the good work of others. For example, in October 1695 he made known in the *Acta Eruditorum* (Ravier 1937, p 101) that a physician of Hildesheim, Konrad Barthold Behrens, was writing a genealogy of extinct families, and at the end of January 1696 he wrote a memorandum requesting support for Johann Andreas Schmid, who wished to investigate a pagan memorial near Helmstedt (*MK*, p137).

In November 1696, Leibniz acquired 3000 books on law and politics for the Library in Hanover. These were purchased, on his advice, from the collection of the late Hanover Secretary, Melchior Ludwig Westenholz, for 1800 taler. Also in the winter of 1696, he designed a medal for the Princess Christine of East Friesland, who at last resigned the administration of the government to her son, composed a verse for the birthday of the Elector (*P*, p 384), and corresponded with Gilbert Burnet, Bishop of Salisbury, on English history and politics. During the years 1696 and 1697, however, his major political work concerned the permanent acquisition for Hanover of the Bishopric of Osnabrück, which at that time was held personally by the Elector Ernst August.

Church reunion

Within a month of receiving Sophie's request to respond to the work of Pellisson on religious controversies and the defence of the Catholic tradition, Leibniz had composed a reply (*K* 7, pp 87–96) in which he was able to cite a number of Catholic theologians in support of the views he held in opposition to Pellisson. Both Protestant and Catholic theologians, he explained, were in agreement that the bases of religious faith were of two kinds; namely persuasion in the form of rational argument on the one hand, and an inner conviction, supposedly a reflection of the divine light, on the other. There was evidently need for some examination of conscience to prevent religion becoming arbitrary, and he invited Pellisson to consider the problem of defining the characteristics by which the true divine light could be distinguished from illusion. Pending clarification of this question, he turned to the problem of persuasion by rational argument. It was, he thought, because argument of this kind in itself could not resolve the controversies of religion that Pellisson supposed infallibility to be necessary. He reconstructed Pellisson's demonstration of the infallibility of the Roman Church on two premises:

1 Infallibility is necessary for the resolution of controversies.
2 If infallibility exists, it can only exist in the Roman Church.

Against the first premise Leibniz argued that it is sufficient to believe the truth of some necessary points without having arrived at a knowledge of the truth by argument. In the case of learned men it is enough for them to see the advantages of Christianity over other religions and in the case of others to believe the word of their pastors. Another argument, which he acknowledged would not be approved by all Protestant theologians but was in conformity with the views of eminent Catholic theologians, was that divine justice would be less than perfect if salvation were to hang on controversies and the chance of receiving good instruction. According to this view, salvation is available in all religions to those who love God. Among the eminent Catholics taking this position was Jacques Paiva Andradius, one of the principal theologians at the Council of Trent, who declared that the redemption of the human species through Jesus Christ was contained implicitly in the general providence of God. Having thus demolished the first premise of Pellisson's demonstration, Leibniz decided that he could dispense with a discussion of the second.

On 1 October 1690 (*K* **7**, pp 97–8) Sophie sent Pellisson's reply to Leibniz from Linsburg, remarking that Pellisson seemed to aspire to lead him into a dispute concerning controversies in which he would find neither pleasure nor profit. For he would not wish to make the travels to discover 'the truth' as he had for the history of the House of Brunswick. In a postscript she informed him that the lady who had sent his memoir to Pellisson was called Marie de Brinon. She had been one of the principals of the school at St Cyr before becoming secretary to Sophie's sister, the Abbess of Maubuisson. Her eloquence, Sophie added, was extraordinary, for she never stopped talking.

In his reply, written from Versailles on 4 September 1690 to Marie de Brinon (*A* I **6**, pp 83–104), Pellisson dismissed Leibniz's appeal to the individual conscience and upheld the infallibility of the Roman Church. Thus he maintained that the Roman Church was the repository of the general truths written in the hearts of men by God and that individual beliefs contrary to these were illusions and imaginations. From this he concluded that the least error of faith, if accompanied by rebellion (that is, refusal to abandon the error), was enough to deprive the individual of salvation. As for the Scholastics cited by Leibniz in support of his views, no Catholic was obliged to defend all they had written. Indeed it was dangerous to rely on such writings and one should rather follow the decisions of Councils and confessions of faith authorised by the Church. He had never heard of the work of Andradius but out of curiosity would search for a copy when he was next in Paris.

Towards the end of October, Leibniz responded to Pellisson's reply with another memoir (*A* I **6**, pp 115–21) in which he complained that Pellisson had not sufficiently examined the views of eminent Catholic theologians concerning the position of material heretics; that is, those who appear to be outside the visible communion of the Church but are judged to be not deserving of condemnation because their fault lies in ignorance or insurmountable error. Leibniz saw this as the central issue, for, in his view, acceptance by the Catholics that the Protestants belonged to this category was a prerequisite for reunion.

In his reply, communicated to Marie de Brinon in December 1690 (*A* I **6**, pp 140–9), Pellisson maintained that material heretics were those who did not know the decision of the Church on the point of doctrine concerned, or who came before the decision had been made. Those who knew the decision, however, and resisted on grounds of conscience (by implication the Protestants) he held to be formal heretics who placed themselves beyond redemption.

Leibniz responded (*A* I **6**, pp 162–71) by claiming that, from the Catholic standpoint, the Protestants should be regarded as only material heretics, for they did not know that the doctrines of faith that they rejected were doctrines of the Catholic Church. According to their view, the Council of Trent was not ecumenical and the doctrines concerned had not therefore been formally decided. In any case, the conclusions of the Council of Trent concerning doctrine were not very different from the Confession of Augsburg. It was the abuses rather than speculative dogmas that prevented reunion. He reminded Pellisson that the protestations of France against the Council had never been retracted. Moreover, disputes concerning the validity of Councils were not new; the Councils of Constance and Basel, for example, had not been accepted in Italy.

At this point Leibniz branched out beyond the theological issues to which the debate had thus far been confined, to introduce the wider political aspects of the reunion problem. As he had just had an audience with Duke Ernst August (on his return from Wolfenbüttel in the middle of January 1691), to present his plans for the history of the House of Brunswick, it seems probable that the subject of the correspondence with Pellisson also arose and that the new approach had the approval of the Duke (*K* **7**, p xliii). Having related the history of the negotiations with Rojas, undertaken with the approval of the late Pope by order of the Emperor and actively supported by the Duke and Duchess, he asked Marie de Brinon to assure Pellisson that there was nothing except consideration of some temporal advantages of princes that prevented the peace of the Church. The misfortunes of the times were not favourable, but perhaps the powers of persuasion of Pellisson and great prelates like the Bishop of Meaux, to whom he had himself sent a letter of Rojas containing details of his plans, could exert some influence on powerful persons of their side for the revival of hopes of reunion. Alluding to Louis, without mentioning his name, Leibniz remarked that this king had it in his power to restore peace and perhaps even the peace of the Church.

Before Pellisson could study this communication, it was lost when the amanuensis to whom he gave it to make a copy for the Abbess of Maubuisson was robbed. Another copy was sent from Hanover (*A* I **6**, pp 190–1) and Pellisson replied on 23 April 1691 (*A* I **6**, pp 192–4). The exchange of views had thus far been conducted through Marie de Brinon with the greatest courtesy and not a little mutual flattery. This may have contributed to Pellisson's failure to recognise the fundamental nature of the narrow gap which separated them, for he expressed to Marie de Brinon his belief that only prayer was needed to bring about Leibniz's conversion, as it had his own. Having been informed by Marie de Brinon that Pellisson was ill, Leibniz addressed a personal letter to Pellisson himself (*A* I **6**, pp 195–7) expressing his good wishes, and also those of the Duchess Sophie, for the restoration of his health, taking the opportunity also to tell Pellisson about his time in Paris and his present work on the history of the House of Brunswick. This brought from Pellisson (*A* I **6**, pp 209–14) the offer of the results of his own historical researches if they would be of use in writing the history of the House of Brunswick and a request for the thoughts of Leibniz and also of his heroine Sophie on the subject of the Eucharist, on which Pellisson was then writing.

After reporting Pellisson's request to Sophie in a letter of 30 June 1691 (*K* **7**, pp 114–17), Leibniz added a comical story he had heard from the Landgrave Ernst, concerning a Capucin of Belgium, who believed that King William was a Catholic at heart and attended mass in secret. The Landgrave was not able to persuade him otherwise, though it was like saying that Innocent XI was a Lutheran or that Louis XIV would receive the Huguenots back in triumph. It was a pity, Leibniz added, that they did not have such a politician at Loccum. Sophie had been depressed; a month earlier she had confided to Leibniz, who was then in Wolfenbüttel, that she was diverted from her sad feelings by listening to the nightingales in her park at Herrenhausen (*K* **7**, p 109). Now taking a health cure, she wittily remarked (*K* **7**, pp 117–18) that Leibniz's letter was more pleasing to read than to answer, because the exercise of the legs was more healthy than that of the head needed to describe her view of the Eucharist. It must be a miracle, she thought, for without divine inspiration it would not be possible to believe something of which one sees the contrary. Not to be outdone, she then told Leibniz a tale of black humour about King William. When a man in a church in Paris asked a priest standing by the holy water where he could find the Prince of Orange, the priest for a joke pointed out a man at prayer. A little later this man was murdered by the questioner in a field near the church.

At the end of July 1691 Leibniz explained more fully to Marie de Brinon (*A* I **6**, pp 235–7) the terms of the Emperor's commission to the Bishop of Neustadt and the agreement he had reached with the theologians of Brunswick-Lüneburg, pointing out that Pellisson, Bossuet and others like them, should take advantage of such a favourable opportunity, which perhaps would present itself only once in a century. In response to a request from Marie de Brinon for the articles agreed with Rojas (*K* **7**, p 123), which she wished to communicate to Pellisson and Bossuet, Leibniz replied that these had already been sent to Bossuet by Molanus some time ago. At this point, Bossuet enters into the correspondence. He had indeed received the articles from Molanus in 1683 but had put them aside and lost them, so he asked for another copy, declaring at the same time that Rome would never give up any decision of the Council of Trent (*K* **7**, pp 134–8). Molanus, the leading theologian of the State as Leibniz described him to Marie de Brinon, composed a new memorandum on the negotiations with Rojas, which was sent to Bossuet in two parts towards the end of 1691 and the beginning of 1692. Meanwhile Leibniz had received from Marie de Brinon the fourth part of Pellisson's *Reflexions* (Ravier 1937, p 125), which included his objections to the first two parts, together with some of the correspondence up to January 1691. Several new editions appeared in 1692 with corrections suggested by Leibniz (*A* I 7, pp 176–7). He regretted that his name had appeared, for his words could be misinterpreted by malicious persons.

Leibniz sent a copy of the fourth part of Pellisson's *Reflexions* he had received from France to his old friend the Landgrave Ernst von Hessen-Rheinfels (*A* I 7, pp 256–62) asking him not to show it to others, however, for he feared that some might not recognise that what Pellisson called his eulogy of Louis XIV was in fact the expression of a wish that this king would use his great power for more worthy ends. In the same letter, he emphasised his view that the plan of Rojas was the only way of raising the schism without the spilling of blood in wars. The Landgrave, however, tended to agree with the jurist Baron Heinrich Julius von Blum of Prague, who regarded the cause of reunion as hopeless in the face of the political realities (*A* I 7, p 300).

On 13 July 1692 (*K* 7, pp 200–1), Leibniz wrote to Marie de Brinon that Bossuet and Pellisson wrote beautifully but when he tested their reasons with logic and cold calculation, they vanished from his grasp. Also, both he and Sophie were disturbed by a letter from the Abbess of Maubuisson which seemed to suggest that there was a plan to publish the letters and writings. This, he explained, would be counter-productive and he stressed the need for secrecy. Bossuet assured him, however, that there was no plan to publish (*K* 7, pp 208–10) and that the misunderstanding probably arose out of a request from Marie de Brinon that he should translate the Latin writings for the princesses.

The long awaited reply of Bossuet to the memorandum of Molanus reached Hanover in September 1692 (*K* 7, pp 213–17). In effect, Bossuet rejected the plan proposed by Rojas, for in his conclusion he declared that the decisions of the Council of Trent were not open to question and that the Protestants were indeed formal heretics whose errors were not excusable. Writing to Pellisson (*K* 7, pp 227–8) in November, Leibniz expressed his disappointment with Bossuet's uncompromising reply, and in a further letter written in December (*K* 7, pp 230–2) complained at the expressions used by Bossuet in a personal sense, for he had declared that Leibniz himself could not be excused from the obstinacy that made him a heretic. Pellisson sent an apology to Leibniz on behalf of Bossuet, who could not bring himself to make it directly, together with New Year greetings (*K* 7, p 232). A month later, on 7 February 1693, Pellisson died at Versailles.

At the end of March 1693, Leibniz informed Bossuet (*K* 7, pp 239–43) that Molanus intended to compose a reply when he returned to his monastery during Lent. Owing to other calls on his time, followed by an illness, however, the document was not ready for despatch until August (*K* 7, p 247). Meanwhile, on 12 May 1693, Leibniz's old friend, the Landgrave Ernst, with whom he had corresponded on matters of philosophy, religion and politics for thirteen years, died in Cologne.

In the autumn of 1693, Leibniz supported the arguments of Molanus by presenting to Bossuet the historical precedents for the position of the Protestants on the Council of Trent (*K* 7, pp 252–60). Disappointed with the response, he accused Bossuet of evading the arguments and closing the door to reunion (*K* 7, pp 266–71). On the same day, 23 October, he wrote to Marie de Brinon, telling her that if they believed they could obtain full agreement on all the decisions of Trent, 'adieu la réunion' (*K* 7, pp 260–4). After a silence of several months, Bossuet announced on 22 April 1694 that his reply to Molanus was nearly ready. At this point, Leibniz decided to bring the Duchess Benedicte (widow of Duke Johann Friedrich), who was then living in Hanover, into the arena. Although Catholic, the Duchess and her daughters, he suggested to Marie de Brinon (*K* 7, pp 276–82), could facilitate reunion as much by the example of their ecumenical attitude as by the penetration of their minds. Marie de Brinon and the Abbess, he added, should themselves take the initiative. Marie de Brinon responded by writing to Bossuet on 18 July 1694 (*K* 7, pp 292–4) telling him that the Abbess, her sister Sophie and the Duchess Benedicte desired him to contribute towards reunion, adding that she herself had consulted a friend at the Sorbonne, who thought that they should agree to the conditions stated by the Lutherans.

While they were awaiting a reply, news reached Hanover of a controversy between the theologians and actors in Paris, which showed just how intolerant the theologians were, even in matters that did not involve articles of faith. One of their number, Franscesa Caffaro, had taken the part of the actors, defending their claims to be admitted to the sacraments, but had to retract when all the theologians of the Sorbonne and many others opposed him. The poets had then supported the actors. Leibniz sent a poem of his own to Sophie (*A* I **10**, pp 70–1), to whom he expressed the belief that the theatre furnished an excellent means of instruction. To Claude Nicaise, canon of Dijon, he sent the same poem, though without admitting that he had composed it himself, and expressed his surprise at the attitude of the theologians to the members of a profession authorised by the king to give public performances (*GP* **2**, pp 549–51).

With growing impatience, Leibniz reminded Marie de Brinon on 24 January 1695 (*K* 7, pp 312–13) that after nine months they had still not received Bossuet's reply to Molanus; they were anxious to know whether there was a chance of reunion or whether the issue would have to be shelved for another century. On 17 March he informed Benedicte (*K* 7, pp 315–17) that Bossuet seemed to have abandoned thoughts of peace. Also he pointedly reminded Marie de Brinon (*K* 7, pp 320–6) that it was at her suggestion he entered into correspondence with Pellisson. She in turn (*K* 7, pp 326–31) was offended by statements of Leibniz which she interpreted as suggesting that she built a tower of Babel and was united to Antichrist. Then on 2 June 1695 (*K* **7**, pp 332–3), he asked her to put to Bossuet the direct question, whether he took the same view as herself, that the demands of the Protestants were impossible and if he would condemn the opinion of those of his own party who believed that they were possible. This evidently brought no response from Bossuet. It would seem that, at a time when the cause of reunion had already suffered a setback in Germany with the death of Rojas on 12 March 1695, the search for agreement with the French Catholics had reached an impasse.

Marie de Brinon enters the scene again two years later, in July 1697, when Sophie showed Leibniz a letter she had received from her sister's secretary, praying for her conversion (*K* **8**, pp 31–2). In his remarks to Sophie (*K* **8**, pp 32–4), Leibniz was rather severe on the good Madame de Brinon in respect of her view that entry to heaven is only by the path of Rome. Sophie herself, however, replied with her usual wit (*K* **8**, pp 34–5). Thanking Marie de Brinon for showing her a better path to heaven than Divine Providence had shown her, she quoted Jesus: 'In my Father's house are many mansions.' Then she added, 'When you will be in yours and I in mine, I shall not fail to make you the first visit.'

Rosamunde von der Asseburg

In the autumn of 1691 Sophie wrote to Leibniz from Ebsdorf asking his opinion concerning a young girl of quality to whom the Saviour appeared and dictated writings in the style of the Revelation of St John (*K* **7**, pp 139–40, 150–1). The young visionary, Rosamunde Juliane von der Asseburg (born in 1672), was the middle of three sisters living in Lüneburg. From infancy she had seen Jesus. Then when she was ten years old, he laid his hand on her head, and being somewhat frightened, she told her mother, who instructed her to ask the Saviour when he appeared again, what commands he had for his maid (*K* **7**, pp 140–2), which she did. Thereafter he appeared to her often and told her what to write. Now her strange powers had become known outside the family circle and she was the subject of some curiosity. Her mother had died recently. If she had still lived, Leibniz confided to the Landgrave Ernst (*A* I 7, pp 188–91), she would have prevented the disclosure. Sophie relates that the girl was given a sealed envelope containing three questions in English, which she answered in German without opening it. Also she prophesied that Christ would appear in glory in 1693 and reign on earth for a thousand years. Sophie quickly corrected this report, however, having heard that the prophetess had not in fact given a date but explained that only God knew the time. Rosamunde had received confirmation of her claims when she saw a vision of her dead sister, who said God had allowed her to appear and tell her she stood next to Christ as she had reported. Sophie regretted that the Superintendent in Lüneburg, Johann Wilhelm Petersen, was in danger of losing his position for accepting Rosamunde's prophecy of the thousand-year reign of Christ on earth. It was her intention, she added, to see Rosamunde incognito at the Superintendent's house in Lüneburg.

Molanus had also been informed and was expected to go to Lüneburg to examine Rosamunde in his capacity as head of the Consistorial Court. Sophie stressed indirectly to Molanus her wish that Rosamunde should not be derided (*K* 7, p 142).

Leibniz and Molanus stated their very different views in letters to each other written on 22 October 1691. Leibniz (*K* 7, p 143) thought Rosamunde might enter into ecclesiastical history and he had no doubt that, if she were in Hanover, Molanus would come to see her and perhaps even invite her to stay with him as his guest, like the Petersens had done in Lüneburg. Molanus (*K* 7, pp 143–4), on the other hand, was far from sympathetic. He thought she should be taken as soon as possible to the waters of Pyrmont to cure her constipation, to which supposed condition he attributed the visions.

In his reply to Sophie, written on 23 October 1691 (*A* I 7, pp 33–7), Leibniz expressed his firm conviction that the visions of Rosamunde happened naturally. He marvelled at the nature of the human mind, of which we only had a limited knowledge, and he regarded Rosamunde's powers as a gift. Such individuals should not be reproached nor coerced to change but should rather be preserved in their elevated state. As he suspected, however, the story about the questions in the sealed envelope had been embellished. Moreover, Rosamunde had never claimed that Christ would answer all questions; it was up to him to decide (*K* 7, pp 150–1).

Leibniz explained to Sophie that there were two ways of distinguishing real perceptions from illusions. First, real perceptions have a relation to the common world in which we live, which is often lacking in dreams. Second, real perceptions are more vivid and distinct than images which only come from the remnants of past impressions. However, a person who has a very strong imagination can have visions so lively that they appear real. When this happens, that which appears has a connection with the things of the world, or things taken as such. This is why young persons brought up in convents where they listen to the old stories of miracles and appearances, if they have a very strong imagination, are susceptible to have such visions, because their head is full of them and their belief that the spirits of the other world often communicate with us does not permit them to have the doubts that the rest of us would have in similar circumstances. Rosamunde saw Christ because the Protestants hardly employed the saints. In the same way, Ezechiel saw visions of buildings, probably because he was an architect at the Court, while Hosea and Amos, living in the country, saw only rustic visions. Having compared Rosamunde with these prophets of Israel, Leibniz remarked to

Sophie that it was not necessary for all the gifts of God to be miraculous. When he employed the natural disposition of the mind and things which surround us to give light to the understanding or uplift to the heart, that was also a gift. He acknowledged, however, that the great prophets had supernatural gifts, for the marvellous connection of all things in the universe, as illustrated for example by the revelations of the microscope, showed that the causes of events were too complex to allow prophecy without the possession of supernatural powers. He feared that if Rosamunde made prophecies of particular events, she would come to harm in the world. Leibniz was angry to hear from Sophie that Petersen was in danger of dismissal for an opinion in such conformity with the Apocalypse. For the Confession of Augsburg was only opposed to millenarianists who disturbed public calm, while the error of those, like Petersen and Rosamunde, who waited patiently for the kingdom of Jesus Christ on earth seemed to him very innocent.

Sophie replied on 25 October (*A* I 7, pp 37–8), saying that she had the same thoughts, as those near her could testify, though she had not explained them so well as Leibniz. His letter, she thought, was much more deserving of publication than the one he had sent for Pellisson on Church reunion. On receipt of this reply, Leibniz immediately wrote to Sophie (*A* I 7, pp 38–40) expressing the hope that the Dukes and the Duchess of Celle, who had been with her at Ebsdorf, held the same view as herself. He thought it best to leave such as Rosamunde alone, provided they didn't interfere in anything that could be of consequence. Sects were born generally by opposition to some particular opinion, under pretext of preventing heresy. In a postscript he expressed his belief that Rosamunde should not be subjected to further tests with sealed envelopes. In her next letter (*A* I 7, pp 43–4) Sophie revealed that Rosamunde's mother had dedicated the child to Jesus while still in the womb, which drew from Leibniz the conclusion that the influence had been partly hereditary. Then on 10 November (*A* I 7, p 53), she reported to Leibniz that Rosamunde had been frightened and distressed when she was made to ask a reply to a sealed letter full of silly questions and found the Saviour angry. Duke Anton Ulrich, Sophie added, had not made up his mind about Rosamunde, though he had remarked that the spirit appearing to her was a liar, since he had told her he was the Saviour.

At this point Sophie returned to Hanover, so that any further discussions with Leibniz about Rosamunde would have taken place face to face in Herrenhausen. The report made by Leibniz on 23 November 1691 to the Landgrave Ernst (*A* I 7, pp 188–91) seems to have benefited from such a direct conversation. For example, he remarks that almost all who approach her believe her to be a demi-prophetess on account of the surprising things she ways. If she were in Italy or Spain, Leibniz confided to his friend, he believed she would be capable of founding a new order. He compared her to St Teresa and St Catherine of Siena.

On 20 February 1692, Leibniz wrote about Rosamunde to Sophie's daughter, the Electress of Brandenburg (*A* I 7, pp 101–4). Sophie Charlotte had already learnt of the affair, as Leibniz knew (*K* 7, pp 164–5), from the Pietist Philipp Jakob Spener, then Provost of St Nicolai in Berlin, who doubted whether Rosamunde had received divine revelation but in the absence of sufficient information did not at that time wish to decide. First Leibniz informed the Electress that, the proceedings against Petersen having resulted in his dismissal, he had retired to Wolfenbüttel. It had been alleged that he contravened an order not to preach his opinion concerning the thousand-year kingdom. But his worst crime had been to publish a book on the visions of Rosamunde,[4] maintaining that Jesus Christ appeared to her in person. Rejecting Petersen's claims, Molanus had argued that, apart from the fact that the expressions used by the Saviour to Rosamunde—my queen, my little dove—were not in conformity with the style of the celestial Chancellery (as far as we know it), there were errors of faith in that the supposed Saviour seemed to assure a universal election; that is, the salvation of every individual. Leibniz told the Electress that he believed Spener, who had defended Petersen, would now give his decision, which he expected to coincide with his own. Then he made the point that, if some persons, badly born and subjected to evil influences in their youth, could imagine that they can conjure up demons, there was no reason not to believe that the opposite causes could have the opposite effect in a girl well born and educated, who perhaps had received at birth the dispositions to have good visions. Finally, Leibniz expressed to Sophie Charlotte his regret that they seemed to live in an age when the external show of piety was the mode, and he commended Spener for his example of moderation and charity, which was the true touchstone of the love of God.

From Celle on 12 March 1692 (*K* 7, p 194), Sophie informed Leibniz that Rosamunde was in Berlin with Frau Sweinitz. Later she became companion to a countess in Saxony, where Petersen visited her in 1708. After this time nothing more is known of her.

Dynamics

Towards the end of 1691 Leibniz received encouragement from two sources to communicate something of his new science of dynamics to his friends in Paris. Earlier in the year he had defended his concept of the measure of force against the attack of Papin (*GM* 6, pp 204–11). One request came from the Abbé Simon Foucher and the Royal Librarian Thevenot, both old friends from his days in Paris, with whom he had renewed contact through

[4][J W Petersen], *A letter to some divines . . . with an exact account of what God hath bestowed upon a noble maid*, translated into English (1695 London).

Brosseau. Writing to Leibniz on 31 December 1691, Foucher remarked that Thevenot was sorry he had not given them a part of the mechanics he had left in Florence (*GP* **1**, pp 400–2). In his reply to Foucher, Leibniz explained that he had left his 'Dynamics' in Florence because his friend Bodenhausen had offered to publish it there. However, he promised that, one of these days, he would send Thevenot a general theorem taken from his dynamics as an illustration (*GP* **1**, pp 402–6). The other request came from Pellisson, to whom he had explained some of the basic ideas during the summer and autumn of 1691. Without having any regard to theology, he explained to Pellisson (*A* I **6**, pp 224–8) he had always judged from natural reasons that the essence of body consisted of something other than extension. For a learned man (the Jansenist Amable de Tourreil), whose name he did not mention to Pellisson, he had written out the simplest demonstration (*GP* **7**, p 447). Through the good offices of Foucher, this appeared in the *Journal des Sçavans* on 18 June 1691 (*GP* **4**, pp 464–6). The new thing he had found in bodies, Leibniz told Pellisson, was force, and on this concept he had established his new science of dynamics. In a further letter (*A* I **7**, pp 191–9) he described the controversy with Catelan and Malebranche, besides mentioning his correspondence with Arnauld on the nature of corporeal substance, and expressed to Pellisson his wish that the dispute should be examined by able geometers. In his response, written on 30 December 1691 (*A* I **7**, pp 225–8), Pellisson pointed out that it would be difficult for Catelan to admit the errors he had evidently made and suggested to Leibniz that he should prepare something on his new dynamics for discussion by the members of the Paris Academy of Sciences.

In response to Pellisson's suggestion, Leibniz composed an *Essay de dynamique*, in which he arranged his thoughts in better order and developed them in more detail than had appeared in the papers relating to the controversy. When he sent the manuscript to Pellisson on 18 January 1692 (*A* I **7**, pp 245–51) asking him to show it first to Malebranche and then to other scholars, including members of the Royal Academy of Sciences, he mentioned that, during his time in Paris, he had been acquainted with Thevenot, Cassini, Gallois and du Hamel, who was then the Secretary. Writing again at the beginning of February (*A* I **7**, pp 263–4) he explained to Pellisson that he had failed to supply a figure for which the copyist had left a space and that he was also sending a correction. He then forgot to enclose either the figure or the correction with the letter and had to send it on a few days later (*A* I **7**, p 280). This may serve to indicate the pressure under which he had to work in dealing with his voluminous correspondence.

Although Pellisson promised in April to send a copy of the *Essay de dynamique* to Malebranche through a mutual friend, Gilles Filleau des Billettes, who had met Leibniz in Paris at the home of Arnauld (*A* I **7**, pp 304–6), he had still not sent it in October, when Malebranche was out of Paris (*A* I **8**, pp 174–9). It seems that Malebranche never actually received a copy, for when he resumed direct correspondence with Leibniz on 8 December 1692 (*GP* **1**, pp 343–4), he made no mention of the *Essay de dynamique*. Under the pretext of waiting for an opportune moment, Pellisson also delayed sending the *Essay de dynamique* to the Royal Academy of Sciences for several months. On 29 June 1692, he informed Leibniz that he had sent a copy to the Abbé Bignon a few days earlier (*A* I **8**, pp 118–21). In fact it was read at a meeting of the Academy on Saturday 28 June and then seems to have been forgotten. Although des Billettes made a copy and entered into correspondence with Leibniz in the same month (*A* I **8**, pp 284–5), he never mentioned the *Essay*. At a meeting on 14 March 1693, Pierre Varignon presented an investigation of Leibniz's claim to have demonstrated that quantity of motion is not conserved, but he referred only to the *Brevis demonstratio* that had appeared in the *Acta Eruditorum* in 1686, making no mention of the *Essay de dynamique*. It would seem that he had either forgotten the *Essay* or believed it to contain nothing new. In any case, the manuscript was inaccessible to him at that time, for as Foucher revealed in a letter to Leibniz on 12 March 1693 (*FCa*, p 104), his *Essay* was among the papers of Thevenot, who had died on 29 October 1692.

Meanwhile, on 6 May 1692 (*A* I **7**, pp 326–8), Leibniz sent to Pellisson for publication in the *Journal des Sçavans* a manuscript of his *Règle générale sur la composition des mouvemens*, a piece in conformity with his dynamics and containing a rule he had discovered several years earlier. Evidently this article came to the attention of the Academy before it was published, for des Billettes made a copy which he filed with his copy of the *Essay de dynamique*. Again Pellisson delayed publication, hoping first to receive a reply from the Academy concerning the *Essay de dynamique*. The article was published in the *Journal des Sçavans* on 7 September 1693 (*GM* **6**, pp 231–3), seven months after Pellisson's death. According to a letter of Leibniz to Bossuet, it was with Pellisson's approval that he allowed the article to appear in the *Journal des Sçavans*, in order to appeal to the public after having failed to obtain any response to his *Essay de dynamique* from the Academy.

Leibniz was himself responsible for holding up the publication of his *Dynamica* by Bodenhausen, because he had delayed again and again sending him the final part. This was still the situation in 1696, when he wrote to L'Hospital (*GM* **2**, p 305) that he had been unable to finish the work, evidently envisaged by him as a definitive treatise that could complete Newton's *Principia*, owing to the new ideas which continually came to him. Pressed by his friends (*GP* **3**, p 162), however, he produced a summary of his new dynamics under the title *Specimen dynamicum* (*Essay on dynamics*). The first part was published in the *Acta Eruditorum* in April 1695, but the second part, intended for the May issue, never appeared.

In his *Essay de dynamique* (Costabel 1960, pp 97–106) Leibniz made clear his desire to avoid a dispute that was merely about words. If others wished to give another meaning to force, he would grant them the same freedom as himself to explain the sense in which they used the term. The real issue on which he sought agreement was that what he himself called force was conserved and not what others had called by that name. Leibniz achieved the greater clarity at which he was aiming by setting out his argument as a formal logical demonstration and avoiding appeal to metaphysical consideration of cause and effect, though he did remind the reader in passing of his principle that the whole effect was equal to the total cause. First, he introduced a definition of equal, lesser and greater force. In effect, this stated that, if in passing from one state to another without external action, mechanical perpetual motion would result, then the force of the first state would be less and that of the second greater; otherwise they would be equal. Secondly, he supposed, as an axiom, that the quantity of force is conserved. Then he postulated that all the force of a body can be transferred to another. Supposing also the law of falling bodies proved by Galileo, Huygens and others, he demonstrated that mv^2 is conserved, so that this quantity is the measure of force. Finally, he demonstrated that the same quantity of motion is not always conserved, otherwise mechanical perpetual motion could be obtained, which is impossible. However, he remarked that the total quantity of progression (in modern terms *momentum*) is conserved.

Among his concluding remarks was one that explains the distinction between static force (*force morte*) and kinetic force (*force vive*). The former consists of a simple endeavour or *conatus*, which has the same relation to the *force vive* (resulting from a summation of *conatuses*) as a point to a line. At the beginning of descent, he explained, when the motion is infinitesimal, the elements of speed are as the distances, but after the summation, when the force has become *vive*, the distances are as the squares of the speeds.

Today the composition of motions would be regarded as a purely kinematic problem. Leibniz, however, could offer his *Regle generale de la composition des mouvemens* as a result in conformity with his dynamics, for he saw the problem essentially as one concerning the effect of forces on a physical body. For simplicity, he supposed that a moving body has four tendencies or motions represented by AB, AC, AD and AE (figure 7.1). Then he imagines the body to be divided into four equal parts, so that each part takes one of the four motions. Each part would need to go four times as far as if the whole body followed that tendency, in order that the quantity of progress (momentum) should be conserved. Not being able to move simultaneously in four directions, the body actually moves in the same way as the centre of gravity of the four parts would move. Consequently, the velocity of the combined motions is 4AG, where G is the centre of gravity of the points B, C, D and E (*GM* **6**, pp 231–3).

A week after the publication of his article on the composition of motions, Leibniz contributed to the *Journal des Sçavans* two examples of problems that could be solved using his rule. Also he further clarified his concept of

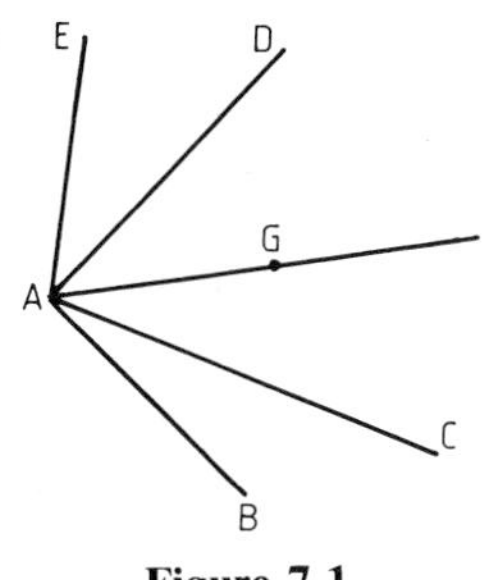

Figure 7.1

force morte, explaining that he used the term 'solicitation' to describe the infinitesimal endeavour or *conatus*, such as the action of gravity or centrifugal tendency, by which a body is solicited or incited to motion, and of which an infinity of impressions is needed to produce a finite motion (*GM* **6**, pp 233–4).

At the beginning of the *Specimen dynamicum* Leibniz declares that he is responding to requests for a fuller explanation of his new dynamics, and having reminded the reader of the demonstration he has given elsewhere that force is a real and essential attribute of bodies, he explains the nature of this force in some detail. First he considers active force, which is of two kinds, primitive and derivative. As the first entelechy, corresponding to soul or substantial form, the primitive force relates only to general causes which cannot suffice to explain phenomena; in other words, primitive force operates entirely on the metaphysical level. The conflict of bodies with each other gives rise to a limitation, as it were, of the primitive force, which manifests itself as derivative force. On the physical plane then, the force of a body to act is this derivative force. Turning now to passive force, Leibniz explains that this is also of two kinds, primitive and derivative. The primitive force of suffering or resisting constitutes what the Scholastics call primary matter, if rightly interpreted. In virtue of this primitive force, bodies are impenetrable and also possessed of a kind of laziness (*ignavia*), so to speak, or a repugnance to motion, so that they do not allow themselves to be set in motion without effect on the force of the bodies striking them. Again, primitive passive force operates only on the metaphysical level, for primary matter (that is, matter without form) is an abstraction. Real corporeal bodies consist of primary matter endowed with form (that is, active force) and these are described by Leibniz as secondary matter. Bodies having been set in motion, their resistance to external influence (primitive passive force on the metaphysical level) manifests itself in various ways in secondary matter (the matter of physical dynamics) as derivative force of suffering or resistance. It is odd that Leibniz does not use here the term 'natural inertia' of bodies, as he had in the *Journal des Sçavans* in 1691, when arguing against the Cartesian concept of bodies (*GP* **4**, pp 464–6).

Like Aristotle, Leibniz thus supposed bodies to consist of matter and form. Every body acted according to its form (entelechy) and suffered according to its matter (materia prima). The primary forces, both active and passive, operated only on the metaphysical level, so that he was able to reject any suggestion that he was returning to the battles of words of the Scholastics by introducing substantial forms as physical causes. On the other hand, he held it to be an abuse of the corpuscular philosophy, excellent in itself, to suppose a crude concept of corporeal substance based only on the appearances without penetrating to the metaphysical foundation. Since, however, the primitive forces manifested their activity and passivity through derivative forces and resistances connected with local motion, it was to these derivative forces, known not only by reason but verified through observation, that the laws of motion applied. Physical dynamics was therefore concerned entirely with derivative forces.

Turning his attention to the doctrine of derivative forces and the problem of how bodies interact with each other, Leibniz first defines the terms *conatus* and *impetus*. Endeavour or *conatus* is measured by velocity (that is, speed combined with direction), while *impetus* is measured by mv, the product of mass (moles)[5] and velocity. Thus the measure of *impetus* is what the Cartesians call quantity of motion. Since motion, however, is a continuous change of position requiring time, Leibniz thought it would be more correct to define quantity of motion as a summation of *impetuses* existing in the moving body multiplied by the corresponding intervals. Thus although he was willing to accept Cartesian terminology, he would have preferred to define quantity of motion as $m \int v \, \mathrm{d}t = ms$, or since s was proportional to v^2, as mv^2; that is, the measure of force he had adopted in the *Brevis demonstratio* of 1686.

Impetus itself, he continues, arises from an infinite succession of impacts on the same moving body, so it too contains a certain element from which it can arise only through infinite repetitions. The infinitesimal impetus (or conatus) he calls solicitation. These mathematical entities, he adds, are not found in nature as such but merely provide the means of making abstract calculations. What really exists in nature is force. He then explains the distinction between static force (*vis mortua*) and kinetic force (*vis viva*). To the examples of *vis mortua* (*force morte*) given in the *Journal des Sçavans*—the centrifugal endeavour and the solicitation of gravity—he adds the force with which a stretched spring begins to restore itself. In impact, however, the force is live and arises from an infinity of impressions of static force. This, he thinks, is what Galileo meant in his enigmatic way when he

[5]In a letter to Johann Bernoulli (*GM* **3**, 2, p 537), Leibniz explains the distinction between *massa* and *moles*: 'matter in itself, or moles, which may be called primary matter, is no substance, not even an aggregate of substances, but something incomplete. Secondary matter, or massa, is not one substance but a plurality of substances.'

described the force of impact to be infinite in comparison with the simple impulsion of gravity.

At this point, Leibniz explains the distinction between absolute, progressive and relative force which he introduced in the second part of his *Dynamica* (see chapter 6) and the various conservation principles which follow.

While remarking that the ancients had only knowledge of static force, Leibniz claimed for himself the explanation of the true concept of force. This, he remarked, he had not possessed when he wrote the *Hypothesis physica nova*, for at that time he had not recognised the internal resistance of bodies. Having discovered that the definition of body was incomplete without the attribution of force, he was able to establish a number of systematic rules of motion; namely, that all change occurs gradually, that every action involves a reaction, that no force is produced without diminishing the earlier force and that there is neither more nor less force in the effect than in the cause. Whether the new principle needed to complete the definition of bodies is called form, entelechy or force, Leibniz regards as immaterial, provided we remember that it can be explained intelligibly only through the concept of forces. He could not agree with the Occasionalists, who had removed all force of action from things themselves and attributed their motion to the direct action of God. For although he admitted that all things arise by a continuous creation of God, he could find no natural truth in things for which the cause had to be found in the divine action and preferred to believe that God had always put into things themselves some properties by which all their predicates could be explained. The hylarchic principle of Henry More he dismissed as a naïve doctrine that had been introduced on the mistaken assumption that there were some things in nature that could not be explained mechanically and that those who undertook a mechanical explanation aimed to deny incorporeal beings. In Leibniz's view the best answer, satisfying piety and science alike, was to acknowledge that all phenomena can be explained by mechanical efficient causes but that these mechanical laws are themselves to be derived in general from higher reasons.

Leibniz holds, however, that there are two ways of explaining phenomena; through a kingdom of power or efficient causes and through a kingdom of wisdom or final causes. For God regulates bodies as machines according to laws of mathematics but does so for the benefit of souls according to laws of goodness or morality. The two kingdoms everywhere permeate each other and their laws are harmonious, so that the maximum in the kingdom of power and the best in the kingdom of wisdom take place together. As an example of the use of final causes in physics, he cited his derivation of the laws of optics from the principle of least resistance, which had been praised by Molyneux.

Turning to the true estimation of forces, he promises an *a priori* method, based on concepts of space, time and action, which will be given elsewhere, and then gives the *a posteriori* method already explained in the *Brevis demonstratio*.

Although the second part of the *Specimen dynamicum* did not appear as promised in the May issue, its contents reveal some of the new ideas that Leibniz was at that time attempting to put in order. First he emphasised that, while force was absolutely real, space and time, like motion, were not real as such but only in so far as they involved the divine attributes of immensity, eternity and activity or the force of created substances. From this he deduced that there could be no vacuum in space and time and that motion, considered apart from force, was relative. Since only force and the effect arising from it at any moment exist, and every effort tends in a straight line, it follows, Leibniz concluded, that all motion is in a straight line, or compounded of motions in straight lines. Secondly, from the principle that all change was continuous, he deduced that there could be no atoms, that is, bodies of maximum hardness. For if such atoms existed, their speeds would change instantaneously in collisions. All rebounds, he held, must arise from elasticity, so that every body, however small, is permeated by a still subtler fluid (the cause of its elasticity), the analysis proceeding to infinity. Another interesting conclusion reached by Leibniz was that, in respect of the motions relative to the common centre of gravity of two bodies, the motion of each after a collision arises from its own internal force. Thus the effect on each body can be derived from the force within itself, without the need of supposing any influence of one upon the other, even though the action of one provides an occasion for the other to produce the change within itself. This idea will find a parallel in his metaphysical doctrine of substances or monads, as he later called them.

In a later *Essay de dynamique* (*GM* **6**, pp 215–31), intended as a further attempt to propagate his ideas on dynamics (which he felt had not been adequately understood), Leibniz explained that, in the case of inelastic collision, the observed loss of *vis viva* was only apparent. For the law of conservation of *vis viva* in the world, which he had demonstrated, he held to be inviolable. Although the parts of the body themselves were perfectly elastic, and hence conserved the whole of the *vis viva* received in the collision, they were not sufficiently organised to transfer this to the whole mass. Consequently some of the *vis viva* received by the body in the collision was distributed among the parts in the form of random motions leading to internal collisions.

Metaphysics

Owing to the close connection between his metaphysics and dynamics—for the one was concerned with the concept of substance in general and the other with corporeal substance in particular—it was natural that Leibniz should wish to clarify and disseminate his ideas in philosophy at the same time as those in his new science of dynamics. Since, in addition, his point of departure had been a rejection of the philosophy of Descartes, he prepared a detailed criticism of the general parts of Descartes' *Principia philosophiae* (*GP* **4**, pp 350–92), which he sent to Basnage de Beauval about the middle of 1692, hoping that he would find a publisher in Holland. Basnage still had the manuscript a year later (*GP* **3**, pp 97), having found no one who wanted to publish it, especially as it was written in Latin (*GP* **3**, pp 81–3). Meanwhile he had shown it to Huygens, who (*GM* **2**, p 139) agreed with Leibniz that the essence of body was not extension, but opposed him in maintaining that resistance to penetration implied the existence of perfectly hard atoms; and also to Bayle, who had been too busy to write any comments (*GP* **3**, p 108). Huygens (*GM* **2**, pp 136–41) suggested to Leibniz the idea of offering his writing as an appendix to the second edition of the *Censura philosophiae Cartesianae* of Pierre Daniel Huet, Bishop of Avranches. The theme of this book, first published in 1689, had been that the Cartesian philosophy was dangerous to religion. This idea, however, came to nothing. It might have been possible, Leibniz remarked to Basnage (*GP* **3**, pp 104–5), if he had been able to obtain for him some comments from learned Cartesians in Holland.

In March 1694 Leibniz published in the *Acta Eruditorum* a short article, *De primae philosophiae emendatione et de notione substantiae* (*On the correction of metaphysics and the concept of substance*) (*GP* **4**, pp 468–70), in which he explained the difference between his concept of 'active force' (*vis activa*) and the 'mere power' (*potentia nuda*) of the Scholastics. While the power of the Scholastics was a potentiality needing an external stimulus, as it were, to be transformed into action, his own concept of active force contained an entelechy and involved endeavour, so that it was carried into action by itself, needing no help but only the removal of impediments. The power of acting inhered in all substance and some action always arose from it, so that corporeal substance, like spiritual substance, never ceased to act. It would be apparent, he added, that one created substance received from another, not the force of acting itself but only the limits and determination of its own power of action. The difficult problem of the mutual action of substances upon each other he reserved for another occasion.

Writing to Bossuet on 12 July 1694 (*A* I **10**, pp 136–44), Leibniz explained that the piece in the *Acta Eruditorum* had been published at the request of a friend in Leipzig. It was just a small illustration of the kind he was content to give until he would have the leisure to order his thoughts on philosophy. But he confided to Bossuet that he was, at that time, writing an essay in which he would give to the public his explanation of the union of mind and body, which he had worked out several years before. The promised essay appeared in June 1695 in the *Journal des Sçavans* (*GP* **4**, pp 477–87). Entitled *Système nouveau de la nature et de la communication des substances, aussi bien que* *de l'union qu'il y a entre l'âme et le corps*, it contained the first published account of Leibniz's metaphysics.

At the beginning of the essay Leibniz tells the reader that, after conceiving the system some years ago, he had discussed it with learned men, especially with one of the greatest theologians and philosophers of the time (Arnauld, who is not mentioned by name), who had been told about his opinions by a person of the highest nobility (the Landgrave Ernst von Hessen-Rheinfels, who again is not mentioned by name). At first this great theologian had found some of the ideas paradoxical but after receiving Leibniz's explanations, he had agreed with some and deferred judgment on others. Leibniz next gives an autobiographical account of his early acceptance of the mechanical philosophy and subsequent recognition of the need to rehabilitate the substantial forms in a way that would make them intelligible. Thus after freeing himself from the bondage to Aristotle, he had at first accepted atoms and the void but having then rejected material atoms—for they were inconsistent with the principle of continuity, as he had explained to Foucher, for example, in a letter published in the *Journal des Sçavans* in June 1692 (*GP* **1**, pp 402–6)—he perceived that it was impossible to find the principles of a true unity in matter which was passive and divisible to infinity. For an aggregate could only derive its reality from true unities, which must be something different from points, since it was clear that the continuum could not be compounded of points. These real unities, he had concluded, could only be found in formal atoms and it was thus necessary to rehabilitate the substantial forms but in a way that rendered them intelligible and avoided the misuse to which they had previously been subjected. Leibniz then explains that the nature of these formal atoms consists of force (that is, activity), so that they possess something analogous to sense and appetite and consequently may be regarded as in some way similar to souls. However, these forms should not be used to explain particular problems of nature, though they are necessary to establish its true general principles. Aristotle called them first entelechies. Leibniz calls them primitive forces, which contain not only the actuality or completion of possibility but an original activity as well.

Leibniz distinguishes minds or rational souls from those which are sunk in matter and can be found everywhere. For minds or rational souls are created in the divine image and follow special moral laws which place them beyond the revolutions of matter. Since every substance which has a true unity can begin and end only by a miracle, he holds that the other forms or material souls were created with the world and will always subsist. The notion of the transmigration of souls from one body to another, however, he dismisses as a fancy far from the nature of things. For the observations of Swammerdam, Malpighi and Leeuwenhoek had led him to believe that the apparent generation of an animal is simply a development and increase of an animal already formed. Likewise, when the animal appears to die owing to the destruction of its grosser parts, the animal in fact survives, its organic body having simply diminished to a size so small that it becomes invisible, just as it was before its birth.

The machines of nature—that is, animals and other corporeal substances depending on mechanical rather than moral laws—thus remain the same machines throughout all transformations. When a corporeal substance is completely contracted, so that it appears to us as a physical point, all its organs are concentrated in this point. Consequently, physical points are indivisible in appearance only. The real unities, which are truly without parts and are the ultimate elements in the analysis of substance, Leibniz calls metaphysical points. These forms or souls have something vital and a kind of perception, mathematical points being the points of view from which they express the universe.

Turning to the problem of the union of the soul with the body, Leibniz begins by saying that he has found no way to explain how body and soul can interact, or how one created substance can communicate with another. Reflection on the impossibility of such interactions had led him to a surprising conclusion; namely, that God had originally created the soul and every other real entity in such a way that all its activity must arise spontaneously from its own nature, yet in perfect conformity with things outside itself. The perceptions of the soul are merely phenomena which arise from its own nature and capacity to represent the universe, though with varying degrees of distinctness, from its point of view. The sequence of representations which the soul produces will naturally correspond to the sequence of changes in the universe itself. The body is expressed or represented more immediately by the soul and acts following the laws of corporeal mechanism at the moment when the soul wills, but without either disturbing the laws of the other. It is this mutual agreement or harmony, regulated in advance for every substance of the universe, which produces what is called their communication and constitutes the union of the soul with the body.

Leibniz points out that his hypothesis has a number of advantages to recommend it. For example, instead of having to say that we are free only in appearance, as some supporters of the mechanical philosophy have thought, it enables us to say that we are determined only in appearance and that in metaphysical strictness we are in a state of perfect independence of the influence of all other created things. Earlier in the year he had written a dialogue *Sur la liberté de l'homme et sur l'origine du mal*, following a conversation with the Brandenburg Counsellor, F von Dobrzensky (Bodemann 1895, p 45). The hypothesis also provides a new proof of the existence of God as the source of the perfect agreement of so many substances which have no communication with each other. Finally, he claims for his metaphysics something more than the status of an hypothesis, for it seems to him hardly possible to explain things intelligibly in any other way.

He adds, however, that ordinary ways of speaking may be preserved, so that when related changes occur, we may think of one substance as acting on another, provided we remember that there is no real interaction.

The publication of the *Système nouveau* in the *Journal des Sçavans* brought an immediate response from the Abbé Foucher (*GP* **4**, pp 487–90), who could not see that the principle of harmony, or concomitance as he called it, had any advantage over the explanations of the Cartesians. Believing Foucher to be away from home, Leibniz sent his reply to the editor of the *Journal des Sçavans* through L'Hospital, hoping thereby to save time (*GM* **2**, pp 297–302). From the reply, he remarked to L'Hospital, he would see in what ways his hypothesis of pre-established harmony—a term he had not in fact used in the published essay—differed from that of Malebranche and the Cartesians. Nevertheless, he hoped that Malebranche would be able to accept his hypothesis, which should be regarded not so much as a rejection of the doctrine of Occasionalism as an improvement and development of it. In his published reply, which appeared in the *Journal des Sçavans* in April 1696 (*GP* **4**, pp 493–8), Leibniz suggested that the repugnance Foucher felt to his hypothesis probably arose because he thought that it was arbitrary and hadn't understood that it was deduced from his concept of unities or substances.

Basnage de Beauval had also evidently made some criticisms, for Leibniz felt the need to clarify the principle of pre-established harmony for him in a letter of 13 January 1696 (*GP* **3**, pp 120–3; **4**, pp 498–500). A month later, Basnage published Leibniz's explanations in his *Histoire des ouvrages des Savans* (*GP* **4**, pp 490–3). In November 1696, Leibniz added further clarifications in a contribution to the *Journal des Sçavans* (*GP* **4**, pp 500–3), where he used the expression 'pre-established harmony' for the first time in print. A long dispute followed the appearance of the objections of Pierre Bayle in the second volume of his *Dictionnaire historique et critique*, published in Rotterdam in 1697 (*GP* **4**, p 418). Another rather unpleasant dispute that Leibniz would have wished to avoid arose out of the appearance in the *Journal des Sçavans* of a letter written in February 1697 to Claude Nicaise (who was a colleague of Foucher) which had not been intended for publication (*GP* **2**, pp 562–5). In this letter, Leibniz had commented on the danger to religion of Descartes' rejection of final causes and the consequent denial of a role for the divine wisdom in the order of things. This drew from the Cartesian Pierre Sylvain Régis (*GP* **2**, p 575) a sharp reply published in the *Journal des Sçavans* in July 1697, in which he suggested that Leibniz should confine himself to mathematics where he excelled and not meddle in philosophy where he didn't have the same advantage. In his defence, published a month later (*GP* **4**, pp 336–42), Leibniz explained that he was only supporting a criticism of Descartes that had already been made more than once by Huet, the Bishop of Avranches, whose views had been the subject of his correspondence with Nicaise at that time. He had no need, as

Régis had rather impertinently suggested, to build his reputation on the ruins of Descartes.

François Mercure van Helmont

An opportunity to explain and discuss his metaphysics with the Electress Sophie presented itself when, in the summer of 1694, she sent him two recently published books she had received from her old friend Baron François Mercure van Helmont and asked for his opinion of them. Concerning the first book, *Verhandeling van de Helle* (*A treatise on Hell*), published anonymously in Groningen, Leibniz was of the opinion that van Helmont would probably not wish to admit his authorship, though his rejection of the concept of eternal damnation had many precedents both in the ancient world and the recent past. The other book, *Het Godlyk Weezen* (*The Divine Being*), contained an orderly exposition of van Helmont's theology by his friend Paulus Buchius, a physician in Amsterdam. While agreeing with van Helmont when he reproved Gassendists and Cartesians for their exclusive attachment to the corpuscular philosophy, which explained all nature by matter or extension, he pointed out that many of van Helmont's ideas were founded on the Jewish Cabbala rather than incontestable reasons. He then explained to Sophie that, by building on certain metaphysical principles, he had formulated the concept of force, which provided the key to the connection between the corporeal and the spiritual. Although he agreed with van Helmont that souls, once created, were indestructible, he could not believe in the idea of transmigration. Leibniz then described for Sophie his theory of the transformation of the same animal, claiming a classical precedent in a work attributed to Hippocrates, and pointing out that he did not apply the theory to man; for being created in the image of God, man was governed by special laws which could only be comprehended in detail by revelation. At the time he wrote out this explanation for Sophie on 13 September 1694 (*A* I **10**, pp 57–65), he was suffering from a feverish cold that had caused him to return home from the fair in Brunswick.

Two years later, van Helmont spent several months in Hanover as a guest of the Electress Sophie. Soon after his arrival in March 1696, Leibniz joined in conversations with him and Sophie for several days on end. He described the circumstances in a letter to Thomas Burnet (*GP* **3**, pp 174–9). At 9 o'clock each morning he would accompany the Baron to Sophie's room. The Baron would sit at the writing table and explain his views. Leibniz listened, but interrupted from time to time when van Helmont failed to express himself clearly.

Although van Helmont had some extraordinary ideas, Leibniz judged him to have a good mind for things of practical utility. An opportunity for him to offer practical advice occurred on 13 August, the day that Leibniz received official confirmation of his appointment as Privy Counsellor of Justice. Having spent the morning discussing his plan for installation of fountains in the park at Herrenhausen with the President of the Chamber, in the afternoon he inspected the region between the town and the park in the company of van Helmont and two advisers, in order to decide the location of the canal that would supply the water for the fountains. During the following week, another series of daily conversations took place in Sophie's room, when the topics of discussion included van Helmont's ideas on immortality and the transmigration of souls. Then after Leibniz's return from a visit to Wolfenbüttel, further exchanges took place before the Baron returned home to Amsterdam on 23 September.

On the departure of van Helmont, Leibniz drafted a report on the conversations, which Sophie sent to her niece, the Duchess Elizabeth Charlotte of Orleans in Paris (*K* **8**, pp 8–11). He described the Baron as an old friend of the Electress of Hanover, who was wont to say, in speaking of him, that he didn't understand himself. Formerly a Roman Catholic, he had become a Quaker and dressed in a habit of drab brown with a cloak of the same colour, so that he looked more like an artisan than a baron. He was very alert and active at 79 years old and accomplished in several fields, including weaving, painting and medicine. The principal opinion he sustained was the immediate transmigration of souls. Leibniz then stated some of the principles of his own metaphysics concerning the nature of souls. In November he sent to Sophie a more detailed account of his metaphysics, explaining particularly his theory of representation (*K* **8**, pp 14–18).

Despite his fundamental disagreements with van Helmont, Leibniz evidently admired the Baron's personal qualities. To Andreas Morell he wrote that van Helmont was one of the few people of extraordinary talent he had met, who shared his view that charity represented the fundamental principle (*MK*, p 143).[6]

Mathematics and logic

When Leibniz challenged the Abbé Catelan in 1687 to find the curve of uniform descent, he introduced a pattern for the promotion of his new calculus that was to be followed in future years. For example, Jakob Bernoulli in 1690 proposed the problem of the catenary and in 1696, his brother Johann proposed that of the brachistochrone or line of quickest descent. In each case contributors were asked not to publish their solutions before the expiry of the time limit but to deposit them with a third party. By this means it was hoped that a number of solutions would be received, all

[6]The relationship between Leibniz and van Helmont is described by Becco (1978).

based on the new analysis and thus demonstrating its superiority over that of Descartes.

It was in fact Leibniz himself who set a time limit of one year for the solution of the problem of the catenary after he had accepted Jakob Bernoulli's invitation to solve it. Both Huygens and Johann Bernoulli sent solutions to the *Acta Eruditorum* before the expiry date. These were published after that of Leibniz, which appeared in June 1691 (*GM* **5**, pp 243–7). He was pleased to see that all three solutions were in agreement (*GM* **5**, pp 255–8). Leibniz published his solution again in March 1692 in the *Journal des Sçavans* (*GM* **5**, pp 258–63), where he emphasised that the solution was made possible by his new infinitesimal analysis, which was entirely different from those of Cavalieri and Wallis and went beyond that of Descartes. At the end of his paper he shows how the catenary could be applied to navigation, if the captain should lose his table of logarithms. A third version of his solution was inserted by Magliabechi in the *Giornale de' Letterati*, published in Modena, as an example of the new analysis (*GM* **5**, pp 263–6).

Another opportunity to show the power of the new analysis presented itself on 27 May 1692 when Leibniz received from Prince Ferdinand of Tuscany a request for a solution of a problem that had been proposed by Viviani under the pseudonym Don Lisci Pusillo. This problem concerned an hemispherical temple with four identical windows, to be designed so that, when the windows were open, the remainder of the surface was quadrable. Having found the solution on the same day, Leibniz sent it to the Prince by the next post. In June, his solution was published in the *Acta Eruditorum* (*GM* **5**, pp 273–8).

In a short article which he sent to the *Journal des Sçavans* (*GM* **5**, pp 278–9) through Nicaise (*GP* **2**, p 537) in July 1692, Leibniz pointed out that Descartes had been obliged to exclude transcendentals from his geometry in order to be able to maintain that all geometrical problems could be solved by his method. Describing the advantages of his own infinitesimal analysis in a further contribution to the *Journal des Sçavans*, published in August 1694 (*GM* **5**, pp 306–8), Leibniz explained that transcendentals, which arose often in applications of mathematics to physics, could always be expressed as series of rationals. Some examples of transcendentals expressed by infinite series had been given a year earlier in a paper published in the *Acta Eruditorum* (*GM* **5**, pp 285–8), where he had obtained infinite series for logarithms, exponentials and sines by assuming a power series and determining the coefficients by repeated differentiation. The Bernoullis, he told readers of the *Journal des Sçavans*, had been the first to apply his new calculus to the solution of problems of physics. They were followed by L'Hospital and his methods had earned the approval of Huygens. Newton, he added, had something similar, though he used an inferior notation.

When Leibniz addressed his first direct letter to Newton on 17 March 1693

(*NC* **3**, pp 257–60), he was hoping to discover what Newton had contributed concerning the inverse method of tangents to the new Latin edition of Wallis' *Algebra*, to be published later in the year as part of the second volume of his *Opera mathematica* (Hofmann 1973, pp 255–66). The first problems of this type encountered by Leibniz had been those proposed by Debeaune, to which he had given attention soon after the discovery of his new calculus in Paris. More recently, Huygens had written to him in the summer of 1690 (*GM* **2**, pp 44–6), proposing the problem of finding the curve whose subtangent is $(2x^2y-x)/(3-2xy)$, having read that the determination of curves from their tangents was one of the types of problem to which the new calculus could be applied. Within two months, Leibniz replied that the equation of the curve (in modern notation) is $x^3y/c=\mathrm{e}^{2xy}$, where c is a constant, and he demonstrated that the curve had the required property, though he did not reveal the method of solution (*GM* **2**, pp 49–55). At the beginning of 1693, Huygens asked Leibniz if he had a method of solving the problems of Debeaune, as L'Hospital had put the question to him, following a renewal of their correspondence, in the course of which L'Hospital had himself supplied the solutions to several inverse tangent problems. Huygens explained that he did not wish to go to the trouble of solving the problems himself, as he was sure that all the difficulties had already been surmounted, either by the Marquis himself, or by Newton (in the work of Wallis, which he believed had been published) or of course by Leibniz, who had thoroughly cultivated this field in which Huygens himself was only a novice (*GM* **2**, pp 148–53). In his reply, Leibniz (*GM* **2**, pp 154–60) claimed to have the idea of a general method but lacked the leisure or patience to work it out, so that he had to be content to consider particular cases. He then gave the solution to one of the problems of Debeaune. Besides inverse tangent problems, Leibniz discussed with Huygens corrections and details concerning his work on resisting media, planetary motion and the explanation of gravity. The correspondence on these matters continued up to Huygens' death in 1695.[7] An idea of the intense pressure under which Leibniz worked may be gleaned from a remark he made to Huygens, that at any given time, he would have up to forty letters needing answers going beyond the normal courtesies.

When L'Hospital (*GM* **2**, pp 249–55), towards the end of 1694, mentioned to Leibniz his plan to write a text-book on the calculus, he included in his letter a sample of the inverse tangent problems he had been able to solve, though he did not expect that it would contain anything new to Leibniz. This required the curve whose subtangent was $\surd(ay+x^2)$, or in other words, the solution of the differential equation $y\,\mathrm{d}x=\mathrm{d}y\surd(ay+x^2)$. First, L'Hospital removed the root by means of the substitution $ay+x^2=u^2$, so that the

[7]Leibniz acquired 121 books from the sale of Huygens' private library. These are now preserved in the Niedersächsische Landesbibliothek, Hanover (Hess 1980).

equation became $2u^2\,\mathrm{d}u = (2ux + u^2 - x^2)\,\mathrm{d}x$. Then using the substitution $u = xz$, he reduced the equation to the form

$$\frac{2z^2\,\mathrm{d}z}{2z - 2z^3 + z^2 - 1} = \frac{\mathrm{d}x}{x}$$

in which the variables were separated, so that the equation of the curve was determined, supposing only the quadratures. The same method, he added, could be applied to any homogeneous equation. It should be mentioned, however, that in an earlier letter (*GM* **2**, p 220), Leibniz had himself claimed to have a general method of solving such equations. In his reply, written on 27 December 1694 (*GM* **2**, pp 255–62), Leibniz remarked that it did not seem necessary to explain anything to L'Hospital, as he had found the method of solving homogeneous differential equations for himself. Encouraging L'Hospital to persevere with the writing of his text-book, Leibniz confided that he was unable to give attention to such matters himself because of the distractions which sometimes overwhelmed him and endangered his health. He then expressed the view that the completion of the inverse method of tangents was in sight. Already he had found a general result that could solve a variety of problems and was capable of further development. This was the general solution of first order linear differential equations. Thus the equation $m + ny + \mathrm{d}y/\mathrm{d}x = 0$, where m and n are functions of x, he explained, reduces to $\int mp\,\mathrm{d}x + py = 0$ on taking $\int \mathrm{d}p/p = \int n\,\mathrm{d}x$. Since p is a function of x, the variables are separated, so that the equation has been reduced to quadratures. Leibniz verified the solution by differentiation. The following year, Jakob Bernoulli proposed the equation $a\,\mathrm{d}y = yp\,\mathrm{d}x + by^nq\,\mathrm{d}x$, where p and q are functions of x. This was evidently one of the further developments to which he had referred, for in March 1696, he noted in the *Acta Eruditorum* (*GM* **5**, pp 329–31), that Bernoulli's equation could be reduced to his own. Johann Bernoulli published such a reduction in 1697, having communicated it to Leibniz in August 1696 (*GM* **3**, 1, pp 323–4).

In 1695 the Dutch mathematician Bernard Nieuwentijt sent Leibniz copies of two works he had recently published in Amsterdam containing criticisms of certain aspects of the new calculus, to which, however, he was not entirely opposed, for as he explained in a letter to Thomas Burnet (*GP* **3**, p 164), Nieuwentijt employed it in part himself. To L'Hospital Leibniz (*GM* **2**, pp 287–9) reported that the objections were directed not only against himself but also against L'Hospital and the Bernoullis. According to Nieuwentijt, all four had employed the new calculus in the solution of problems without having demonstrated the principles. After mentioning to L'Hospital that he would reply to Nieuwentijt's objections in the *Acta Eruditorum*, Leibniz expressed the hope that L'Hospital's book would soon appear, for then he expected such complaints to cease.

Besides raising objections against the concept of infinitesimals, especially those of higher orders, which he completely rejected, Nieuwentijt also claimed that the method was not applicable to curves whose equations involved exponentials. Leibniz published his reply in the *Acta Eruditorum* in July 1695, attempting to resolve the difficulties concerning infinitesimals and illustrating the applicability of the new analysis to transcendental curves by showing how to differentiate exponentials (*GM* **5**, pp 320–7). In the following month he published an addendum, where he mentioned the death of Huygens with some dismay and pressed his brother to publish his manuscripts (*GM* **5**, pp 327–8). The controversy with Nieuwentijt was taken up by Jakob Hermann of Basel, a student of Jakob Bernoulli, who published a detailed reply to the objections in 1697 (Ravier 1937, p 70).

In September 1696, soon after Johann Bernoulli, then Professor of Mathematics in Groningen, had proposed the problem of the brachistochrone or curve of quickest descent, in a letter to Leibniz (*GM* **3**, 1, p 283) and in the *Acta Eruditorum*, Leibniz advertised the challenge in the *Journal des Sçavans* (*GP* **4**, pp 501–2), inviting solutions also from those who followed methods other than the new analysis employed by Bernoulli and himself to arrive at their independent solutions (*GM* **3**, 1, pp 290–5, 302–9), which they had found to be in agreement. Bernoulli, he informed the reader, would receive solutions up to Easter of the following year. As the purpose of the challenge was to demonstrate the superiority of the new calculus over the methods of Descartes, Leibniz could be pleased with the result. For besides Johann Bernoulli and himself, the only other mathematicians who solved the problem and discovered the curve to be a cycloid were Jakob Bernoulli, L'Hospital and Newton.

In the spring of 1696, during one of his visits to Wolfenbüttel, Leibniz described his binary system of arithmetic, invented years before, in conversation with Duke Rudolf August. Evidently the Duke was struck by the way in which the numbers were generated out of 0 and 1; this suggested to him an analogy with the Biblical creation of the world out of nothing by God. As a New Year present, he sent to the Duke at the beginning of 1697 a design for a medal illustrating the binary system and the analogy with the creation (figure 7.2) (Loosen and Vonessen 1968, pp 19–23).[8] The reverse side of the medal shows a picture of darkness over the water, with light streaming from the top; this illustrates the creation story. In the middle of the picture, a table of binary numbers with denary equivalents illustrates the meaning of the system, while examples of binary addition and multiplication are given at the sides. Along the rim is written '2 3 4 5 etc. Omnibus ex nihilo ducendis. Sufficit unum.' Underneath the picture is written, 'Imago Creationis Ann Chr. Inven G. G. L. MDCXCVII.' On the front of the

[8]Although the letter itself contains only a verbal description of the proposed medal, Leibniz also made three sketches (Zacher 1973, pp 51–2). The figure shows a design, based on one of these sketches, published by Ludovici (1737).

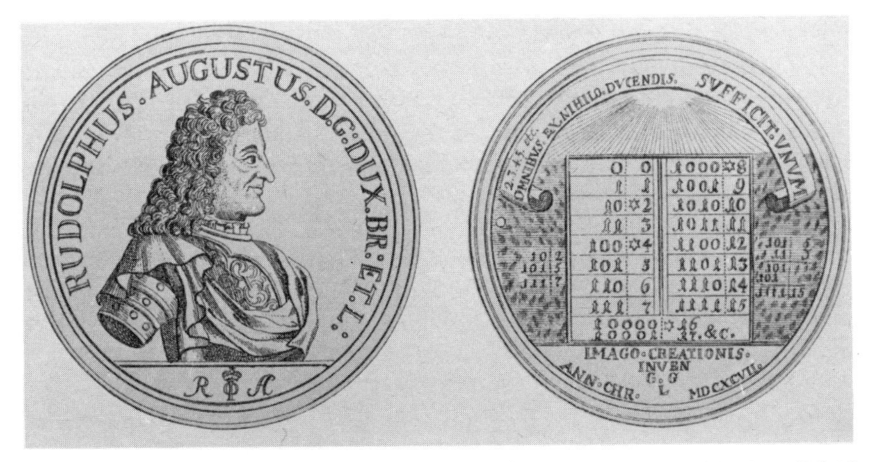

Figure 7.2 Medal illustrating the analogy of the binary system with the Biblical creation. Altered arrangement of the illustration in C G Ludovici, *Ausführlicher Entwurf einer vollständigen Historie der Leibnitzischen Philosophie* (Leipzig 1737). Courtesy of the British Library.

medal, we find 'R ♕Φ A.' In his letter Leibniz explains that the symbol Φ with the crown above, surrounded by the Duke's initials, represents 0 with a 1 through it. Remembering, however, that Φ in Greek is one symbol, it reminds us that 'unum necessarium' (one thing is needed): a Biblical text (Luke 10:42).[9]

Although the Duke had suggested the idea of the image of the creation with the account of the Bible in mind, Leibniz himself adopted the idea to find in the binary system an image of his own philosophical system. For Leibniz, creation was continuous, in the sense that God conserves created monads and produces them continuously by a kind of emanation, as we produce our thoughts. Thus to Johann Christian Schulenburg (*GM* **7**, pp 238–40) in Bremen he remarks that, in the binary system, he had found a very beautiful picture of the continuous creation of things out of nothing and their dependence on God for their continued existence.

Leibniz recalled his early interest in logic and defended its value in a letter written in 1696 to Gabriel Wagner (*GP* **7**, pp 514–27), who had published a fierce attack on it in the German weekly he had founded in Hamburg. Since logic or the art of reasoning embraced not only the art of judging but also the art of discovery, Leibniz held that it should be considered as the key to all the arts and sciences. Although, as he explained to Wagner, he had found some basis for the art of reasoning when he was still a novice in mathematics, and indeed had published something on it in his twentieth year—*De arte*

[9]This may be taken to symbolise the necessary existence of God.

combinatoria—he had finally come to see that the real foundation was to be found in mathematics.

At some time during the years following his return from Italy, Leibniz extended his algebraic logical calculus in various ways. For example, he introduced a predicate constant *ens*, which may be seen as a precursor of the existential quantifier. Also he developed an interpretation of predicates as propositions instead of concepts, so that inclusion between concepts became implication between propositions, and the predicate constant *ens* appeared as the truth value (*verum*). The most complete formulations of a logical calculus achieved by Leibniz are contained in two manuscripts, the first of which originally carried the title, later deleted, *Non inelegans specimen demonstrandi in abstractis* (*A not inelegant example of demonstrating in the abstract*) (*GP* 7, pp 228–35). Here he made the transition from an intensional logic of concepts to a pure calculus of classes. A new predicate constant *nihil* (for *non-ens*) was introduced and also the relation of incompatibility and its negation. Besides disjunction, symbolised by +, this calculus also contained a relation symbolised by −. For example, man − rational = beast. One of the propositions of this calculus was that $A-B=C$ if and only if $A=B+C$ and B and C are incompatible. The calculus developed in the second manuscript (*GP* 7, pp 236–47) lacks the relation of subtraction, while conjunction as well as disjunction is symbolised by + (or rather ⊕), so that the meaning has to be determined from the context. In this calculus, an extensional as well as an intensional interpretation is expressly given. Leibniz did not make sufficient distinction between the formal structure of his logical calculus and the interpretation of its content, immediately treating the beginnings as axiomatic and the rules of transformation as principles of deduction. Yet his work represents an approximation to the elements of a logical calculus or formal language and hence an example of a *characteristica universalis* (Mittelstrass 1970, pp 444–8).

Geology

Leibniz had taken an interest in geology from the time of his first visits to the Harz mountains and he never missed an opportunity to study fossils and formations in the course of his journey to Italy. In the summer of 1692, the discovery of a prehistoric tooth in the vicinity of Wolfenbüttel was brought to his notice and he was asked for an opinion on it. Describing the find to Sophie (*K* 7, pp 201–2), he explained that, if it belonged to a giant, as some supposed, his calculations showed that the giant would have to be as tall as a house. On the basis of descriptions he had read, it seemed possible that the tooth belonged to an elephant, but as elephants did not inhabit cold countries, at least in modern times, he thought it more likely that the find was a relic of some great sea monster, since the discovery of fossils of sea animals indicated that the earth had once been covered by the ocean. To provide evidence for a more reliable interpretation, he had asked that as many pieces as possible of the skeleton should be collected.

In April 1691 Leibniz mentioned to Huldreich von Eyben (*A* I **6**, pp 436–43) that he was writing up an account of the geological history of the earth. This work, entitled *Protogaea*, was not published during his lifetime, but a brief announcement appeared in the *Acta Eruditorum* in 1693 and a popular account of his theory of the earth's history was included in his *Théodicée*, published in 1710 (*GP* **6**, pp 262–3). The earth, he supposed, was originally molten, so that the rocks which form the base of the earth's crust are the slag remaining from a great fusion. Evidence for this theory of the formation of the rocks through fire was provided by the presence in their interior of products of metals and minerals similar to those found in furnaces. As the surface of the earth cooled, the vapour that the fire had driven into the air fell back, dissolving the salt remaining in the cinders, and formed the oceans. Violent changes continued to occur in the crust, like the collapse into cavities described by Thomas Burnet[10] in his *Telluris theoria sacra*, and the effects of several deluges and inundations were to be seen in the existence of marine fossils in places now far removed from the sea. These violent upheavals finally ceased, however, and the globe took the form that it has today. At the end of his brief account in the *Théodicée*, Leibniz remarked that Moses had indicated the history of the earth in a very few words; the separation of light and darkness was a description of the fusion caused by the fire, and the separation of water and dry land related to the effects of the inundations.

In the *Protogaea* itself, Leibniz refers briefly to the opinions expressed by some, that the animals which inhabit the earth today were originally aquatic, then became amphibious as the water receded and finally abandoned the water for the dry land. Besides being contrary to Scripture, this speculation, he remarks, is subject to other difficulties (Saint Germain 1859, p 16). Yet in response to the objection that the sea does not now contain some of the animals revealed in the fossils, he replies that, on the one hand, the depths of the ocean have not been searched, and on the other, it is to be presumed that, in the course of so many geological revolutions, a great number of animal forms have been transformed (Saint Germain 1859, pp 65–8).

Medicine

Whenever an opportunity presented itself, Leibniz sought to bring about an improvement in the primitive state of the science of medicine and the

[10]This Thomas Burnet was an Anglican cleric who had worked with Cudworth and Henry More during several years he spent in Cambridge.

effectiveness of medical practice by making recommendations to the appropriate authorities. After meeting Leibniz in Modena, the physician Bernardino Ramazzini, known as the father of industrial medicine, had written to Magliabechi to say that he had never before met anyone who was so accomplished in all the fields of knowledge (*A* I **5**, p 685). Evidently one of the subjects they discussed was the value of medical statistics. For Ramazzini produced such a record for 1690, having been encouraged, it seems, by Leibniz (*GM* **4**, p 519), who then recommended that the authorities in Germany and France should follow Ramazzini's example. First, he persuaded the Nuremberg physician Georg Volckamer (Bodemann 1895, p 366), who was then President of the Academia Leopoldina in Vienna, founded in 1672 in imitation of the Royal Society of London, to include Ramazzini's statistics as an appendix to the *Ephemerides* of the Academy for 1691 (Ravier 1937, pp 123–4). Then, in 1694, he contributed a letter to the *Journal des Sçavans* asking for an annual publication of medical statistics for Paris, the Isle de France and the other provinces, following the pattern established by Ramazzini (*D* **2**, 2, pp 162–3). This was a theme to which he frequently returned.

When news of the discovery in France of a cure for dysentery came to Leibniz in 1691 in a letter from Justel in England, he took steps to secure its introduction in Germany. For example, he sent a report to the President of the Academia Leopoldina, who published his letter, though at that time he was not able to give the name of the plant (*K* **7**, p 125). From Magliabechi in Florence, to whom he mentioned the discovery of the cure, he received the additional information that the Marquis de Louvoy had ordered the army physicians to acquire supplies of the plant. Reporting to Sophie (*K* **7**, pp 123–4), Leibniz remarked that he didn't know what the cure was, but that it seemed rather important and he suggested that the Duke should himself endeavour to discover the details of the cure. Four years later, having discovered that the cure was the ipecacuanha root from South America, Leibniz sent a memorandum on its application to Sophie and also asked the Countess Maria Aurora von Königsmarck (*A* I **11**, pp 453–5), sister of the murdered Count, to promote its use in Saxony. He prepared a detailed account for the Academia Leopoldina, which was published as a separate report in January 1696 (Ravier 1937, p 24).

When the question of the primitive state of the science of medicine arose in correspondence with the Electress Sophie in 1693, Leibniz (*A* I **9**, pp 38–41) expressed the view that, in order to make advances in knowledge and the effectiveness of treatment, the physicians needed facilities for research and the dissemination of its results which only governments could provide. Moreover, he believed that governments had a moral duty to give this support in aid of the welfare of those whom they governed. Medicine, he remarked, was only in its infancy. For example, the circulation of the blood had been discovered only sixty years before. While he agreed with Sophie that some physicians were charlatans—like the metal diviner Jacques Aymar, who had been exposed in the *Journal des Sçavans*—there were others, he pointed out, who had discovered excellent remedies, and such responsible physicians should not be discouraged from making their researches by an attitude of contempt for medicine by the public in general. Ordinary physicians, preoccupied with making a living by their practice, were sometimes hindered by the unreasonable demands of patients. It was essential that diagnosis should precede treatment. Symptoms should be carefully observed and the course of the illness, including the response to treatment, written up. Accounts of the most interesting cases should then be published, so that the knowledge gained could be applied as widely as possible. Large hospitals, he suggested, provided the greatest scope for systematic observation and research but this would require the provision of adequate finance and personnel. The importance of a supply of good physicians and surgeons for armies and navies was fairly obvious, but Leibniz went further in advocating the idea of preventive medicine. To Sophie he recommended the setting up of a permanent Council of Health, consisting jointly of political Counsellors and physicians, which would take steps to curb the effects of the epidemics which occurred almost every year and resulted in numerous deaths, especially among the poor.

For the treatment of his own ailments, frequent heavy colds and hoarseness, Leibniz sought the help of Count Francesco Palmieri, who he believed, as leader of the Opera, would know of the old and tried remedies used by the singers (Bodemann 1895, p 213).

Family affairs

Towards the end of 1690, Leibniz's nephew, Friedrich Simon Löffler, sought advice from his famous uncle concerning his further studies (*A* I **6**, pp 605–7). After obtaining the master's degree in theology at the University of Leipzig in 1689, he had transferred to the University of Wittenberg. Leibniz advised his nephew to concentrate on the Scriptures and Church history (*A* I **6**, p 610). It was no doubt for the purpose of gaining further insight into the Scriptures that the young Löffler spent the years 1692 and 1693 in Hamburg, studying Hebrew and Greek. Meanwhile he kept his uncle informed on events concerning Church politics in Saxony, including the opposition to Pietism in Leipzig and the movement of the Pietist Philipp Jakob Spener from Dresden to Berlin (*A* I **6**, pp 615–16). On completing his language studies in Hamburg, Löffler made a tour of north Germany and Holland, seeking out the leading theologians and historians, before becoming a Lutheran pastor in Probstheida, near Leipzig. On his travels he helped his uncle by distributing the printed title page of the *Codex juris gentium diplomaticus* and soliciting information on additional documents for inclusion in the second volume (*A* I **9**, p lxviii). Leibniz, in return, advised his nephew on the preparation of his dissertation on the Trinity (*A* I **11**, pp 222–34).

When Anna Elizabeth, the daughter of his half-brother Johann Friedrich, married in December 1690, Leibniz congratulated his niece with a couplet (*A* I **6**, pp 607–9). Johann Friedrich was a teacher at the Thomas School in Leipzig and a member of Spener's circle. He was highly critical of the edict issued in 1692 by the Dukes of Wolfenbüttel against Pietists and Chiliasts after Johann Wilhelm Petersen, on retiring from Lüneburg to Wolfenbüttel, had supported the claims of Rosamunde von der Asseburg in his sermons (*A* I **8**, pp 605–6), and expected his brother to take a similar view. Although Leibniz had been sympathetic to Rosamunde and also to Petersen, he defended the action of the Dukes, pointing out to his brother that this had to be taken in order to prevent the disturbance of public Church discipline by fanatics. He added, however, that it should not prevent moderates like Spener from promoting piety in animated private meetings (*A* I **8**, p 614). In fact Spener was welcomed as Court Chaplain in Berlin.

The production of an Opera in Leipzig at the beginning of 1694 by the Italian Girolamo Sartori prompted Leibniz (*A* I **10**, pp 673–4) to express his views on theatrical performances to his brother. Having informed him that Sartori was once in the service of Duke Johann Friedrich, before going to Brunswick and Hamburg, he went on to say that he did not decry this kind of entertainment and that it was a much better form of relaxation than drinking and gambling. Although Johann Friedrich was less enthusiastic about the moral values of the theatre, there can be little likelihood that he indulged in the alternative diversions suggested by his famous brother. This unpretentious, pious man died on 19 March 1696, at the age of 64 years, 2 months, 2 days, 3 hours, as Leibniz noted for his nephew, Friedrich Simon Löffler.

In the autumn of 1697, Leibniz was able to send a poem of congratulation to his nephew Johann Friedrich Freiesleben, the son of his half-sister Anna Rosina, on his graduation in Halle as a doctor of law (Bodemann 1895, p 63).

According to his first biographer, Johann Georg Eckhart, Leibniz thought of marriage in the year 1696, at the age of fifty, but as the lady took time to consider the proposal, his inclination waned and the opportunity passed. After this he tended to say that he always thought there was enough time but now realised it was too late (Eckhart 1779, p 198). It seems that he was fond of children, for he was known to have often distributed cake to those playing in the neighbourhood (Guhrauer 1846 **2**, p 364).

The last days of the Elector Ernst August

The Elector had been ill at the beginning of 1697 when Leibniz sent good wishes for his restoration to good health. Thanking him, Sophie (*K* **8**, pp 21–2) quoted a remark she had heard in a similar case, that it was a pity good wishes did not help; she added, however, that her husband was better. By the beginning of August, his condition called for another consultation with his doctors. On this occasion, Leibniz expressed his surprise to Sophie (*K* **8**, pp 35–6) that one of the most able physicians of the time had not been consulted, although he was near at hand. This was Heinrich Meibom of Helmstedt. If he were ill himself, Leibniz declared, he knew no-one to whom he would rather go for treatment and that was the best recommendation he could give. From this time, the Elector's health began to deteriorate more rapidly.

The most important political event of 1697 was the conclusion of the Treaty of Ryswick, bringing to an end the conflict between France and the League of Augsburg. In Leibniz's view, however, the Treaty was the most disgraceful the Empire had ever made (*K* **8**, p 40; **6**, pp 162–70)—he could not find words to describe his pain at the permanent loss of Strasbourg—and he saw in it a danger to the Protestant religion. Since the conversion of the Elector of Saxony to Catholicism, the defence of the Protestant cause in Germany needed more than ever the cooperation of the Houses of Brunswick and Brandenburg.

Even against this background of uncertainty, Leibniz could look to the future with optimism. Writing to Magliabechi on 30 September 1697 (*D* **5**, p 118), he gave an outline of the work he still wished to accomplish. This included the writing of a *Theodicy* and completion of his scientific and technical researches and inventions. Also there were many problems on mathematical, historical and philosophical themes that he wished to resolve, but success in these fields, he added, would depend on the help he could receive from younger men or other scholars of diligence and sagacity.

Already the principles of his dynamics and metaphysics had been established and published, while his calculus had been accepted and successfully applied by some of the most able mathematicians of the time. He had just reaped the benefit of one of his technical inventions when he accepted delivery of a new carriage made to his own design (Gerland 1906, pp 236–41). In some of his other concerns, Leibniz reached a stage of significant achievement or promise at this time. For example, in April he edited a collection of letters and essays by members of the Jesuit Mission in China, entitled *Novissima Sinica* (*Latest news from China*). This was the result of the links he established with members of the Mission following his meeting in Rome with Grimaldi. Through a diplomat in Paris, he received a present from the leader of the Mission, P Verjus, in 1692, and Carlo Mauritio Vota supplied much information after Leibniz established a correspondence with him in 1694. One of the copies of the *Novissima Sinica* that Leibniz sent to Verjus came into the hands of Joachim Bouvet, a member of the Mission who had just returned home to Paris on leave. Bouvet wrote to Leibniz on 18 October 1697 expressing his commendation of the *Novissima Sinica* and giving him more recent news from China. With his letter Bouvet sent a copy of his own recently published *Portrait historique de l'Empereur de la Chine*. Recognising that this biography of the Emperor would make a very suitable addition to his own book, Leibniz sought permission to include it in a new edition, and this being granted, he prepared a Latin version, which appeared in the second edition of the *Novissima Sinica* in 1699. To Andreas Morell (Bodemann 1895, p 190) Leibniz expressed the view that the Protestants should also send missionaries to China. The conversion of a single man of great influence, such as the Emperor of China, he believed, would achieve more than the winning of a hundred battles. In the years that followed, the correspondence with Bouvet proved to be of great importance in relation to the dissemination of Leibniz's binary arithmetic.

For many years Leibniz had been interested in the study of national languages. Accepting the single origin of mankind, he supposed that there must have been an original language, whose remnants are widely scattered among existing languages, for he did not believe that any of the languages then in use could have been the original. This led him to search for linguistic examples from all parts of the world. He saw the study of national languages as complementary to the study of history, providing a source of knowledge concerning the origin and migrations of nations. Assuming that the principle of continuity operates in the natural development of language, the existence of a discontinuity points to a migration. Following his return from Italy, he wrote to Huldreich von Eyben (*A* I **6**, p 442) raising objections against the influential Swedish thesis, according to which the Germanic nations came from Scandinavia (Aarsleff 1975, p 125). The resolution of this problem, he believed, would depend on a better knowledge of the languages of Asia and those of the regions of south-east Europe in contact with the Greek world. At this stage, he therefore preferred to collect material rather than make premature generalisations (Schulenburg 1973, p119). After an interval of over twenty years, he renewed his correspondence with the Jesuit Adam Adamandus Kochanski in Warsaw, partly because he hoped the Polish king might still secure permission for the missionaries to travel through Russia on their way to China—this would enable them to collect language specimens—and partly for the information Kochanski could himself supply (*A* I **8**, pp 596–7). The study of the German language was something that Leibniz wished to promote and at the end of 1696 he proposed for this purpose the establishment of a Wolfenbüttel German Society under the leadership of Duke Anton Ulrich.

In 1691 Leibniz had received from Georg Friedrich Mithoff, the magistrate in Lüchow (*A* I **6**, pp 513–19), information which enabled him to study the Slavonic influences in the language spoken by the 'Wends' then living in the neighbourhood of Lüneburg. For an opportunity to obtain specimens of the languages of Russia and Siberia he had to wait until August 1697, when Czar Peter stayed incognito in Coppenbrügge with the Russian legation on its way from Berlin to the Netherlands. Leibniz hoped for an introduction through Count Palmieri to the Russian general Franz Lefort and through him an audience with the Czar. While the Electress Sophie and her daughter Sophie Charlotte, the Electress of Brandenburg, visited the Czar in Coppenbrügge, Leibniz went to Minden, where he enlisted the help of the General's nephew, Peter Lefort, in securing specimens of the languages of Russia and Siberia. He was only able to see the Russian legation pass by at a distance, but on returning to Hanover, he corresponded with Peter Lefort for about two months concerning the language specimens. Also in the year of the Elector's illness, Leibniz published a small pamphlet on the origin of the German nation, drawn chiefly from the evidence of national languages (*D* **4**, 2, pp 198–205), in which he refuted the Swedish thesis.

In October 1697, Leibniz received news from Johann Jakob Chuno, a Brandenburg diplomat he had met two years earlier, that an Observatory was to be constructed in Berlin. This unexpected development seemed to him to present the possibility of achieving one of his principal ambitions, the establishment of an Academy of Sciences, if not in Hanover, then in Berlin. The proposal concerning the Observatory had come from Sophie Charlotte herself and had been taken up by the Prime Minister Eberhard von Danckelmann before his sudden fall from grace. Sophie Charlotte, it seems, had complained that in Berlin, the meeting place of so many scholars, no calendar was produced and neither astronomer nor observatory existed. Leibniz informed Chuno that he would be willing to help with advice and he also wrote to Sophie Charlotte (*K* **8**, pp 47–50), quoting exactly the report he had received from Berlin (in order to avoid any misunderstanding) and with much flattery suggested extending the plan so as to establish an Academy of Sciences comparable to those of Paris and London.

Although Leibniz was anxious to take advantage of Sophie Charlotte's proposal, he was conscious of the need for circumspect diplomacy, in order to avoid any offence to the Elector of Brandenburg, who was jealous of his authority. He formed a plan based on two circumstances: first, the need for closer agreement and mutual understanding between the Houses of Brunswick and Brandenburg in view of the danger to the Protestant religion posed by the Treaty of Ryswick, and second, the fact that Sophie Charlotte was a link between the two Houses. In a memorandum addressed to Sophie Charlotte and her mother (*K* **8**, pp 53–5) he proposed that he should act as a confidential messenger between them, thus facilitating their close cooperation and mutual help in pursuing the common interests of the two Houses without risking the accidents of letters. In his covering letter to

Sophie Charlotte (*K* **8**, pp 50–3), he stressed the need for the two Houses to work together for the preservation of the Protestant religion and explained that his visits to Berlin would seem quite natural. For as he had to visit the Library in Wolfenbüttel, an occasional extension of his journey to Berlin, in order to give advice on scientific matters, concerning which he was obviously well qualified, should not provoke suspicion of an intention to interfere in the affairs of Brandenburg. Before the end of the year, the Brandenburg diplomat Ezechiel Spanheim brought Leibniz the news that Sophie Charlotte wished to receive him in Berlin. At the end of his letter of thanks (*K* **10**, pp 40–2), he mentioned his correspondence with the Jesuit missionaries in China and offered to send her information concerning the great philosopher Confucius or the ancient Chinese kings, who, he added (surely in jest), were close to the Flood and consequently among the first descendants of Noah.

On 15 November Leibniz was called to Herrenhausen, where the Elector's condition gave cause for concern. At 8 PM he reported to Bartolomeo Ortensio Mauro that he had been in Herrenhausen an hour before, talking with Sophie and the doctors in the ante-chamber. The Elector was much better and had taken solid food that day. He had in fact been gravely ill for several months, and according to Leibniz's report to Sophie Charlotte (*K* **10**, pp 45–8), had been tended with unfailing devotion and danger to her own health by the Electress Sophie. Apart from a short visit to Brunswick about the middle of December, Leibniz spent the next two months in Hanover.

For some time the Court of Brandenburg had taken to heart the union of the Protestants and secret negotiations between the theologians in Berlin and Hanover, belonging respectively to the Reformed (Calvinist) and Lutheran sects, had been in progress with Leibniz as the intermediary. When Ezechiel Spanheim, late in December, brought him Sophie Charlotte's invitation in principle to an audience in Berlin, the official purpose of his visit to Hanover over difficult winter roads was to present to the sick Elector a memorandum of the Court Chaplain, Daniel Ernst Jablonski, on the agreements and differences of the Lutheran and Reformed sects. In practice, the Lutherans had not been shown much favour at the Brandenburg Court, for appointments were generally given to members of the Reformed sect in preference to Lutherans of greater merit (*K* **10**, p 38). After the meeting with Spanheim, Leibniz returned home with a headache. Just over three weeks later, on 23 January 1698, the Elector Ernst August died in Herrenhausen, after a reign which had seen Hanover raised to Electoral status but had also witnessed tragedies whose causes were at least partly rooted in the Elector's ambition for the aggrandisement of his House.

8
Hanover and Berlin
(1698–1705)

When the Elector Ernst August was followed by his eldest son, Georg Ludwig, Leibniz's position remained outwardly unchanged, but in reality he no longer enjoyed the same confidence of the ruling prince or the support for his multitudinous activities that he had received from Ernst August. Indeed, the new Elector seems to have had little interest in his activities or understanding of their value, so that, while he had to agree with the universal belief that his Privy Counsellor of Justice was a very able man, he tended to find fault with him instead of recognising that his free spirit was always usefully employed.

One manifestation of the critical attitude of the new Elector to Leibniz was a series of complaints, sustained over many years, concerning his undertaking to write the history of the House of Brunswick-Lüneburg. For example, in October 1698, the yearly allowance he had been granted in 1696 (on the occasion of his promotion), for an assistant to help with his historical work, was cancelled because he had not yet produced any draft. At the beginning of 1699, he sought help from Eleonore d'Olbreuse, the Duchess of Celle, to have the allowance restored (*MK*, p 157), pointing out that he had never intended writing a Brunswick history that could be read for amusement, but rather a history without precedent in Germany or Italy for the soundness of its scholarship. No-one, he added, should expect him to devote all his time to the work, for most men were allowed leisure. He had chosen to use his leisure for work on the progress of science and other problems, and the approbation he had received for this work brought prestige to Brunswick-Lüneburg as well as himself. Two years later, at the same time as criticisms were made of his service of so many masters, Leibniz was provided with an amanuensis, who would help with the historical work and also have the duty of reporting on its progress (*MK*, p 173). Leibniz was persistent and firm, however, in stating his case. In April 1702, he presented a memorandum claiming the payment of expenses he had incurred in the historical researches and proposed further, that on completion of the Brunswick history, he should receive an annual pension of 2000 taler, so that in old age, if God left him life and capacity, he might have means, opportunity and freedom to complete his own researches.

Clear evidence of the Elector Georg Ludwig's hostile attitude to Leibniz is provided by a letter that he wrote to his mother, the dowager Electress Sophie, in October 1703 from the hunting lodge at Linsburg (*MK*, p 186). Although he had arranged accommodation for him, the Elector informed his mother, Leibniz, after whom his sister Sophie Charlotte languished so much, was not there. It seems that Leibniz actually arrived at Linsburg on the same day (*K* **9**, p 46), having at first been refused permission to go by the Elector, who then changed his mind. Continuing his complaints, Georg Ludwig remarked that he never knew where Leibniz was, and that whenever he asked him why he was so elusive, he always had the excuse that he was working at his invisible books. To prove the existence of these, the Elector added, would require as much trouble as Jaquelot (the chaplain to the French colony in Berlin) devoted to the books of Moses.

Among the books that the Elector could not see were two volumes of unpublished chronicles of German history with the title *Accessiones historicae*, which he edited and published in 1698 and 1700 respectively, besides a supplement, published in 1700, to his earlier collection of constitutional documents, the *Codex juris gentium diplomaticus*, and several political tracts. Contributions to the learned journals continued to flow from his pen. Concerned mainly with his responses to criticisms raised by correspondents, these included important clarifications of his philosophy and mathematics. As an example of his concern for more practical things may be mentioned his criticism of those who opposed the use of a threshing machine invented by the Bailiff of Oertzen. Against the claim that such machines deprived the poor of livelihood, he argued that there was always opportunity for employment on other useful work, though at first this might be unfamiliar (Bodemann 1895, p 66).

Early in 1700 Leibniz was elected a foreign member of the reconstituted Royal Academy of Sciences in Paris. This brought him into correspondence with Fontenelle and a year later with Pierre Varignon, who defended the differential calculus against the attacks of Michel Rolle in the Academy and himself applied the new techniques to the problems of motion under central forces. Leibniz was prompted to offer a justification of the logical basis of the calculus by some remarks he had read in the *Mémoires de Trevoux* (*GM* **4**, p 95). In return for his election to the Academy, he contributed papers on the binary system of arithmetic.

In November 1698, the Abbé le Thorel, whom he had met in Hanover, conveyed to Leibniz the ardent wish of Antoine Verjus, the leader of the

Jesuit Mission to China, that Leibniz should become Royal Librarian in Paris (*MK*, p 155). Recounting how he had declined an invitation to become Curator of the Vatican Library (with the prospect of becoming a cardinal), rather than change his religion, he explained that the same obstacle would prevent him from going to Paris. Two years later (Bodemann 1895, p 360) he offered Verjus some advice on the missionary art, advocating the method employed by St Paul at Athens on seeing an altar to an unknown god. According to this accommodationist view of missionary work, the practices and doctrines of the Chinese should, as far as possible, be given a Christian interpretation. In return he asked Verjus if he could obtain from China the Hebrew Old Testament of the Jews there,[1] so that alterations made by the European Jews out of hate of the Christians could be detected. The high point of his relationship with the Jesuit Mission to China occurred on 1 April 1703 when he received from Peking a letter of Joachim Bouvet, claiming to have found in his binary arithmetic the key to the interpretation of the hexagrams of the *I ching*, and, consequently, to the origins of Chinese philosophy.

At the beginning of 1704 Leibniz received from Lady Damaris Masham, a daughter of Ralph Cudworth, a copy of her father's principal work, *The true intellectual system of the universe* (1678). The book was not unknown to him, for he had seen a copy in Rome. Owing especially to her connection with Locke (who lived with her family in his last years), Leibniz took the opportunity to correspond with her, taking care to give her a clear understanding of his metaphysical system.

On 28 May 1704, Johann Friedrich Hodann, who had worked as Leibniz's secretary for two years, completed the alphabetical lexicon for the *Encyclopedia* (Couturat 1903, p 510). The five tables of definitions, of which this was the last, constituted the most complete draft of a major project which he had seen as a stepping-stone to the achievement of his universal characteristic. As with previous lexicographers, like George Dalgarno,[2] the classification system adopted by Leibniz was based on the Scholastic distinction between substances (things and beings) and accidents (attributes or qualities) (Couturat 1901, p 171).

On all political matters, the advice of Leibniz continued to be sought and valued. For Hanover, the most important political question of current concern was the promotion of the claim of the Electress Sophie to the English succession for herself and her descendants. This required delicate diplomatic approaches in which Leibniz played a leading role, sometimes even taking the initiative. Another political problem, which perennially occupied a great deal of Leibniz's time but held little prospect of success, was

[1] In his *China illustrata*, Athanasius Kircher wrote that descendants of one of the sons of Noah had settled in Bactria, from where colonists had reached China.

[2] Leibniz was influenced by the writings of Dalgarno (Couturat 1901, pp 544–8).

that of the reunion of the Churches. In accordance with the wishes of the Elector, he spent the first week of September 1698 with Molanus in Loccum, drafting a memorandum for Count Franz von Buchhaim, the new Bishop of Wiener-Neustadt, who had been commissioned by the Emperor to continue the negotiations of his predecessor on the reunion of the Protestants and the Catholics. This document consisted of a series of propositions concerning the concessions each side would have to make before reunion could take place, followed by a catalogue of thirty vacuous controversies between the Catholics and the Protestants (*FC* **2**, pp 172–93).

In the spring of 1700, the Elector received a request from the Emperor Leopold I (*K* **8**, p xxx) asking him to send Leibniz to Vienna for deliberations on reunion matters with Bishop von Buchhaim, because he was the most experienced, discreet and qualified man to facilitate the work. The visit took place in the autumn, and during his stay in Vienna, Leibniz also visited the Abbey in Melk, had several conversations with the Papal Nuncio Da Via and searched through the papers of Rojas, making numerous extracts. In particular, he transcribed a summary by Rojas on the negotiations between 1661 and 1693 (*FC* **1**, pp 1–16). The Emperor expressed his satisfaction with the work of Leibniz, who had displayed rational thought, untiring industry and uncommon knowledge (*K* **8**, p xxxi), in a letter that he gave him for delivery to the Elector.

At the request of Duke Anton Ulrich (*FC* **2**, pp 205–6) Leibniz resumed his correspondence with Bossuet in the spring of 1699, first having sought the permission of the Elector, to whom he expressed the view that, although there was little hope of reunion between the Catholics and the Protestants in their day, it would be to the credit of the Protestant princes to show their good intentions by not abandoning a realisable project (*FC* **2**, pp 247–50). Instead of responding to Leibniz's request for replies to the precise questions that had remained unanswered in 1695, Bossuet presented a new exposition of the principles of the Catholics with an application to the problems of the canonical books. In collaboration with Duke Anton Ulrich, whose Library in Wolfenbüttel contained a fine collection relating to the history of the Reformation, Leibniz responded with one hundred and twenty-four arguments (*FC* **2**, pp 318–73). In particular, he claimed that the decree of the Council of Trent concerning the canonicity of the books of the Bible was demonstrably false (*FC* **2**, p 384). This, he believed, was a sufficient reason (*FC* **2**, p 372) for rejecting the validity of the Council and thus removing the chief obstacle to reunion. Hoping to rescue Leibniz from what he believed to be a position of error, Bossuet defended the Council of Trent with sixty-two arguments. Leibniz replied, point for point, in his letter of 5 February 1702, which brought the correspondence to an end.

In September 1698, when Leibniz and Molanus were discussing the reunion of the Catholics and Protestants in Loccum, the Brandenburg Court Chaplain Jablonski arrived in Hanover for a personal conference with them on the reunion of the Lutheran and Reformed sects. Since the prospect of harmony on the Eucharist and Predestination seemed slender, Leibniz would have been satisfied with civil and ecclesiastical tolerance. The Elector of Brandenburg, however, wished for conformity of belief and complete union in a single Church to be called Evangelical (*DS* **2**, pp 63, 233–5). Just before Jablonski returned to continue the conversations on 4 October, the Electoral Library and Leibniz's home were moved into 10 Schmiedestrasse, the house of the widowed Sophie Elizabeth von Lüde.[3] This remained his home in Hanover for the rest of his life. The following month, Leibniz made his first short visit to Berlin, spending two days in conversations with Jablonski on the reunion question. A year later, in October 1699, he expressed to Jablonski his fear that the proposals for reunion were premature, for if the right time were not chosen, the result could be disastrous (*DS* **2**, pp 109–14).

Another form of harassment suffered by Leibniz in Hanover was the repeated refusal of the new Elector to grant him permission to visit Berlin in response to invitations from Sophie Charlotte. The first refusal followed a visit that she made to Hanover in August 1698 when she repeated to Leibniz in person the invitation she had sent through Ezechiel Spanheim at the end of the previous year. When, two months later, permission was again refused, Leibniz pointed out to the Elector (*K* **10**, pp 52–3) that Sophie Charlotte had sent the invitation through her mother, no doubt believed that his permission had already been given and awaited his arrival with impatience. Unless the Elector changed his mind, Leibniz added, he would have to make some excuse to Sophie Charlotte, such as illness caused by the bad weather. This was tantamount to telling the Elector that there was no real justification for preventing his visit to Berlin.

Another year went by before Leibniz received from the Court Chaplain Jablonski (*DS* **2**, pp 151–2) an invitation in the name of the Elector of Brandenburg himself to visit Berlin in order to give advice on the establishment of a Society of Sciences. A few days later, on 28 March 1700, Leibniz (*K* **8**, pp 150–1) wrote to his master saying that he wished to inform him of the invitation before accepting it. Like the honour bestowed on him the previous month by his election as a foreign member of the reorganised Royal Academy of Sciences in Paris (*K* **8**, pp 149–50), the decision of the Elector of Brandenburg to give him direction of the similar Society he had resolved to found in Berlin (which he would exercise from his home in Hanover) could not be displeasing, Leibniz thought, to the Court in Hanover. He then explained that he needed to make small journeys for the sake of his health, since he led a very sedentary life, working all day for the service of the Elector and his House or the reputation of his Court. For

[3]This house was destroyed in World War II. The façade has now been reconstructed on a new site.

Plate 5 The house in which Leibniz lived in Hanover. Drawing made in 1828. Courtesy of the Historisches Museum, Hanover.

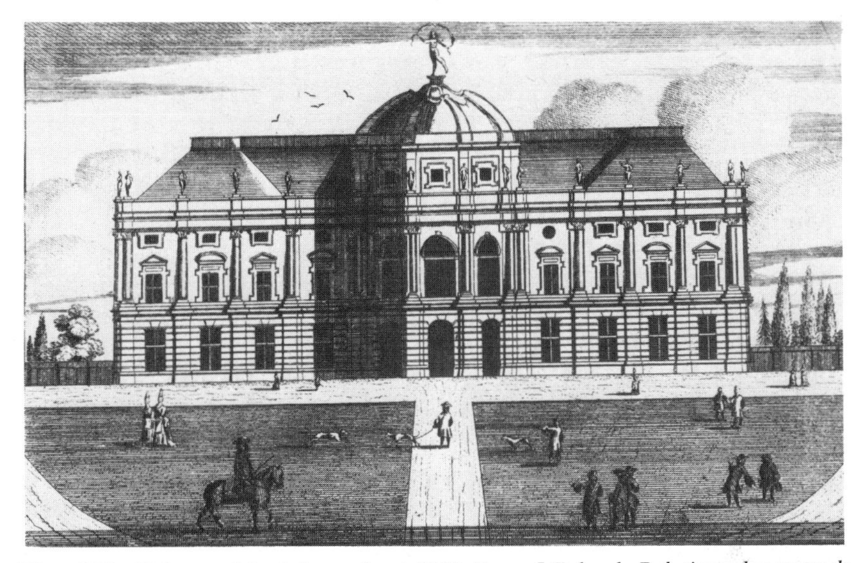

Plate 6 The Palace at Lützenburg about 1702. From J Toland, *Relations des cours de Prusse et de Hanovre* (The Hague 1706). Courtesy of the Bodleian Library, Oxford.

example, he would have to visit a spa for treatment in the spring, so why (he seems to imply) should he not also visit Berlin. The work on the history, he added, would proceed in his absence, since it involved the ordering of material already prepared.

At the end of April, Leibniz set out on the journey to Berlin, visiting Celle and Wolfenbüttel on the way. From Wolfenbüttel his progress was slow owing to illness, so that it was 11 May before he arrived in Berlin, where he had difficulty in finding accommodation, because the city was crowded with visitors for the forthcoming wedding of Princess Luise Dorothea Sophie, the daughter of the Elector's first wife, to Prince Friedrich von Hessen-Kassel. He lost no time in making a first visit to Lützenburg (later Charlottenburg), where the Electress Sophie Charlotte received him kindly and assigned him a room in her delightful palace, which was still under construction (*K* **8**, pp 151–5). A few days later he confided to Sophie (*K* **8**, pp 151–5) that he was out of his element, for the demands of the social life at Lützenburg left him with only four hours sleep a day, and in consequence, he was rather deranged, but being able to see Sophie Charlotte was worth the discomfort. By the end of the wedding festivities early in June, he was feverish and in need of a rest cure (*K* **8**, pp 167–9).

The deed of foundation of the Berlin Society of Sciences having been ratified and his own position as President of the Society secured, Leibniz took leave of the Elector of Brandenburg in Berlin on 20 August and visited Lützenburg to take leave of the Electress on the evening of the same day. She invited him to stay the night. It is perhaps an indication of the intensity of his feelings towards the Electress that Leibniz was upset when, having taken leave of him, medals were presented to him on her behalf by her lady in waiting, Fräulein von Pöllnitz. Evidently, he thought she should have presented them in person. In a letter to Sophie (*K* **8**, pp 207–8), he complained that it was as if the Electress had taken him for a stranger. The following day he wrote to Sophie Charlotte from Berlin, making the same complaint (*K* **10**, p 80). He also remarked that he had a headache caused by the heat and requested another audience just before noon the next morning, when he would be passing by Lützenburg, for he didn't know when another opportunity to see her would occur.

Early in September, the two Electresses, Sophie and Sophie Charlotte, sent an invitation to Leibniz to accompany them on a visit to Aachen, where they intended to solicit support for the elevation of the Elector of Brandenburg to become King in Prussia among the royalty staying at the international spa and in the neighbouring Netherlands. At this time Leibniz was in Wolfenbüttel planning a visit to the spa at Töplitz in northern Bohemia, and according to a letter he wrote to George Stepney (*K* **8**, p 239), regretted receiving the invitation too late to accept. Having taken treatment in Töplitz during the last week in September, he travelled from there to Prague and Vienna, where he arrived at the end of October for the conversations with Buchhaim that the Emperor had requested. At the beginning of this journey, he was still not well, and for several days suffered from a heavy cold with diarrhoea, sweating at night and hoarseness in the morning.

Leibniz was back in Hanover when, on 18 January 1701, the Elector of Brandenburg crowned himself King Friedrich I in Prussia, an unusual form of title necessitated by the fact that part of the Duchy was Polish (Schnath 1978, p 557). Leibniz designed a commemorative medal and offered congratulations on behalf of the Berlin Society of Sciences (*K* **10**, pp 338–46). Writing to the Queen in July (*K* **10**, p 81), he explained that he had not been able to visit her because he was being pressed to advance the history. However, he hoped it would be permitted for him to breathe a little and that the summer would not pass before he could satisfy his desire to see her again. But the summer passed, and at the end of August, he had to explain that he was still unable to leave Hanover (*K* **10**, pp 82–4), for Lord Macclesfield had brought to the Electress Sophie the Act of Settlement concluded on 12 June, which decreed in favour of Hanover the succession to the English Crown. As soon as the English party had gone, he promised, he would make haste to Berlin. The English, he remarked, wished the Electress and Electoral Prince to go to England, a possibility he viewed with dismay, since it would mean that, besides only being able to visit Sophie Charlotte so infrequently, he would also be completely deprived of the company of Sophie.

From Wolfenbüttel Leibniz wrote to Sophie on 23 September (*K* **8**, pp 287–8), telling her he intended to visit her daughter in Lützenburg, but that he would soon return, because he wished to spend the winter shut up in his shell in Hanover in order to get on with his own work. When Sophie showed this letter to her son, he complained that Leibniz should at least tell him where he was going, for he never knew where to find him (*K* **8**, pp 290–1).

Leibniz arrived in Berlin at the beginning of October, where he divided his time between conversations on philosophy and politics with Sophie Charlotte in Lützenburg and consultations with the Court officials concerning the organisation and work of the Society of Sciences. On 2 December the Queen gave him authority to act as her political agent with Elector Georg Ludwig (*K* **10**, pp 91–2). A few days later, he returned to Hanover to report on political matters (*K* **10**, pp 95–102). Anxiety had arisen both in Hanover and Celle as well as Brandenburg and Prussia on account of the army Duke Anton Ulrich, with French help, had raised in Wolfenbüttel. Leibniz returned to Berlin on 23 December with a letter to the Queen from her brother (*K* **10**, p 114) and instructions (*K* **10**, pp 103–5) for the diplomatic approaches he should make at the Court. For the last three days of the year he scarcely left his room owing to a headache and after this he lost his voice (*K* **10**, p 130) before accompanying the Queen on a visit to Hanover in the second half of January 1702 (*GP* **3**, p 287).

Scarcely had Sophie Charlotte returned to Berlin before she invited Leibniz to visit her at Easter (*K* **10**, p 136). Her lady-in-waiting, she told him, had bought a book on mathematics but the terms and ideas were so difficult that they needed him to explain them. A little later Fräulein von Pöllnitz (the lady-in-waiting) told him he should come at once, because as it said in the German proverb, 'Wann die Katze nicht zu Haus ist danzen die Meuse auf den Bänken', and besides, the Queen was without company (Guhrauer 1846 **2**, app, p 24). In response to repeated pleas of this kind, he could only reply that he would go to Berlin as soon as the Elector gave him permission. Soon after presenting a memorandum to the Elector on the enlargement of the Library on 22 May (*MK*, p 179), he evidently secured his permission to undertake the journey, for he arrived in Lützenburg on 11 June. For his visits to the Queen, she provided him with a travel document, which entitled him to receive from the representatives of Prussia in towns on the way, such as Distorf, Neuendorf and Tangermunde, fresh relays of six horses free of charge (*K* **10**, p 218). On this occasion he spent a whole year in Berlin, although he would have returned earlier, if illness had not prevented him from travelling.

When Sophie Charlotte visited Hanover for the annual carnival in January 1703, Leibniz decided that, instead of accompanying her, he would remain in Berlin to learn the King's intentions concerning his proposals for the funding of the Society of Sciences (*K* **10**, pp 381–3). These were communicated to him by the King's Minister Friedrich von Hamrath on 5 February (*K* **10**, pp 383–4). He then expressed to Sophie (*K* **9**, pp 8–10) his desire to return home in order to attend to his affairs and also because, on account of his long stay and the friction that existed between the two Houses at that time (focused on the Duke of Celle), he was suspected of being a spy. In reply (*K* **9**, pp 4–5) Sophie told him that the Queen was returning by order of the King and that she expected he would stay some time with her before returning home. Writing on 6 March to the Minister of State, Baron von Goertz (*K* **9**, pp 12–13) in Hanover, he expressed his wish to return as soon as his health permitted. Lack of nursing had allowed him only slowly to recover, and although his appetite had returned, his insomnia remained. He assured the Baron that his work on the history of Brunswick had progressed almost as much as if he had been in Hanover, since he had been able to use the books in the Royal Library. Towards the end of March, there was a recurrence of a swelling in his toe (not gout) from which he had suffered the previous December, and this prevented him from following the Queen to Lützenburg and Potsdam, where she was going to attend a wedding. In May, when he was with the Queen in Lützenburg, she wrote to Hans Caspar von Bothmer in Hanover saying that she was reluctant to let him go (*MK*, p 184), adding that he was very vexed by the rumours in Hanover that he wished to leave the service in order to live in Berlin. This, Leibniz had told her, would be to exchange freedom for slavery. After a year in Berlin, he returned to Hanover on 3 June 1703.

Two months later, Leibniz wished to accompany Sophie on a visit to Berlin, but the Elector refused to give him permission. On this occasion, Sophie left him an apartment in Herrenhausen, where he could work in peace, and it was at this time that he started his serious study of Locke. Writing to Sophie on 22 August (*K* **9**, pp 34–7) he reported that, soon after her departure, he had been attacked by a fever on account of the heat. When the fever abated he had not been able to work at anything and so had decided to visit Duke Rudolf August in Brunswick. He was returning to Hanover the next day, and if the Elector was away from Herrenhausen, he intended to make a short visit to Berlin. This, however, did not take place.

Leibniz next visited Berlin in January 1704 when he held discussions with Jablonski on the reunion of the Protestant Churches. Sophie (*K* **9**, pp 71–2) expected him to return with her daughter for the Carnival, but following political discussion in Wolfenbüttel with Duke Anton Ulrich, whose brother Duke Rudolf August died on 22 January, he made a brief visit to Dresden to promote the idea of a Society of Sciences there, before returning to Brunswick for the fair. To Sophie (*K* **9**, pp 72–3) he reported that, as the fair was not very interesting, he had intended to return to Hanover after the first week, but Duke Anton Ulrich had asked him to stay a few days, in order to console him following the sudden death of his wife, the Duchess Elizabeth Juliane. By the middle of February, he was again in Hanover, where he soon received another invitation from Sophie Charlotte to visit her in Berlin. Writing to her in March (*K* **10**, pp 224–6) he explained that he had to stay in

Hanover until his portrait, commissioned by the Grand Duke Cosimo III of Tuscany, had been painted. She hoped (*K* **10**, p 140) he would not delay when the portrait was finished and would visit her in the company of the Countess von Kielmansegg—the natural daughter of Sophie Charlotte's father and his mistress, the Countess von Platen—for she wished to hear his views on Locke (*K* **10**, pp 230–1). Meanwhile Leibniz appealed to her to give the impoverished philosopher, Gabriel Wagner, a small employment as a microscopist, for the observation of the wonders of nature in miniature that would throw light on the composition of larger bodies (*K* **10**, pp 224–6). Her brother, the Elector, had made no response to a similar appeal. When the destitute philosopher had been taken ill with a fever in Hanover, Leibniz had given him 2 taler to tide him over his illness. Praising the Countess of Kielmansegg for her patience in trying to help Wagner, who preferred to starve rather than sign his name for alms (Bodemann 1895, p 373), he remarked that, when he had gone—and he seemed to be anxious to send him to Berlin as soon as possible—they would only need to recommend him to Providence (*K* **10**, p 228).

Leibniz set out for Berlin in June but had to return from Brunswick when Sophie needed his advice on a letter she received from England. Then in July he was delayed by pain in his ankle which prevented him from walking (*K* **9**, pp 89–90). On 9 August he set off again, and after some historical research in Wolfenbüttel, arrived in Lützenburg at the end of August (*K* **9**, pp 92–5).

When the Vice-Chancellor, Ludolf Hugo, died in September, Leibniz asked the Queen to recommend him to her brother for the office, for which he was next in line of seniority (*K* **9**, pp 95–6). As a good friend, she agreed to put his interests above her own, for if he succeeded to this office, she would no longer have the pleasure of seeing him in Lützenburg (*K* **9**, p 96). Leibniz also applied to the Elector (*K* **9**, p 101) for the Provostship of Ilefeld, another office held by Hugo, so that he could bring to completion Hugo's work on the re-establishment of a school there. While Sophie could think of no-one more suitable in qualities of equity and justice for the office of Vice-Chancellor (*K* **9**, pp 96–7), both she and Molanus (*K* **9**, pp 99–100) were surprised that he wished to concern himself with so many tedious affairs. The Elector, she told him (*K* **9**, pp 101–2), seems to complain 'that your merit, which he esteems infinitely, is of no use to him, that he sees you rarely, and of the history you have undertaken to write, he sees nothing at all'. Moreover, he did not think Leibniz would find the duties of the office agreeable. The office of Chancellor had lapsed in Hanover in 1669 and the Elector Georg Ludwig wished to abolish that of Vice-Chancellor. By allowing this office to lapse, he probably rendered Leibniz a service.

History and politics

Towards the end of 1698, when his assistant Joachim Friedrich Feller was dismissed from his service owing to a breach of confidence, Leibniz acquired two new helpers for his historical work. One of these was Johann Georg Eckhart (Eckhart 1779, p 170), who at that time lived in Hanover but in 1706 became Professor of History at Helmstedt. With the cooperation of Leibniz, Eckhart also edited a German language review journal, *Monatlicher Auszug*, which appeared in Hanover between 1700 and 1702.[4] The other helper was Friedrich August Hackmann, who also became a professor in Helmstedt. In 1698 Hackmann received a commission from the Duke of Celle to search for historical documents in London (Guhrauer 1846 **2**, app, p 98). Then, in 1699, ten years after Leibniz's own journey to Italy, he was commissioned by the Court in Hanover to undertake a similar journey in order to gather more material for the Brunswick history. In the spring of 1699, Leibniz attempted to obtain official recognition for another helper in the collection of historical documents, the French genealogist and historian Charles René d'Hozier, by proposing the award of a medal to him. At the same time he suggested that copper engravings of seals and medals should be prepared in Hanover for inclusion in the finished work (*MK*, p 157). The following year, on his first extended visit to Berlin, he requested permission to use the Brandenburg archives, so that he could continue his historical work there.

One of Leibniz's first political tasks under the new Elector had been the composition of the oration to be read at the funeral of the Elector Ernst August on 2 April 1698 (*P*, pp 45–82). At this time, the question of the English succession became for Hanover a political problem requiring delicate diplomacy. When King William III of England stayed with the Duke of Celle at his hunting lodge in Göhrde, near Lüneburg, in October 1698, Leibniz took the initiative in suggesting to the Duchess of Celle that she should approach the King on the subject (*K* **11**, pp 15–17). She proposed to the King that he should name the Electress Sophie and her descendants as being directly in the line of succession and agree to a marriage (when he would be old enough) of the delicate nine-year-old Duke of Gloucester (the next in line after his mother Princess Anne) with the young Princess of Hanover. To these proposals the King responded favourably, as the Duchess informed Leibniz, who meanwhile had conversed on the subject with the English Ambassador, George Stepney. Sophie herself evidently took no part in these approaches and the King made no declaration to her.

Two years later, on 20 August 1700, when he was taking leave of the Electress Sophie Charlotte in Berlin before returning to Hanover, news reached Leibniz of the death of the Duke of Gloucester. The following day,

[4]The contributions of Leibniz to this journal are reprinted in *DS* **2**, pp 313–438.

in a letter to Sophie (*K* **8**, pp 207–8), he remarked that it was now more than ever necessary to think about the succession. In a letter to Leibniz, written at the same time, Sophie (*K* **8**, pp 205–7) remarked that, if she were young, she might be flattered by a crown, but now, if she had the choice, she would prefer to prolong her life rather than increase her grandeur. Although the King, following the death of the Duke of Gloucester, kept his ministers in ignorance of his views (*K* **8**, p xxxii), it seems that he took Stepney into his confidence, and sought indirectly through him to elicit a response from Sophie. For Stepney explained in a letter to Sophie (*K* **8**, pp 208–13) that, if she would indicate her consent by a single word through Leibniz, after she had communicated with the King, he would work for her cause in England. Leibniz was thus officially drawn into the affair by the King's permission to Stepney to mention him in his letter to Sophie (*K* **8**, p xxxv). In her reply to Stepney, Sophie (*K* **8**, pp 214–15) was less than enthusiastic (she was then 71 years old) and pointed to the right of succession of the young Prince of Wales (the son of James II), who she believed would be pleased to become a Protestant in order to regain the throne his father had lost. Remarking that she was in Aachen with her daughter, who had been taking the waters, she informed Stepney that she would soon have an opportunity to discuss the matter with the King himself, for on leaving Aachen, they intended to visit the King of Holland.

By the beginning of 1701, both Sophie and Leibniz had returned home and on 18 January they conferred in Celle with the Duke Georg Wilhelm and the English Ambassador John Cresset on the question of the English succession. Following this meeting, Sophie wrote to the King asking his advice on what she should do (*K* **8**, p 240). He evidently regarded this as an indication that she would be prepared to agree with his wishes. On 21 February, he made a speech to Parliament emphasising the need to ensure the succession in the Protestant line after himself and the Princess Anne (*K* **8**, p lii), though he did not name Sophie at this stage. Meanwhile, Leibniz (*K* **8**, pp 239–43) wrote in confidence to Stepney, pointing out that among the English only Cresset besides himself was in full knowledge of the deliberations, and suggesting that the cause of Hanover should be promoted by the publication of pamphlets appearing to come from elsewhere. On 1 May, Stepney replied from Vienna, where he had been sent as ambassador, saying that the English were well disposed to the succession on the King's recommendation and that there was no need of pamphlets to prepare men's minds (*K* **8**, pp 257–8). On 14 August, Lord Macclesfield brought to Sophie the Act of Settlement, which decreed the succession in the Protestant line in favour of Hanover. He also carried a letter of recommendation from Gilbert Burnet, Bishop of Salisbury, introducing him, in the Bishop's words, to a man whose fame not only filled the House of Brunswick but the whole of Germany (*MK*, p 173).

While the reluctant Sophie and the Elector (who was admitted to the Order of the Garter) accepted the succession in principle, they took no steps to assert their claim. Leibniz, on the other hand, lost no opportunity to promote their cause, not least in his correspondence with diplomats and in conversations with the many influential English visitors who came to Hanover. Evidently concerned at the lack of liaison between England and the Elector, Leibniz employed one of his favourite literary devices—in this case, an anonymous piece supposedly written by an Englishman—to show that, apart from the tie of the Act of Settlement, it was in the best interests of England to act in close alliance with Hanover (*K* **9**, pp 61–6). The danger he foresaw was that a continuation of the apparent coldness between the two Courts might give the enemies of Hanover an opportunity to overturn the Act of Settlement.

Another major political problem in which Leibniz was involved concerned the relationship of Hanover and Brandenburg with Wolfenbüttel, which was in an alliance of neutrality with France at the beginning of the War of the Spanish Succession. When Charles II of Spain died in November 1700, he left a will offering the undivided succession to the Dauphin's second son Philip, the sixteen-year-old Duke of Anjou, thus setting aside the previously concluded Partition Treaty and disregarding altogether the rival claim of the Emperor Leopold I. By September 1701, England and Holland had signed a secret treaty with Leopold, out of which developed the Grand Alliance against France, comprising England, Holland, Austria, Denmark, Prussia, Hanover, the Palatinate and later Savoy, which changed sides in 1703. Soon after the death of King William III in the spring of 1702, the allies began fighting in western Europe by besieging Kaiserswerth on the Rhine.

Early in 1701, Leibniz argued the legality of the Austrian succession in Spain in a letter supposedly written by a citizen of Amsterdam in reply to a letter supposedly written by a citizen of Antwerp defending the action of France in supporting the Duke of Anjou (*FC* **3**, pp 308–44). Both letters, which first appeared as pamphlets, were provided with German translations and published, with introduction and extracts from relevant constitutional documents, under the title, *La justice encouragée*. In a conversation with Sophie on 8 April 1701, Leibniz looked further into the future, speculating on the consequences that would follow if, as seemed likely, the Elector of Bavaria, who was allied with France, should inherit the throne of Austria. Although Prussia could divide the spoils of the Emperor with Bavaria, it would not be to their ultimate advantage, for their relationship with France would be like that of the animals with the lion in the fable (Guhrauer 1846 **2**, app, p 32).

Shortly before he received Queen Sophie Charlotte's authority to act as her political agent in relations with her brother the Elector Georg Ludwig early in December 1701, Leibniz presented to her a memorandum on the efforts of France to divide the Empire by subsidising Cologne, Bavaria and Wolfenbüttel, enabling Wolfenbüttel in particular to raise an army out of all proportion to the needs of defence, in return for treaties of neutrality (*K* **10**, pp 88–91). Then, on behalf of the Queen, he advised the Elector concerning the steps he should take, in collaboration with Prussia, to avert the danger (*K* **10**, pp 95–100). Having returned to Berlin with instructions from the Elector (*K* **10**, pp 103–5) he reported to him late in December on the negotiations that were taking place between Prussia and Wolfenbüttel (*K* **10**, pp 114–17, 119–21). Both Leibniz and the Queen were agreed that persuasion rather than threats of force should be used to bring about the disengagement of Wolfenbüttel from the alliance with France. In the end, the matter was settled by negotiation after troops of Celle and Hanover had crossed the frontier into Wolfenbüttel on the night of 19 March 1702 and seized the towns of Peine and Goslar. Leibniz sent a full account of the reasons for this action to Count von Buchhaim, the Bishop of Wiener-Neustadt, in a letter of 30 March (*K* **10**, pp 133–6). The mediators, he explained to Sophie Charlotte in a letter of 22 April, had testified to the moderation of Celle and Hanover, in demanding only what was necessary for their security and the general good, forbearing to take advantage of their position to require recognition of the ninth Electorate by Wolfenbüttel (*K* **10**, pp 143–5).

By the end of 1703, the conduct of the war appeared to Leibniz to have reached a stage of mortal lethargy. In a memorandum (*K* **9**, pp 51–61) presumably written for Sophie and through her for the Elector, he pointed to the need for greater coordination of strategy and resources. To this end, he recommended the setting up of two general headquarters, one in Holland and the other in the Empire. Among a number of other suggestions he made was a plea for better care of the soldiers, not only by the provision of medical aid but also by prevention of unnecessary casualties. This could be achieved by the employment of better strategy and good command; in other words, by the application of science rather than brute force. In another memorandum (*K* **9**, pp 66–70), evidently intended for Vienna, he recommended the appointment of the Elector, who had proved himself to be a very able general, as Commander-in-Chief of the Imperial forces. This came to pass a few years later under Emperor Joseph I. At the same time as he sought, by such memoranda, to influence the conduct of the war, he also contributed propaganda designed to gain popular support in England and Holland (whose rulers favoured the partition of the Spanish Empire) for the rights of Austria to the undivided succession. In characteristic fashion he passed off his writing (*FC* **3**, pp 360–431) as a piece originally written in Spanish and sent to him by a person of quality, which would merit publication in England and Holland. Several editions, in French and Dutch, were published in Holland during 1703 and 1704.

Leibniz was well received on his early visits to Berlin, not only by Sophie Charlotte but also by the Court officials and the Elector of Brandenburg himself. After visiting Sophie Charlotte in Lützenburg in the summer of 1700, he spent a week in Oranienburg with the Prussian Privy Counsellor Heinrich Rüdiger von Ilgen, who had been authorised by the Elector to explain to him in confidence his views concerning the relations of Brandenburg with Hanover and Celle, which, on account of differing political interests, had sometimes been somewhat cool and even strained. After Leibniz had discussed the various points with Sophie Charlotte (*K* **10**, pp 70–6), she commissioned him to communicate to von Ilgen her own observations on what he had said (*K* **10**, pp 76–9). With the acquisition of a more informed view of the policies of Brandenburg, Leibniz was in a better position to advise Sophie and the Court in Hanover on related problems. Through von Ilgen, the Elector sought the advice of Leibniz on the administration of justice, though he does not appear to have acted on the recommendations. These Leibniz communicated to von Ilgen in two memoranda (*K* **10**, pp 331–6), explaining that his ideas had been worked out at the age of 21 by command of the Elector of Mainz and had later been put into practice in Saxony. The particular recommendations, concerning official proceedings, the needs of the defence and uncertainties in the law, all follow from the general principle that justice is the charity of the wise man and the judge should help those who cannot help themselves.

In the autumn of 1704, when for the last time he was able to walk in the park at Lützenburg with Sophie Charlotte and her mother, Leibniz met the eighteen-year-old Princess Caroline of Ansbach, whose presence gave much pleasure to the Queen (*K* **9**, p 93). The Elector Johann Wilhelm of the Palatinate wished Caroline to marry his nephew, the Austrian claimant to the Spanish throne, and had sent his Confessor Ferdinand Orban to Lützenburg in order to seek her agreement and conversion. Having resolved to keep her religion, she found Leibniz on hand to draft the official letter declining the marriage (*K* **9**, pp 108–9). Her marriage in the following year to Sophie's grandson, the Electoral Prince, was a love match (*K* **11**, p 3). Later she became Princess of Wales and eventually Queen of England.

Philosophy

In the second volume of his *Dictionnaire historique et critique*, Pierre Bayle included an article on Jerome Rorarius, Papal Nuncio of Clement VII at the Court of Ferdinand, King of Hungary in the sixteenth century, who had written a book claiming to show that animals not only have rational souls but can reason better than men. There was much in the description of animal behaviour by Rorarius, Bayle explained, that would embarrass the followers of both Aristotle and Descartes. He then went on to describe the new theory of Leibniz, whom he regarded as one of the greatest intellects of Europe, remarking however that he had encountered difficulties in this theory. These were concerned especially with the explanation of the relationship between soul and body. Leibniz sent a reply to Basnage de Beauval for insertion in the

Histoire des ouvrages des savans. This was published in July 1698 (*GP* **4**, pp 517–24).

Bayle could not understand how the activity of the soul arises spontaneously from its nature; for example, how the soul of a dog could spontaneously feel pain immediately after feeling pleasure. Leibniz surmised that Bayle had in mind the axiom that a thing remains in the state in which it is, unless it is compelled to change. This axiom was, in fact, one of the foundations of his system. For while a body at rest remains at rest, a body in motion retains its motion, or its process of change, if nothing impedes it. Now, in his view, it was in the nature of created substance to change spontaneously following a certain order which constituted its individuality and was in exact agreement with what happens to every other substance and throughout the universe. Although he could claim to have demonstrated all this, Leibniz asked for the present only that it should be granted as a hypothesis suitable for explaining phenomena. Thus the law of change in the soul of the dog carries it from pleasure to pain at the moment it is hit with a stick, because this law of change was designed to represent what occurs in the body in the way that the animal experiences it, and even to represent, through its relation to the body, everything that happens in the universe.

Another idea that caused Bayle difficulty was the supposed implication that the soul of the dog, constructed so as to feel pain at the moment when it is hit, would still feel pain if no-one were to strike it. This, Leibniz replies, is a situation that could not arise in the natural order, for the soul of the dog has been constructed to represent successively the changes of matter which actually occur in the universe. God could have given each substance its own phenomena independent of those of others (in which case the situation envisaged by Bayle might occur) but then God would have created as many worlds without connection, so to speak, as there are substances.

Finally, Bayle claimed that Leibniz's dislike of the Cartesian system of occasional causes was based on a false assumption; for, as God's intervention followed only natural laws, it did not involve the perpetual miracle supposed by Leibniz. In the system of occasionalism, Leibniz replied, God was a perpetual supervisor, like a man charged with constantly synchronising two bad clocks, which were themselves incapable of agreeing, whereas his own system assumed a natural agreement. Occasionalism implied miracle, because God had decreed rules without providing a natural means of carrying them out. Similarly miraculous, he added, was Newton's gravitation. For it was not enough to give a primitive gravity to bodies; a natural means by which it could act was also needed.

Replying to Leibniz's clarifications in the second edition of the *Dictionnaire*, published in 1702 (*GP* **4**, pp 524–54), Bayle declared that, if the possibility of the hypothesis of the pre-established harmony were well founded, he would prefer this hypothesis to the scholastic interaction of mind and body or the assistance hypothesis of the Occasionalists, because it gives a noble idea of the Creator and removes every trace of miraculous guidance from the ordinary course of nature. Leibniz attempted to satisfy Bayle's reservations in an article, *Réponse aux réflexions contenues dans la seconde édition du Dictionnaire Critique de M. Bayle, article Rorarius, sur le système de l'Harmonie preétablie,* which was published in Masson's *Histoire critique de la république des lettres* in 1702 (*GP* **4**, pp 554–71).

Bayle believed it to be improbable that an automaton, such as Leibniz—from the point of view of physics—conceived the body to be, could be designed. For it would be like designing a ship to sail into port without a pilot. Leibniz was surprised that Bayle should deny this to be possible for God, but surmised that he had misunderstood the hypothesis to require the impression on the body of a faculty in the Scholastic sense, like those ascribed to bodies in the Middle Ages to draw them towards the centre. If, however, he had in mind a faculty that could be explained by the laws of mechanics, then there would be no difficulty from God's point of view. As bodies were divisible and actually divided to infinity, the smallest corpuscle received some impression from the slightest change in all others and must therefore be an exact mirror of the universe, so that an infinitely penetrating mind could see in that corpuscle everything that had happened and would happen not only to it but to all bodies everywhere.

Leibniz had failed to satisfy Bayle in regard to his earlier objection that he could not understand how the activity of the soul arose spontaneously from its nature. Returning to this difficulty, Bayle now claimed that the soul had no instruments for the execution of its own laws. Comparing the soul (conceived by Leibniz as a monad or atom of substance) to an atom of Epicurus surrounded by the void, Bayle held that the soul, having once had a feeling of pleasure, should always retain it, in the same way that the atom retained its motion. To this Leibniz replied that the atom (though in reality there is no such thing) has a simple tendency to move, whereas the soul, though indivisible, involves a composite tendency, each of the multitude of present thoughts tending to a particular change according to its relationship to all the other things in the universe. The soul should not therefore be compared with an atom but with the universe it represents. Changes in the universe are mirrored in the thoughts of the soul and this accounts for the variety of their modifications. The instruments by which the soul moves from one perception to the next in the series, Leibniz explains, are the present thoughts (both clear and confused) from which the future thoughts are born.

In the course of his article, Leibniz takes the opportunity to point out that the pre-established harmony is a good interpreter between soul and body and this success, he claims, shows that his system combines what is good in the hypotheses of Epicurus and Plato, the greatest of materialists and idealists.

A controversy between Johann Christoph Sturm, professor of mathematics in Altdorf, and the physician Günther Christoph Schelhammer in

Kiel, provided for Leibniz an opportunity to expound the basic ideas of his philosophy in the course of a critical review of the dispute. This was published in the *Acta Eruditorum* in 1698 (*GP* **4**, pp 504–16) under the title *De ipsa natura* (*Concerning nature itself*). Sturm accepted the position of Robert Boyle, who proposed that the term 'nature' should be replaced by 'mechanism', while Schelhammer defended the concept of nature as something more than material. In his review, Leibniz not only answered Sturm, who had corresponded with him following the publication of the *Système nouveau* in 1695, but directed his criticism more generally against the Occasionalists and others who would deny the reality of the physical world. The doctrine of occasionalism, he thought, had dangerous consequences which its defenders did not intend; namely, the implication of pantheism; for, since that which does not act can in no way be a substance, the doctrine implies created things to be mere modifications of the one divine substance and so, like Spinoza, makes out of God the nature of the world itself.

On a visit to Leiden in 1698, Johann Bernoulli defended Leibniz's dynamics in conversation with the professor of mathematics and philosophy, Burchard De Volder, who had taken the side of Papin in the dispute with Leibniz. There ensued a correspondence between De Volder and Leibniz, conducted through Bernoulli, in which Leibniz first defended the principles of his dynamics and then proceeded to expound and clarify the metaphysical foundation in the idea of substance, so that, in a sense, the correspondence forms a continuation of that with Arnauld.

In one of his letters to De Volder, Leibniz explains his reason for starting with dynamics and then going on to metaphysics (*GP* **2**, pp 192–5). The estimation of forces he describes as the gateway to the true metaphysics; for by these considerations, the mind is gradually freed from the false notions of matter, motion and corporeal substance commonly held, and in particular by the Cartesians, when it comes to understand that the rules of force and action cannot be derived from these notions, so that there is something higher in bodies themselves. Believing De Volder, whom Huygens had chosen as his literary executor, to be an able searcher after truth, Leibniz resolved to answer his objections as fully as possible. Repeated efforts to explain his concept of substance to De Volder, however, were of no avail, for De Volder continued to reject this concept as complex and to prefer what he called a logical approach, in which substance was defined as something having one simple attribute. In the case of corporeal substance, this simple attribute was extension, so that De Volder in effect adhered to the Cartesian view. Leibniz patiently pointed to the logical inconsistencies of De Volder's concept of substance and then emphasised the need of his metaphysics in order to give reality to the physical world. As the correspondence continued into its sixth year, 1704, Leibniz became increasingly irritated by what he regarded as De Volder's failure to understand his arguments. This correspondence, like that with Arnauld, remained unpublished, but in the course of trying to convince De Volder, Leibniz had succeeded in clarifying in his own mind some of his own ideas.

A particularly clear statement of Leibniz's views is to be found in a letter of 20 June 1703 (*GP* **2**, pp 248–53), where he expressed to De Volder the hope that his reply to Bayle (who, he remarks, has now accepted his position, though with reservations about the spontaneous thought of the soul) will have made his ideas clearer to him as well. Leibniz explained that he regarded substance, being endowed with primary active and passive power, as an individual monad. The monads were real, while extended matter was only a well-founded phenomenon—that is, an appearance of an aggregate of monads, strictly derivative from monadic characteristics. The forces arising from mass and velocity, belonging to aggregates or phenomena, were therefore derivative, and were found to be continually changing. When he spoke of a primitive force as enduring, he did not mean the conservation of the total motive power but an entelechy which expressed this total force as well as other things. Derivative forces were nothing but modifications and echoes of primitive forces. Since any modification presupposed something permanent, he rejected De Volder's hypothesis that there was nothing in bodies apart from derivative or mechanical forces.

Another clarification contained in the same letter concerns the concept of organic corporeal substance. Aggregates of monads were of two kinds; inanimate objects consisting of mere aggregates on which only the mind conferred unity, and organic objects consisting of aggregates united by substantial forms. To De Volder Leibniz explains that, if mass is thought of as an aggregate containing many substances, it is still possible to conceive of a single pre-eminent substance or primary entelechy in it. This pre-eminent monad, or dominant monad as he calls it, in some unspecified way makes the aggregate into a real unity. In the letters to Arnauld, Leibniz had explained that one monad may be said to act on another when the one expresses more distinctly than the other the cause or reason for the changes in both. This suggests that, in a sense, the dominant monad is able to use the others as organs of perception and activity, but further clarification had to wait for a later stage. Summarising his concept of organic corporeal substance for De Volder, Leibniz distinguishes (i) the primitive entelechy or soul, (ii) primary matter or primary passive power, (iii) the complete monad formed by these two, (iv) mass or secondary matter (the organs in which innumerable subordinate monads concur), and (v) the plant or animal which the dominant monad makes into one machine.

Writing to the Electress Sophie from Berlin on 11 November 1702 (*K* **8**, pp 385–6), Leibniz told her he had received a visit from Isaak Jaquelot to seek his advice concerning a new edition of his *Dissertation sur l'existence de Dieu*, which had been published at The Hague in 1697. After leaving France following the revocation of the Edict of Nantes, Jaquelot had spent time in

Heidelberg and The Hague before becoming Court Chaplain to the French colony in Berlin. Following their meeting, Leibniz wrote out for Jaquelot his criticism of the Cartesian demonstration of the existence of God; namely, that Descartes had not shown the idea of God to be possible (*GP* **3**, pp 442–7). In the second attempt, Jaquelot argued that, as there are contingent beings and these must have a cause, a necessary being exists. While expressing his approval of this argument, which he had himself used in a criticism of François Lamy published in the *Mémoires de Trevoux* in 1701 (*GP* **4**, pp 405–6), Leibniz pointed out that a complete demonstration must show that this necessary being has all perfections (*GP* **3**, pp 448–54).

Early in February 1704, Jaquelot (*GP* **3**, pp 462–4) asked Leibniz for an explanation of certain difficulties he had found in the system of pre-established harmony, on which he had made some reflections while an illness kept him by the fire. Leibniz responded immediately with a letter from Wolfenbüttel (*GP* **3**, pp 464–6) giving a brief exposition of the principal ideas of his system. But Jaquelot failed to understand it. He concluded that the differences between Leibniz's system and the Cartesian system of occasional causes were insignificant, except in one respect, which was not favourable to the system of pre-established harmony. In his view this system entirely destroyed the freedom of the will (*GP* **3**, pp 466–7). It appears that Jaquelot sent Leibniz a pre-publication copy of his *Conformité de la foy avec la raison* (1705), which contained an appendix criticising the principle of pre-established harmony. In a letter of September 1704, Leibniz reacted sharply to what he considered to be an unjust accusation of having proposed a system with dangerous consequences for religion (*GP* **6**, pp 558–60). Concerning the freedom of the will, Leibniz explained that our future actions are in us, but only in a manner of inclination which does not carry any necessity with it, although they have certitude for God. In other words, God knows the free choices we shall make. In a further letter (*GP* **6**, pp 567–73), Leibniz remarked that there was neither pleasure nor profit in receiving a communication which imputed to him views he did not hold and was written without any attention to what he said or the arguments he advanced.

Leibniz found greater pleasure in his correspondence with Lady Masham, for, as he remarks in a letter to Thomas Burnet (*GP* **3**, pp 297–9), he believed that her friend, the great English philosopher John Locke, had some part in it. Thanking her for sending him a copy of her father's book, Leibniz (*GP* **3**, pp 336–7) remarked that he had added to the intelligible world described in it a small system of pre-established harmony between substances, which Bayle referred to in the two editions of his *Dictionnaire*, more amply in the second, where he introduced new objections. Having consulted the article on Rorarius in the first edition of the *Dictionnaire* (she did not have the second by her) and also Leibniz's article, *Système nouveau*, cited by Bayle, she wrote to Leibniz (*GP* **3**, pp 337–8) seeking clarification of his forms, on which he seemed to build his hypothesis. For he called them primitive forces, souls, substantial forms, even substances themselves but such as are neither spirit nor matter, so that she had no clear idea of them. She also asked him to send her a summary of his response to Bayle's new objections. In reply (*GP* **3**, pp 338–43) Leibniz described the main principles of his philosophy, adopting an approach based on observation and the uniformity of nature rather than the analysis of matter as infinitely divisible. For example, the observation that there is in us a simple being endowed with action and perception leads us to believe that there are such beings active throughout in matter, differing only in the manner of perception. To confirm her understanding, Lady Masham repeated the principles of Leibniz's metaphysics in her own words, requesting his corrections where necessary (*GP* **3**, pp 348–52). Throughout the summer and autumn of 1704, Leibniz continued to answer Lady Masham's queries and endeavoured to convince her that his system of pre-established harmony was more natural than the Cartesian system of occasional causes. One of her questions concerned the need for bodies, for it seemed to her that the body served no purpose, if the mind sufficed. In reply, Leibniz pointed out that God wished an infinity of other beings. Our body is a kind of world full of an infinity of creatures which also merit existence, and if our body were not organised, our microcosm would not have all the perfections it should have and the macrocosm would not be as rich as it is (*GP* **3**, pp 352–7). With regard to her fear that the freedom of the will was destroyed in his system, he assured her that this was not so (*GP* **3**, pp 361–4). At the time of this correspondence, Locke was too ill to pursue further studies in philosophy but he was pleased that Leibniz would contribute some corrections to his *Essay concerning human understanding* (*GP* **3**, p 351). On 24 November 1704, Lady Masham (*GP* **3**, pp 364–6) wrote to Leibniz telling him of the death of Locke, who had been her friend for half a lifetime.

Mathematics

In October 1699 John Wallis published the third volume of his *Opera mathematica*, which provided some of the background to the invention of the calculus, for besides the correspondence of Leibniz and Wallis, it contained the letters of Leibniz to Oldenburg and the two famous letters of Newton—'Epistola prior' and 'Epistola posterior'—to Leibniz. To Magliabechi (Bodemann 1895, p 163) Leibniz explained that, when Wallis asked his permission to publish these items, he had been content to leave the selection to him, as he would otherwise have had to seek his autographs out of the mass of papers from which a great part had been lost, and he had not been dissatisfied with the result. He contributed a review of the work to the *Acta Eruditorum* in May 1700. A publication less pleasing to Leibniz was a work of Fatio de Duillier on the brachistochrone, in which Leibniz was named as the second inventor of the calculus, after Newton, and accused of arrogance in claiming all for himself and failing to recognise the contributions of others. Whether Leibniz had borrowed anything from Newton Fatio would leave to the judgment of those who, like himself, had seen Newton's letters and other manuscript papers (*NC* **5**, p 98). Leibniz published a reply in the *Acta Eruditorum* (*GM* **5**, pp 340–50) and he was confident that readers of the correspondence in the work of Wallis would recognise that the insinuations of plagiarism had no basis, but he was concerned that the Royal Society had apparently sanctioned Fatio's attack by giving its *imprimatur* to his work. On complaining to Wallis, Leibniz was assured that the discourteous imputations of Fatio did not have the support of the Royal Society, though a Vice-Prèsident had inadvertently approved the book for publication, believing it to be a harmless mathematical treatise (*GM* **4**, pp 71–3). Leibniz also received an apology from the Secretary of the Royal Society, Hans Sloane (*NC* **5**, p 96).

An elementary exposition of integration using Newton's method of fluxions was published in the summer of 1703 by George Cheyne, under the title *Fluxionum methodus inversa* (*On the inverse method of fluxions*). In this work, as Johann Bernoulli pointed out to Leibniz in a letter at the end of September (*GM* **3**, 2, pp 722–5), Cheyne would make them all 'Newton's apes, uselessly retracing his steps of long before', and, in particular, claimed that the method of series using undetermined coefficients and comparison of terms which Leibniz had published in 1693 was in fact the method of series discovered by Newton at least seventeen years earlier. In reply (*GM* **3**, 2, pp 726–30), Leibniz described this claim as inept, for when he published the method in 1693, which he had shown to Huygens and Tschirnhaus in Paris in 1675, it was not apparent to him or to anyone else (at least in the public domain) that Newton too possessed such a thing, and in a letter of 25 March 1704, he commented further that he had encountered no indication that the differential calculus or an equivalent to it was known to Newton before it was known to him (*GM* **3**, 2, pp 744–5). Irritation with Cheyne had evidently hardened his attitude towards Newton, for in 1676 he had been satisfied that Newton had something similar to his own calculus.

The many errors in Cheyne's book seem to have prompted Newton at least to publish his own account of the method of fluxions. Hooke having died in 1703, Newton was in a position to publish his *Opticks* without risk of a hostile reaction from his former rival. When the first edition appeared in 1704, he took the opportunity to append two mathematical treatises, concerning respectively the method of fluxions and the classification of cubic curves. Leibniz was impressed by the *Opticks* but in his review of the mathematical appendices, published in the *Acta Eruditorum* in January 1705, he expressed polite approval without displaying great enthusiasm. Having described the concept of fluxions, contrasting Newton's kinematic view of the variable with the static view of his concept of differences, Leibniz refers the reader to the works of Cheyne and Craig for further information. To Johann Bernoulli he remarked that there was nothing in the work on fluxions that was new or difficult for them (*GM* **3**, 2, p 760). Both in his correspondence with Johann Bernoulli (*GM* **3**, 2, pp 702–8) and in papers contributed to the *Acta Eruditorum* (*GM* **5**, pp 350–66) Leibniz developed further techniques for the integration of rational functions by resolution into partial fractions. He also explained to Bernoulli, in the context of a problem proposed in the *Journal des Sçavans*, the application of the calculus to caustic curves that he had first introduced in his paper on optic lines published in the *Acta Eruditorum* in 1689 (*GM* **3**, 2, pp 732–6), and to which L'Hospital had devoted particular attention in his *Analyse des infiniment petits*.

An attack on the principles of the differential calculus was sustained in the reconstituted Académie des Sciences in Paris during the years 1700 and 1701 by Michel Rolle, a self-educated mathematician who had gained a reputation for his skill in diophantine analysis and algebra. It is said that Rolle, one of the salaried mathematicians in the Academy—the others were Jean Gallois and Pierre Varignon—had been incited to make the attack by influential persons. An account of the lecture in which Rolle first presented his objections to the Academy was sent by Varignon to Johann Bernoulli, who passed it on to Leibniz (*GM* **3**, 2, pp 635–42). Rolle's principal objection concerned the lack of clarity in the concept of the differential and he rejected completely, though without reason, the concept of infinities and infinitesimals of higher orders. As L'Hospital, an honorary member of the Academy and the acknowledged expert in the differential calculus, was not present, the company looked to Varignon to make a response. For Varignon, the first Professor of Mathematics at the Collège Mazarin, had been active in applying the differential calculus to the problems of motion. In the absence of the President, the Abbé Bignon, he found no support for his proposal that Rolle should first publish his objections in the journals, so that the wider public outside the Academy would also have an opportunity to judge their validity. Varignon read his reply to the Academy at the beginning of August 1700, claiming that Rolle had no understanding of the calculus that he attacked and demonstrating the orders of infinities and infinitesimals in the manner of the ancients. When Rolle then attempted to show, taking as examples the curves $a(y-b)^3=(x^2-2ax+a^2-b^2)^2$ and $y=2+\sqrt{(4+2x)}+\sqrt{(4x)}$, that the determination of maxima and minima by means of the differential calculus did not give the same results as the method of Jan Hudde (*GM* **3**, 2, pp 662–4), Leibniz expressed his satisfaction with Varignon's response, which exposed the errors in Rolle's reasoning. From 1702, the debate moved outside the Academy, when Rolle found himself in conflict with Joseph Saurin, who was then mathematical editor of the *Journal des Sçavans* and a firm supporter of the differential calculus.

Towards the end of 1701 Varignon (*GM* **4**, pp 89–90) expressed his concern to Leibniz that a writing of his in the *Journal de Trevoux* (*GM* **5**, p 350) exposed his concept of the infinitesimal to misinterpretation and asked for clarification, addressed either to Bernoulli or himself, that would help them in dealings with opponents of the calculus. The Abbé Gallois had evidently been misled by the short article into believing that Leibniz understood the differential or infinitesimal to be only a very small magnitude but fixed and determinate, like the earth in relation to the heavens. Leibniz (*GM* **4**, pp 91–7) replied as soon as he received the letter, which had taken two months to reach him since it had arrived in Berlin from Groningen only after he had left for Hanover with the Queen. The letter in the *Journal de Trevoux*, he tells Varignon, was sent in response to some objections made in the journal concerning infinitesimals, and his aim had been to explain infinity simply by incomparables; that is, to conceive the different orders of infinity as quantities incomparably greater or smaller than other quantities, and that there was no need to make mathematical analysis depend on metaphysical controversies. The infinitesimals were not real lines but ideal notions which expressed analytically real magnitudes, in the same way, for example, as imaginary numbers in algebra. The calculation with infinitesimals, however, was rigorous, for he took infinitesimals as small as he pleased and his method only differed from the style of Archimedes in being more direct in expression and more in conformity with the art of invention.

Leibniz offered public clarification of his views in another article in the *Journal de Trevoux*, entitled *Justification du calcul des infinitésimales par celuy de l'algèbre ordinaire* (*GM* **4**, pp 104–6). Here again he emphasises that his calculus is rigorous, since the error can be taken less than any assignable quantity. Moreover, the infinities and infinitesimals are so founded that all happens in geometry, and even in nature, as if they were perfectly real. As an illustration he takes his law of continuity, in virtue of which it is permitted to consider rest as an infinitesimal motion, coincidence as an infinitesimal distance and equality as a limiting case of inequality. Although the limits are not rigorously included in the series—for example, the circle is not a polygon—nevertheless they have properties as if they were included, for otherwise the law of continuity would be violated.

When Varignon (*GM* **4**, pp 99–104) told Leibniz that Fontenelle planned to write a work on the metaphysical elements of the calculus, he again emphasised (*GM* **4**, pp 106–10) that, in his view, it was not necessary to consider infinities or infinitesimals as other than ideal things or well founded fictions. He then mentioned his own metaphysical idea of simple substances as truly indivisible, but these, he added, were immaterial and only principles of activity.

Although Varignon (Aiton 1964) applied the differential calculus to the problems of planetary motion, he supposed only that these bodies were subject to centripetal forces and never mentioned the harmonic vortex of Leibniz. In effect, he adopted the Newtonian position and developed a descriptive or mathematical theory on the hypothesis of the gravitational attraction, without offering any speculation concerning the physical cause of this force.[5] Starting with two general rules for rectilinear motion, $v = dx/dt$ and $f = dv/dt = ddx/dt^2$ (where x represents distance, v speed, f force and t time) and eliminating dt, he obtained the result $f dx = v dv$, which he recognised as the equivalent of proposition 39 of Book I of Newton's *Principia*. Then for motion in a curve under the action of a centripetal force, he established the result $f = -v dv/dr$ (where r is the distance from the centre of force), though he failed to recognise its equivalence to Newton's next proposition. From this result he deduced Newton's propositions on planetary motion. For example, beginning with the polar equation of the ellipse and assuming Kepler's law of areas, he calculated $v dv/dr$, and hence the centripetal force, by differentiation. While the hypothesis of Kepler concerning time, adopted by Newton and Leibniz, was 'la plus physique', he showed that his method could also be applied to other hypotheses such as those of Seth Ward and Cassini. He recognised that, in the absence of resistance, a centripetal force implied Kepler's law of areas but he was content to quote Newton's proof of this. These results were presented to the Academy in 1700 and supplemented in the following year by a paper relating the centripetal force and radius of curvature. Leibniz encouraged him to pursue his investigations, suggesting that he should consider problems involving more than one centre of attraction. For in David Gregory's book on astronomy, he remarked, the action of the sun and planet on a satellite was considered but not the action of the principal planets on each other, which would also merit attention. In 1703, Varignon presented to the Academy a paper treating some rather artificial cases. For application to the real planets, what was needed was a solution of the general inverse problem; that is, the determination of the orbit, given several forces directed to moving centres. Even the simplest case of the inverse problem, relating to two bodies, was beyond Varignon's reach, for his expertise was limited to the differential calculus. Although he had sought instruction in the integral calculus from Johann Bernoulli, this had been withheld in consequence of an agreement between Bernoulli and L'Hospital. In April of the following year, 1704, Varignon informed Leibniz that L'Hospital had died after a short illness, leaving only an unfinished work on conic sections (rather than the textbook on the integral calculus that had been expected) and in reply Leibniz informed Varignon that Wallis was also dead. Towards the end of

[5]Varignon had earlier defended the Cartesian explanation of terrestrial gravity in his *Nouvelles conjectures sur la pesanteur* (1690 Paris). In this work, written before he had any knowledge of the infinitesimal calculus, he applied mathematics to determine the speed and acceleration of bodies falling through the air.

the year, Varignon (*GM* **4**, pp 113–27) sent the letter seeking advice concerning the calculation of centrifugal force that led Leibniz to recognise the error he had made in his own paper on the planetary motions.

During his visit to Berlin in the summer of 1700, Leibniz evidently sought the collaboration of the Court mathematician Philippe Naudé in further researches on the binary system. For on his return from the conversations on Church reunion with Buchhaim in Vienna, he received a letter from Naudé containing tables of series of numbers in binary notation, including the natural numbers up to 1023 (Zacher 1973, pp 237–8). Thanking Naudé for the pains he had taken to compile these tables, Leibniz (Zacher 1973, pp 239–42) explained his intention to investigate the periods in the columns of the various series of binary numbers. For it was remarkable that series—such as the natural numbers, triples, squares and figurate numbers generally—not only have periods in the columns but that in every case the intervals are the same, namely 2 in the units column, 4 in the twos column, 8 in the fours column and so on. In the case of the triples (table 8.1), for example, the periods in the last three columns were 01, 0110 and 00101101. Already he had noticed a good theorem: that the periods consisted of two halves in which the 0s and 1s were interchanged; but the general rule for the periods in successive columns had thus far eluded him.

Table 8.1 Triples in binary notation.

00000	0
00011	3
00110	6
01001	9
01100	12
01111	15
10010	18
10101	21

Leibniz considered the periods as objects of algebraic operations. This may be illustrated by the deduction of the periods for the triples from those of the natural numbers. Let $a, b, c, d, e, \ldots$ represent the periods of the units, twos, fours, eights, . . . columns in the sequence of natural numbers. Thus $a=01$, $b=0011$, $c=00001111$, and so on. Then multiplying *edcba* by 11 (that is, 3), we get the periods for the triples.

The product is

$$\begin{array}{r} edcba \\ edcba\ . \\ \hline \end{array}$$

To perform the addition we note that in the last column we have $a=01$. Then for $b+a$ Leibniz writes

b	0011	
a	0101	
	0110	period
	0001	reserve.

This is really

b	0	0	1	1
a	0	1	0	1
	00	01	01	10

The units are taken for the period 0110 and the twos for the reserve 0001. Thus 0110 will be the period of the second column and the 0001 is reserved or carried to the calculation of the third column. The third column will therefore be $c+b+0001$, that is

c	00001111	
b	00110011	
	00010001	reserve
	00101101	period
	00010011	new reserve.

The periods for succeeding columns are calculated in the same way.

The possession of Naudé's tables enabled Leibniz to compose his *Essay d'une nouvelle science des nombres*, which he sent from Wolfenbüttel on 26 February 1701 to the Paris Academy of Sciences to mark his election as a foreign member. In the essay (Zacher 1973, pp 251–61) and also in his letter to Fontenelle, he explained that the new system of arithmetic was not intended for practical calculation but rather for the development of number theory. To Fontenelle, he remarked that, before publication, there was perhaps a need to add something more profound and he hoped that some young scholar might be stimulated to collaborate with him to this end. L'Hospital (*GM* **2**, pp 339–41), to whom Leibniz confided his belief that one of the principal applications would be a better representation of transcendental numbers than that in terms of infinite series of rational fractions, recommended the young Academician Antoine Parent, who would be willing to work for him in return for a salaried position in the new Academy of which Leibniz was the President. Without mentioning Parent, Leibniz (*GM* **2**, pp 341–3) remarked in his reply that, although he was President of the Berlin Society of Sciences, his powers were limited, and finance was at that time available only for essentials. Concerning his decision to communicate his binary system, although the applications had not yet been achieved, Leibniz explained that, in view of his many commitments that prevented him from bringing his researches to completion, he feared that his continued silence might lead to the loss of an idea which seemed worthy of conservation.

China

In his reply of 2 December 1697 to Bouvet's first letter, Leibniz described the nature of his own researches, in which he had shown by mathematics that the Cartesians did not have the true laws of nature. To arrive at these, he explained, it was necessary to suppose in nature not only matter but also force, and the forms or entelechies of the ancients were nothing other than forces (Aiton and Shimao 1981, p 73). Bouvet, in his letter of 28 February 1698, written before his return to Peking, expressed the view that the ancient Chinese philosophy did not differ from that of Leibniz, for it supposed in nature only matter and movement, which was the same as form, or what Leibniz called force. The ancient Chinese philosophy, he added, was embodied in the hexagrams of the *I ching*, of which he had found the true meaning. In his view they represented in a very simple and natural manner the principles of all the sciences, or rather a complete system of a perfect metaphysics, of which the Chinese had lost the knowledge a long time before Confucius. It is in the 'Great appendix' of the *I ching* that the words 'yin' and 'yang' make their first appearance as philosophical terms, used to describe the fundamental forces of the universe, symbolising the broken and full lines of the trigrams and hexagrams.

Leibniz wrote to Bouvet on 15 February 1701, at the time he was composing his essay for the Paris Academy, and it was therefore natural that he should describe for his correspondent the principles of his binary arithmetic, including the analogy of the formation of all the numbers from 0 and 1 with the creation of the world by God out of nothing. Bouvet immediately recognised the relationship between the hexagrams and the binary numbers and he communicated his discovery in a letter written in Peking on 4 November 1701. This reached Leibniz in Berlin, after a detour through England, on 1 April 1703. With the letter, Bouvet enclosed a woodcut of the arrangement of the hexagrams attributed to Fu-Hsi, the mythical founder of Chinese culture, which holds the key to the identification (figure 8.1).

Two arrangements of the hexagrams in the Fu-Hsi order are illustrated in the woodcut, one in the form of a circle and the other in the form of a square. The Greek words, written by Bouvet, indicate the top and bottom of the square arrangement. In the bottom left-hand corner, underneath the diagram, Leibniz has written: 'Ut apparet conferendo characteres circuli et quadrati, respectu circuli superius est quod remotius a centro' (By

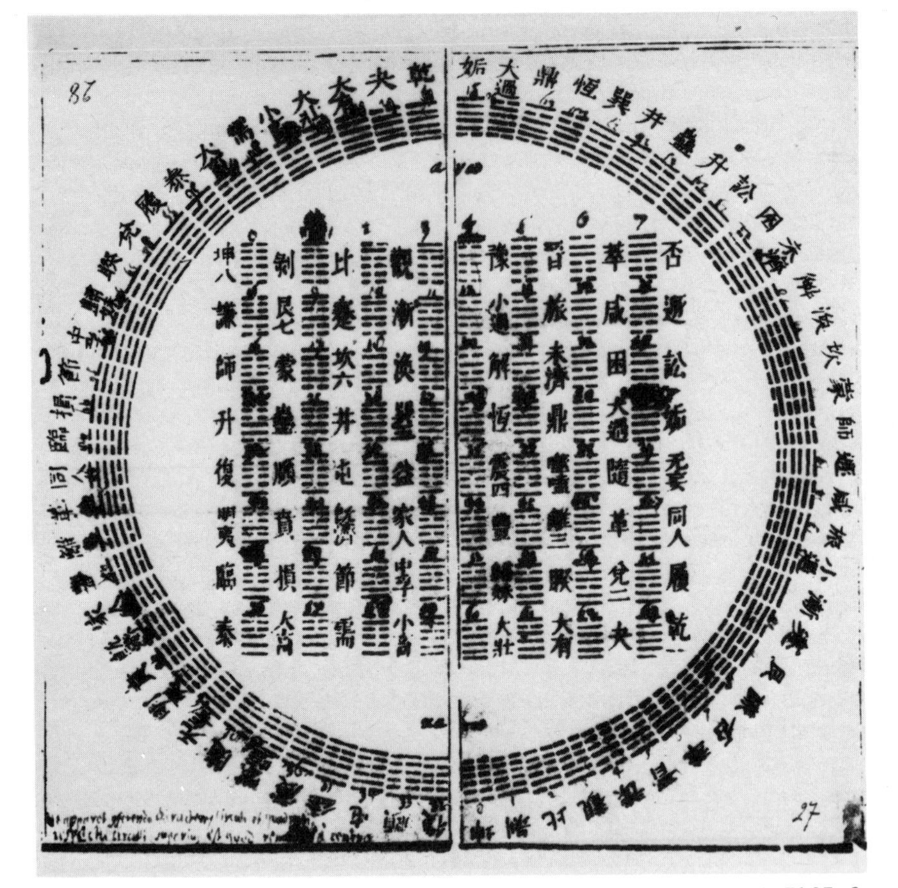

Figure 8.1 The Fu-Hsi order of the hexagrams. Leibniz-Briefe 105 (Bouvet), Bl 27–8. Courtesy of the Niedersächsische Landesbibliothek, Hanover.

comparing the characters in the circle and square, it is clear that, in the circle, the top is furthest from the centre). Leibniz surmised that Fu-Hsi modelled his circular arrangement on the earth's globe, where the most distant parts from the centre are taken as the highest. In both the circular and square arrangements he read the hexagram from bottom to top as a number in binary notation from left to right, taking a broken line to represent 0 and a full line to represent 1. Against each hexagram he has written in red the corresponding number in denary notation. In the circle, the numbers go from 0 to 31, starting at the bottom and proceeding anticlockwise to the top. Then there is a jump across the circle to 32 after which the numbers go to 63 by proceeding clockwise to the top. A curious feature of the square arrangement, he remarked, was that the numbers followed the pattern of European writing.

Leibniz accepted Bouvet's discovery with great enthusiasm. Having no reason to doubt the antiquity of the Fu-Hsi arrangement of the hexagrams that Bouvet had sent him, he was evidently delighted that this figure—'one of the most ancient monuments of science', as he described it—should have been found to be in agreement with his own binary arithmetic. In his reply to Bouvet (Zacher 1973, pp 275–86) he suggested that the interpretation of this ancient document, with the aid of a new discovery coming from Europe, should increase the respect of the Chinese for European science and in consequence for the Christian religion. He even suggested to Bouvet that one could believe Fu-Hsi himself to have had in mind the Biblical creation when he invented the trigrams, from which the hexagrams were supposed to have been formed by combinations in pairs. For taking 0 to symbolise the void, which preceded the creation of the heaven and earth, there existed at the beginning of the first day only God. At the beginning of the seventh day, however, everything existed, and 7 is written in binary notation 111 without 0. It is only in this way of writing with 0 and 1, he added, that we see the perfection of the seventh day which passes for holy, where it was again remarkable that its character has relation to the Trinity.

Within a week of receiving Bouvet's letter, Leibniz had communicated the discovery to his friend Carlo Mauritio Vota, the Confessor of the King of Poland, and sent to the Abbé Bignon for publication in the *Mémoires* of the Paris Academy his *Explication de l'arithmétique binaire, qui se sert des seuls caractères 0 et 1, avec des remarques sur son utilité et sur ce qu'elle donne le sens des anciennes figures Chinois de FOHY* (Zacher 1973, pp 292–301). Ten days later he sent a brief account to Hans Sloane, the Secretary of the Royal Society (Aiton 1981).

Owing to his separation from the real scholars of the time, who for political reasons shunned the Court circles on which he had to rely for his information, Bouvet had been mistaken in his belief in the antiquity of the Fu-Hsi arrangement of the hexagrams. For this order was the creation of Shao Yung (also known as Shao K'ang-chie), who lived in the eleventh century. Moreover, it is clear that his derivation of the order from the segregation table by continued dichotomy of white and shaded rectangles (figure 8.2) makes no appeal to an arithmetical interpretation of the hexagrams. The circular arrangement is obtained by dividing the table into two halves, placing these side by side in two columns and then opening them out to form a circle.

In the *I ching* the hexagrams are arranged in a different order, attributed to King Wen (*ca* 1050 BC). In this arrangement, the asymmetrical hexagrams are placed next to their mirror images and the symmetrical ones next to their inverses (that is, the hexagrams obtained by interchanging the full and broken lines). This order lacks even a superficial resemblance to a number system.

Bouvet's great discovery, to which Leibniz gave his enthusiastic support,

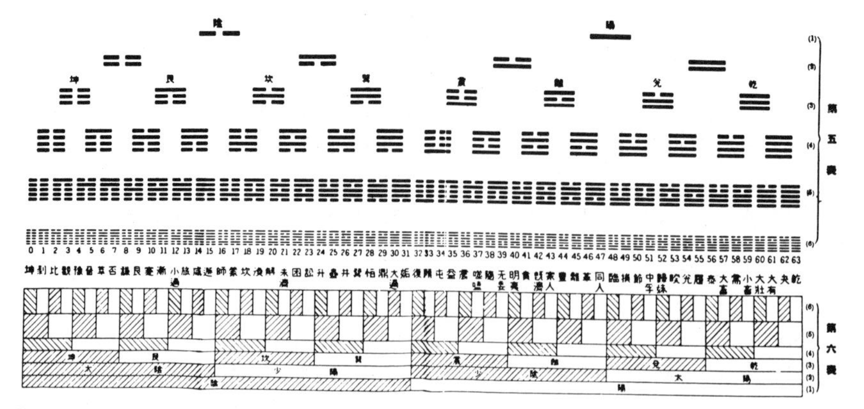

Figure 8.2 Shao Yung's segregation table. From Gorai Kinzō, *The influence of Confucianism on German political thought* (Tokyo 1929) (in Japanese). Courtesy of Eikoh Shimao, Doshisha University, Kyoto.

was therefore a misinterpretation based on bad Sinology. Generously but mistakenly, Leibniz had been willing to follow Bouvet in attributing his own invention to Fu-Hsi, thereby giving support to the myth that the ancient Chinese possessed advanced scientific knowledge which later generations had lost.

Correspondence with Fontenelle

When Leibniz received from Bernard Le Bovier de Fontenelle, the Secretary of the Paris Academy of Sciences, the diploma, dated 13 March 1700, recording his election as a foreign member, he took the opportunity afforded by this new honour to seek information concerning the scientific work then in progress in France. Thanking Fontenelle, in a letter of 3 September (Cohen 1962, p 76) for sending the diploma and expressing his hopes for the new Society of Sciences in Berlin, Leibniz asks if Cassini, De la Hire and others are still satisfied with the ellipses of Kepler or if they believe other hypotheses concerning planetary orbits to be in better agreement with observations. He also enclosed corrections of the *Rudolphine Tables* of Kepler, relating to the elements of the sun and moon, on which he sought the opinion of the French astronomers. Fontenelle's reply of 8 December (*FCa*, pp 198–203) contains information on a variety of scientific matters. Having summarised the technical arguments of the astronomers relating to the proposed corrections of the *Rudolphine Tables*, Fontenelle informed Leibniz that, for a long time, Cassini had abandoned the ellipse of Kepler, preferring an 'ellipse' in which the product of the 'focal distances' is constant. De la Hire did not appear to favour any particular curve and

Fontenelle suspected he would find in the end that the planets moved within certain limits but did not describe any curve that was regular and exact. Fontenelle then described the work of Varignon, who had given a general method of determining the central force (centripetal and centrifugal) which moves the planets. At the beginning of his research, Fontenelle remarks, Varignon did not fail to acknowledge that Leibniz and Newton had been the first and only ones to apply geometry to the discovery of these gravities of the planets towards the sun. In these investigations, Fontenelle added, Varignon employed only the differential calculus, to whose inventor he was pleased to express his obligation.

Turning to more personal matters, Fontenelle tells Leibniz that the Academicians have applauded the choice of the Elector of Brandenburg for the President of his new Society of Sciences. Pointing out that Leibniz was also regarded as one of the principal members of the Paris Academy, Fontenelle invited him to communicate one of his discoveries, which he would be pleased to include in the *Histoire*. This was an annual publication, required by the new rules of 1699, containing the most important communications of the Academicians during the previous year. It was in response to this invitation that Leibniz sent his *Essay* on the binary system of arithmetic. He must have been disappointed with the response of the Academy, as communicated to him by Fontenelle on 30 April 1701 (*FCa*, pp 204–7), for it reveals a complete misunderstanding of his intentions. First there is a reference to the inconvenience of the large number of figures, but this would only arise in practical calculations, which Leibniz had expressly excluded. Secondly, the Academicians waited with impatience for the applications he had promised, whereas he had hoped to find among their number an able collaborator with whom to pursue his investigations. Finally, his request concerning publication had been disregarded, for Fontenelle promised that his paper would appear in the *Histoire* for 1701, although Leibniz had indicated that such publication would be premature. Over a year later, in a letter written from Lützenburg on 12 July 1702 (*FCa*, pp 207–16), Leibniz again asked Fontenelle to withhold publication until he could himself provide some better illustrations, since no assistance had been forthcoming from the Academy. On this occasion he asked Fontenelle if the comet observed in Berlin had also been observed in France and on behalf of an astronomer enquired whether the zodiacal light of Cassini had been observed south of the equator. Also in response to some remarks of Fontenelle, he gave a history of the discovery of phosphorus, including his own dealings with Crafft and Brand. Criticising some propositions of Tschirnhaus Fontenelle had sent him, he defended the superiority of his own methods over those of Descartes and Tschirnhaus. Remarking on Fontenelle's interest in his infinitesimals, Leibniz asked him for an opinion on his philosophical essays, and especially on his explanation of the union of the mind and body and the communication of substances. For the concept of the infinite entered into these considerations, though in a manner other than it was taken in the infinitesimals, which he regarded as something ideal. He had sent Bayle a reply to the objections he had made in the second edition of his *Dictionary*, but he would be pleased to profit from the views of Fontenelle before his reply was printed, so he promised to send him a copy.

Having waited in vain for the arrival of the reply to Bayle, Fontenelle wrote to Leibniz on 18 November 1702 (*FCa*, pp 216–21), telling him that he already knew of his system of the mind but that, in his view, the nature of the human mind was incomprehensible to the human mind, for it could only know that which was of an inferior order; namely, extension and its properties. The zodiacal light, he informed Leibniz, had been seen in Goa, South America and at the Cape of Good Hope, while the comet seen in Berlin had also been observed in Rome and Paris in the same month of April. Again he pressed Leibniz for the applications of his binary arithmetic, so that the *Essay* could be published. In accordance with his wish, he had withheld it from the *Histoire* for 1701.

When early in April 1703 Leibniz sent to the Abbé Bignon his account of the discovery by Bouvet of the relationship between his binary arithmetic and the hexagrams of the *I ching*, he sent a letter to Fontenelle by the next post (*FCa*, pp 224–8) explaining that this discovery would justify the publication of his binary system, but he asked that his new paper and not the old one should be inserted in the *Mémoires* of the Academy. This was shorter and included an explanation of Bouvet's discovery. Then he made another attempt to explain his metaphysics. Starting with some questions raised by Fontenelle concerning the laws of motion, he pointed out that he had shown repeatedly in one of the learned journals that the laws based on the assumption that body consists only of extension and impenetrability (such as those of Descartes and Malebranche) result in absurdities and in particular violate his principle of continuity. The laws of motion, he explained, were a consequence of the entelechy or primitive force God had put in corporeal substance, for if they were not, then they would need a continual miracle (as he had alleged to be implied in the doctrine of Occasionalism). Rejecting Fontenelle's view that God must necessarily create all that is possible, he points out that, because of the relations of all creatures, there are incompatibilities between the possibles. Appropriately, as Fontenelle is a writer, he illustrates his own concept of God's choice of the best or most perfect by comparing God with a poet or an architect (which he truly is) who selects the good in preference to the bad.

On 6 July 1703 Fontenelle (*FCa*, pp 229–31) informed Leibniz in a letter that his ingenious explanation of the hexagrams by means of binary arithmetic had been read to the Academy and that it would be published in the *Histoire* for 1703. It seems that Leibniz was impatient to have his memoir published, for in a letter of 9 September 1704 (*FCa*, pp 231–2), Fontenelle had to explain that the memoir belonged to 1703 and that they were only at the stage of printing the memoirs for 1702. Concerning Leibniz's metaphysics, Fontenelle made some polite remarks but in effect closed the discussion. His memoir appeared in print in 1705 when the volume for 1703 was published. Relations between Leibniz and Fontenelle became strained owing to Fontenelle's neglect of a paper left in his care for publication in the *Journal des Sçavans*, in which Leibniz replied to some criticisms of Lamy concerning the pre-established harmony (*FCa*, pp 233–5). In consequence, the correspondence came to an end.

The Berlin Society of Sciences

On 19 March 1700, the Elector of Brandenburg gave his approval for the foundation of an Observatory and Society of Sciences in Berlin. Before making his journey to Berlin at the invitation of the Elector, Leibniz composed two memoranda setting out his recommendations concerning the project (*FC* 7, pp 599–618). These make clear his utilitarian and philanthropic aims for the Society. By contrast with the Societies of Paris, London and Florence, which were devoted to the satisfaction of simple curiosity and the promotion of purely scientific discoveries without application, the Berlin Society, he recommends, should combine theory with practice for the benefit not only of the arts and sciences but also for the country and its inhabitants through the promotion of manufacturing industry and commerce. Above all, the Society should be concerned with the real sciences, mathematics and physics, each of which he divided into four classes. Mathematics consisted of (i) geometry, including analysis, (ii) astronomy and its related fields, such as geography, chronology and optics, which would be supported by an Observatory provided with all the necessary instruments, (iii) civil, military and naval architecture, together with painting and sculpture, and (iv) mechanics with its applications to technology. Physics consisted of chemistry and the three kingdoms; that is, the mineral, vegetable and animal kingdoms. While the mineral kingdom was concerned mainly with mining and smelting of metals, the vegetable kingdom embraced agriculture, horticulture and forestry, and the animal kingdom included in its domain the study of anatomy, animal husbandry and the science of hunting, to say nothing of the higher science of medicine. The study of all sciences, Leibniz added, would be greatly facilitated by the provision of libraries and what he called a theatre of nature and the arts, which would include museums together with botanical and zoological gardens. Lastly, Leibniz recommended that advantage should be taken of the good relations which existed with Moscow to establish through the Society a Protestant mission to China. From this would follow not only a commerce of merchandise and manufactured objects but also of knowledge and wisdom with this ancient civilisation.

Even before his arrival in Berlin on 11 May, Leibniz had been in correspondence with the Court Chaplain Jablonski concerning the Society of Sciences. Explaining to Jablonski his preference for the term 'Society', he pointed out that, in Germany the term 'Academy' was generally associated with teaching (*DS* **2**, pp 153–61).[6] For the important post of Secretary, he stated his preference for a young physician, knowledgeable in mathematics, mechanics and chemistry, with the ability to understand French and English and to write Latin and German. The Court Chaplain's elder brother, Johann Theodor, was chosen for this position. The first appointment to be made, however, was that of Gottfried Kirch as astronomer. Well known for his calendars and ephemerides, as well as his observations of comets, and recommended by Leibniz in a letter of 26 March to Jablonski (*DS* **2**, p 155), Kirch was immediately invited to Berlin to take charge of the Observatory (*DS* **2**, p 167), whose construction Jablonski rather too optimistically expected to proceed quickly.

Having arrived in Berlin, Leibniz was commissioned to compose the charter for the Society (*K* **8**, p 172), which the Elector ratified on his birthday, 11 July. On the following day, Leibniz was formally appointed as President (*K* **10**, pp 328–30). In the official document, he is referred to as Gottfried Wilhelm von Leibniz, though there is no record in the state papers of his elevation to the nobility.[7] As President, he received an annual allowance of 600 taler to cover the costs of travel and correspondence (*K* **10**, p 331). It was expected that he would visit Berlin once a year, as Jablonski explained to a correspondent in Danzig (*MK*, p 167).

Early in 1701, Leibniz entered into correspondence with the Prime Minister of Brandenburg, Johann Casimir Kolbe von Wartenberg, and besides offering his congratulations on the crowning of the King in Königsberg, sought his help in obtaining further support for the Society of Sciences (Bodemann 1895, p 379). Then, on 18 March, he sent to the Secretary, Jablonski, a design for the membership diploma and some details concerning proposed members (*DS* **2**, pp 188–94). Soon after his arrival in Berlin in October, he reported to Sophie that he had been working for the Society (*K* **8**, pp 291–3) and, in the following month, he composed a memorandum in which he again pointed to the opportunities that existed for

[6]Friedrick the Great, in a statute dated 24 January 1744, renamed the Society 'Die Königliche Akademie der Wissenschaften' (Harnack **2**, p 263).

[7]Leibniz himself added the 'von' and this was accepted by the officials in Berlin. Leibniz's great-great-grandfather had a nephew who was ennobled in 1600 by the Emperor Rudolf. The coat-of-arms (with added embellishment) (Eckhart 1779, p 191) of this Paul von Leibniz, who died childless, was used by Johann Friedrich Leibniz in letters to his half-brother, and was also adopted by Leibniz when he entered the service of the Guelfs. Apart from the first letters to Bossuet, where he signed himself 'de Leibniz' (*A* I 7, p 272), his use (or rather misuse) of the title seems to have been confined to his letters in German (Müller 1966, pp 8–10).

the establishment of a Protestant mission to China under the direction of the Society (*K* **10**, pp 353–66). In a later memorandum, evidently written after 1 April 1703 (for it contains a reference to Bouvet's discovery) (*K* **10**, pp 366–71),[8] he returns to this theme, mentioning his belief that the King would already have sent an Evangelical Mission if the success of Sweden in the war against Russia and Poland had not blocked the route to China.

Leibniz had always recognised that his dream of a learned Society could not be achieved without an adequate source of revenue. It was for the purpose of financing such a Society that he had attempted to improve the output of the Harz mines by the introduction of his windmills. For the realisation of his far-reaching plans concerning the new Berlin Society of Sciences, a great deal of money would be needed, and if this could not be found, there was a danger that the Society would continue to exist only on paper. He proposed many schemes for the provision of income, including lotteries (*FC* **7**, p 626) and the revenue from the checking of standards of weights and measures, in which connection he suggested the introduction of a metric system (*FC* **7**, pp 635, 642). In the early years, income was in fact provided by the profit from the printing and sale of calendars, for which the Society had been granted a monopoly (*FC* **7**, p 619). Following the introduction of the Gregorian calendar at the beginning of 1700, Erhard Weigel had suggested this means of providing money for an Observatory. On 9 May 1704, Leibniz had to complain to von Wartenberg concerning an infringement of the calendar monopoly by a publisher in Berlin itself (*K* **10**, pp 387–8). Leibniz took a very serious view of the affair and demanded punishment, for the calendar monopoly, as he pointed out to von Wartenberg, had been the only source of revenue of the Society up to that time.

In the autumn of 1702, Leibniz proposed the introduction of silk-culture as a major source of income for the Society (*FC* **7**, pp 287–97). He asked for the planting of mulberry trees in the Royal Gardens at Potsdam and elsewhere to provide nourishment for the silkworms. Many years earlier, he noted, Johann Philipp von Schönborn, the Elector of Mainz, had begun a plantation of mulberry trees in the neighbourhood of Würzburg. The trees, he explained, were not difficult to cultivate and required little care, for only the leaves and not the fruit were needed. Many people would be employed in new work; for old people, children and others without much employment could be engaged to feed and tend the silkworms. In order to increase the chance of success of the project, he sought the patronage and authority of the Queen herself. On 8 January 1703, Sophie Charlotte granted him a patent for silk-culture throughout Prussia for the benefit of the Society (*K* **10**, p 372). The circumstances of the war of the Spanish succession increased the attractiveness of transplanting to Prussia an industry for which

[8]Klopp dates it incorrectly.

Germany had been dependent on France. Quite clearly, this was a long-term project, for the mulberry trees would have to grow before any profit could be earned. When Sophie Charlotte visited Hanover in January 1703, Leibniz remained in Berlin to await the King's intentions on the planting of the mulberry trees at Keppenich and Potsdam. He was no doubt pleased to learn from the Privy Counsellor Friedrich von Hamrath on 5 February that the King approved of the proposal. As he had stressed to von Wartenberg his desire to avoid delay, he must, however have been disappointed with the King's decision to defer the establishment of the plantations until the following year, on the grounds that the season was too advanced for the necessary arrangements to be made in time for planting in 1703 (*K* **10**, pp 383–4). In the spring of 1704, Leibniz himself took the initiative. Writing to the Queen on 18 May (*K* **10**, pp 245–8), he told her he was sending some seeds of mulberry trees he had received from Italy and asked her to allot a place in her garden for them to grow, until the trees were ready to be transplanted into their proper places. For her gardener he enclosed instructions for the planting and care of the seeds and saplings (*K* **10**, pp 247–8). He made a small experiment himself in his garden in Hanover. The project never prospered but he persevered to the end of his life. For, as his biographer Eckhart remarked, it was his disposition never to yield to difficulties (Eckhart 1779, p 174).

Plan for a Society of Sciences in Dresden

Writing from Lützenburg on 29 September 1702, Leibniz (*K* **8**, p 370) mentioned to Sophie that he had often spoken with Count Jakob Heinrich von Fleming, the Saxon ambassador in Berlin. One subject of conversation was the manufacture of silk, for the Count sought for himself and Leibniz a patent for silk-culture in Saxony, which was granted by the Elector Friedrich August of Saxony (the King of Poland) on 11 May 1703. The idea of founding a Society of Sciences in Dresden was probably born about this time or shortly after. It is mentioned by Leibniz in a letter to Carlo Mauritio Vota written on 4 September 1703 (Bodemann 1895, p 368). Having been well received by the Queen in Berlin during the first months of 1703 and with equal warmth in Hanover, so that he came to admire both Sophie Charlotte and her mother, Vota was well disposed to grant a reasonable request of their friend Leibniz. As confessor of the Elector of Saxony, he was in an ideal position to present to him the idea of a Society of Sciences in Dresden. This was to be modelled on the Berlin Society (*FC* **7**, pp 218–29) with Leibniz as President. The income would be derived mainly from a calendar monopoly and the profits from the manufacture of silk. He prepared a memorandum for the Elector and early in 1704 visited Dresden for a few days almost incognito to promote the idea among the Court officials. On 18 August he sent his secretary Johann Georg Eckhart to Saxony to continue the negotiations (Eckhart 1779, p 174). Then in December he spent three weeks in Dresden, where an audience with the Elector was arranged by Count von Fleming, General von der Schulenburg and his old friend Ehrenfried Walther von Tschirnhaus. Leibniz proposed that Tschirnhaus should collaborate with him in the establishment of the Society. To this the Elector agreed and even showed enthusiasm for the project but in the circumstances of the war with Sweden—the Elector had already been driven out of his Polish kingdom—the Society of Sciences in fact remained a pious dream.

Conversations with Sophie Charlotte

When Sophie Charlotte was born in 1668, she was named after her mother Sophie and the Princess Elizabeth Charlotte (later Duchess of Orleans), Sophie's niece, who had spent a great part of her youth with her aunt. After her marriage to the Elector of Brandenburg in 1684, at the age of sixteen, Sophie Charlotte had been without political influence until the fall of Danckelmann in 1697. Until this time she had seen Leibniz as her mother's friend, but in the changed circumstances of her own position, she took the initiative in inviting him to Berlin as her own friend and teacher. She delivered the invitation personally in August 1698 when she spent three weeks in Linsburg with Leibniz and her mother. Soon after this summer holiday at the hunting lodge, Elizabeth Charlotte of Orleans wrote to Sophie (who had always shared with her niece the substance of her correspondence and conversations with Leibniz) that she could not really be sad when she had Leibniz by her, for she had seen from all he had written that he must be good company. Towards the end of the year, Sophie and Leibniz lost an old and colourful friend with the death of Baron François Mercure van Helmont in Amsterdam at the age of 81 (Bodemann 1895, p 86). In writing to Sophie Charlotte (*K* **10**, p 144), Leibniz assumed that she was familiar with van Helmont's vivid description of the other world.

In a letter of 1 September 1699, Sophie Charlotte told Leibniz that he could henceforth regard her as his disciple and one who appreciated his merit (*K* **10**, p 54). After many delays, Leibniz made his first visit to Sophie Charlotte in May 1700, where he was warmly received as a guest in her delightful palace. At that time the palace was still under construction and, apart from himself, there were only three other guests, though ambassadors and diplomats came to visit every day and the social obligations involved the keeping of very late hours. In a letter to Sophie (*K* **8**, pp 151–5), Leibniz said he had been charmed by her grandson, the young Electoral prince, who asked questions about Hanover and the Observatory. Someone had sent the

Plate 7 Queen Sophie Charlotte. Painting by F W Weidemann, in the Schloss Charlottenburg, Berlin. Courtesy of the Verwaltung der Staatlichen Schlösser und Gärten.

Electress a proposal for a perpetual motion machine. Something of this kind, Leibniz remarked to Sophie, was needed for the fountains, as there was not enough current in the river, but he would not count on it.

Preparations were in progress for the marriage of Princess Luise Dorothea Sophie, the daughter of the Elector's first wife, to Prince Friedrich von

Hessen-Kassel, so that it might seem unlikely that Sophie Charlotte had time for philosophical conversations. Leibniz described the festivities in letters to Sophie. On 28 May the bridegroom made his entrance with a splendid retinue of carriages, horses and men. This day of gold, as Leibniz described it to Sophie (*K* **10**, p 61), was followed by a day of diamonds, on which the wedding ceremony took place. For the prince and princess he wrote an epigram in German, Latin and French, wishing them happiness as radiant as gold and as lasting as diamonds (*K* **10**, p 62). The festivities continued until 10 June and included ballet, opera, masquerade, bear hunt and firework display (*K* **8**, pp 157–8), besides a fête with a flotilla of gondolas on a canal near Lützenburg (*K* **8**, p 154). When the party was in Oranienburg on 4 June, Leibniz had a day of porcelain, for the Elector showed him his collection in company with the Landgrave of Hessen (*K* **8**, pp 159–61). In the midst of all these activities, Sophie Charlotte found an opportunity to speak to him about news from France concerning the Spanish succession that she had received from her mother.

A few days after the end of the wedding festivities, Leibniz sent to Fräulein von Pöllnitz some thoughts on the real distinction between the mind and the body (*K* **10**, pp 62–70). These were intended for the edification of Sophie Charlotte but Leibniz had doubts about the suitability of the last part, in which he used a mathematical argument. For he remarked to Fräulein von Pöllnitz that it seemed to him inappropriate to present thorny arguments involving numbers and figures to the Electress unless she raised such questions herself. In the other parts of the writing he demonstrated the real distinction between the mind and body using simple philosophical arguments, adding quotations from the Bible in confirmation.

The theme of Leibniz's first philosophical writing for Sophie Charlotte was suggested by a request he had received from her mother for his views on a dispute between the Elector Georg Ludwig and Molanus (*K* **8**, pp 162–4). The Elector maintained that thought is material, since it is composed of things which enter in us through the senses and we can think of nothing except what we have seen, heard or tasted. Sophie agreed with her son, for she had not been convinced by the counter-arguments of Molanus. In reply, Leibniz (*K* **8**, pp 173–8) informed Sophie that he approved of the sentiment of Molanus, although he took another route to achieve it, since Molanus was a Cartesian and he had found difficulties in the philosophy of Descartes. Thought, he declared, was not confined to what comes through the senses, for we can think of force, action, time, unity, true, good and a thousand other things of this kind. Moreover, it was not material which enters the mind through the senses, but its idea or representation, which is not a body but an effort or reaction. While he believed that this should suffice for those who do not love a great discussion (the Elector), he explained that he was adding something for those who wished to penetrate the question (Sophie). Then follows an exposition of his theories of substance and representation, which, though expressed in popular terms, is more demanding than the document he composed for Sophie Charlotte. This was only to be expected, as his discussions with Sophie concerning philosophy had been going on for many years.

Leibniz first demonstrates for Sophie that the soul is not material. Since matter is divisible to infinity, so that its parts, however small, are multitudes of substances, the true unities, which these multitudes presuppose, cannot be matter. There must be force and perception in these unities themselves, for otherwise there would not be force and perception in what is formed, which can only contain repetitions and relations of what is already in the unities. The unities or monads, he explains, are not all equally noble and in all organic bodies there is only one dominant or principal monad, which is its soul. It is the 'me' in us, which is superior to most other souls, because it is a mind and reasons by means of universal, necessary and eternal truths, founded not on the senses, nor on induction of examples, but on the internal and divine light of ideas, which constitutes right reason. As an illustration he takes the property that the differences of successive squares form the series of odd numbers. Induction from particular examples always leaves the question open, but the result can be demonstrated by mathematical reasoning based on the internal light independent of the senses.

To show that thought is not material, Leibniz explains that matter cannot enter a true unity, for otherwise it would not be a unity but a multitude. Consequently, that which is in the mind is not material but a representation of material: a representation, without extension, of that which is extended. As an illustration—it was this that he hesitated to show to Sophie Charlotte—he takes the case of a right angle (figure 8.3). It is clear that the angle is measured not only by the large arc BCD, but also by the small arc EFG, however small it may be. These arcs, therefore, are represented in the centre by the relation of inclination at the centre, which is in the lines when they go out from A. Minds, which are as centres, represent in themselves what happens in the multitudes they perceive, according to their point of view. Finally he remarks that the problem of the union of mind and body has

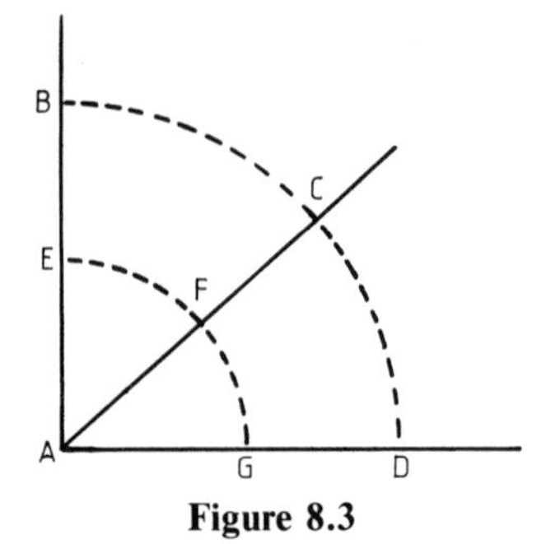

Figure 8.3

been resolved in a system he has explained elsewhere; the pre-established harmony.

Sophie replied briefly (*K* **8**, pp 178–9), raising the question whether, if his unity were alone, would it not have in common with the divinity the property of acting always on diverse things? The remainder of her letter deals with a miscellany of political matters and the charming observation that the swans had produced three cygnets, which they carried on their backs and wings when they became tired of swimming.

Sophie's question brought from Leibniz (*K* **8**, pp 180–2) an answer in which he clarified the distinction between God and the monads in his theory of representation. The unities, he explains, are never alone and without company, for otherwise they would be without function and have nothing to represent. God represents the universe distinctly and perfectly from the source, so that the universe is as he makes it. But minds represent the universe after the event and accommodate to what is outside themselves. Hence God is entirely free, but we are in part in bondage, insofar as we depend on other things and our perceptions or representations are confused. God is the universal centre, we particular centres. In other words, God doesn't have a point of view; he sees the universe as it really is.

Just one month after the wedding festivities, Lützenburg was again preoccupied with frivolities, when a masquerade was arranged in the theatre in honour of the Elector's birthday. This took the form of a village fair with a variety of stalls selling hams, smoked sausage, ox-tongues, wine, lemonade, tea, coffee, chocolate and similar items. The entertainment started with a procession in which a quack doctor entered on a kind of elephant, followed by his wife (played by Sophie Charlotte) in a sedan carried by Turks. Then followed a clown, dancers and a dentist. After a variety of performances, including a small ballet and comic extractions of teeth, the real doctor came to challenge the quack and finally the Elector entered in the disguise of a Dutch sailor to make a number of purchases from the stalls. The part allotted to Leibniz was that of an astrologer with telescope in hand, but the Duke of Wittgenstein kindly stood in for him, making advantageous predictions for the Elector (*K* **8**, pp 196–9). This self-made entertainment, Leibniz reported to Sophie, gave as much pleasure as a costly grand opera would have done.

In September, Sophie Charlotte and Leibniz went their separate ways, she to Aachen with her mother and he to Vienna. Over a year was to pass before they were again together in Lützenburg in October 1701. Earlier in the year, the Queen remarked to Fräulein von Pöllnitz, 'Here is a letter of Leibniz that I send you. I love this man; but I am angry that he treats everything so superficially with me' (Guhrauer 1846 **2**, p 248). Opportunities for more sustained philosophical discussions presented themselves over the next eighteen months. Leibniz spent the winter of 1701 mainly at Lützenburg and accompanied the Queen on her visit to the Carnival in Hanover at the beginning of 1702. Then in June of that year he was again in Berlin, where he stayed until June 1703, often as a guest of the Queen in Lützenburg.

Among the Queen's guests in Lützenburg was often to be found the English freethinker John Toland, who had accompanied Lord Macclesfield to Hanover in August 1701 as a member of the delegation bringing the Act of Settlement before making his first visit to Berlin (*K* **10**, p 82). Although his opposition to the Catholics and Stuarts made him a declared partisan of the House of Hanover, Toland was a dangerous ally, as Baron Schütz warned Leibniz in a letter sent from London on 11 July 1702 (*K* **8**, pp 356–7). Toland, the ambassador explained, was disliked by the bishops, clergy and ministers of state, and he hoped he would have enough discretion not to appear in the same place as the Electress Sophie. Writing to the Prime Minister Count von Platen (*K* **8**, pp 357–9) on 26 July, Leibniz explained that, despite Sophie's efforts to keep him away during her summer visit to Lützenburg, Toland had arrived in Berlin. The Queen was not prepared to forbid him to come to Lützenburg and Sophie was reluctant to interfere in a foreign Court. Soon after Sophie's return to Hanover, Toland read to the Queen a discourse on the mind, following closely the materialist doctrine of Lucretius. With biting sarcasm, Leibniz remarked in a letter to Sophie that, instead of playing with philosophy, which was not his metier, Toland would be better employed researching facts (*K* **8**, pp 362–3).

Living in a beautifully situated nobleman's house at Buckow, not far from Berlin, was an old friend of Leibniz, the crippled General Heinrich Heino von Fleming, who was evidently much admired by the Queen. Leibniz visited him for a few days in October 1701, reporting to Sophie that his friend had abandoned the corpuscular philosophy in favour of his own (*K* **8**, pp 299–300). In September 1702, the Saxon General Jakob Heinrich von Fleming (no relation of Leibniz's friend) visited Berlin with his wife and joined in the philosophical conversations which took place in Lützenburg. Also present in Lützenburg was the daughter of Leibniz's friend (*K* **8**, p 370) and the guests from Saxony took the opportunity to accompany her on a visit to her father in Buckow, whose mind, Leibniz remarked to Sophie, was as free as his body was not. When the visitors from Saxony departed, Sophie Charlotte complained that she had no company apart from Leibniz and she encouraged him to dispute with Toland for her diversion (*MK*, p 181). In the summer of the following year, when Leibniz had returned to Hanover, Sophie Charlotte confided to him as a friend (asking him not to show the letter to others) that the duties she had to endure in Berlin gave her no joy. She wished that she were paralysed like General von Fleming in Buckow—but only, of course, if his patience and intelligence were included in the bargain (*K* **10**, pp 210–12).

The summer of 1702 was a time of great happiness and contentment for both Leibniz (Bodemann 1895, p 122) and Sophie, who, after her return to Hanover, thanked Leibniz for news of Lützenburg, 'where my heart remains and where I believe I have spent the best days of my life' (*K* **8**, p 359). Sophie Charlotte had herself introduced one of the themes of the summer conversations during her visit to the Carnival in Hanover, when she asked Leibniz to read in her presence part of a letter supposedly written for Sophie by a friend of the late M de Guenebat in Paris (*K* **10**, p 141). When Sophie Charlotte returned to Berlin, there ensued a correspondence with Leibniz on the theme. If the friend of Guenebat, whom he tentatively identified as Monte-jean (*K* **10**, pp 140–1) had said that there would be a time when the mind was without body, then, Leibniz explained, he was in error and opposed to the ancients, including the Church Fathers, who supposed that only God was incorporeal. Although he could not go into such detail concerning the other life as van Helmont would have done, he assured the Queen that things could be known without seeing them, as mathematics could demonstrate; a clear reference to his view that there is something in our thought that does not come from sense. The exchange of views by letter is mostly on this rather superficial level. For example, Leibniz and Sophie Charlotte exchange pleasantries about King William and the possibility of a transmigration of his mind into a healthy body. On a more serious note, the Queen stated her view, applauded by Leibniz, that the search for truth should be disinterested and not motivated by fear or hope. Her tranquil temperament, she remarked (*K* **10**, pp 136–7), inspired her to believe that she had less to fear from the future than the present, while all she had heard about the devil had not made her fear death. Mathematics, she told Leibniz, was all Greek to her, apart from unity, of which, thanks to him, she had a little understanding. With his letter of 22 April, evidently the last before his visit to Berlin, in which he explained the distinction between the unity of mathematics and that of philosophy—the former had parts and the latter did not—he sent for Fräulein von Pöllnitz, who wished to study mathematics, an illustration of his binary system, which he described as a pleasant method of arithmetic he had invented one day, where all the numbers are written with 0 and 1 (*K* **10**, p 145). Meanwhile Sophie had received another letter from Paris; this time from her niece the Duchess of Orleans, who had presented a beautiful parrot to her milliner. This made her think of Leibniz, who says animals have understanding, are not machines as Descartes supposed, and have immortal souls (*MK*, p 178).

Following the conversations in the Queen's park, Leibniz set down his thoughts on the two basic questions in an ordered exposition entitled *Lettre touchant ce qui est indépendant des sens et de la matière* (*K* **10**, pp 154–67). The two questions were:

1 Whether there is something in our thoughts which does not come from sense.
2 Whether there is something in nature which is not material.

Leibniz begins his answer to the first question by comparing our use of the external senses to the way in which a blind man uses his stick. They help us to know their particular objects, such as colours and sounds, but they do not help us to know what these sensible qualities are. They are in fact occult qualities, so that, far from understanding sensible things only, it is these that we understand least. Besides these occult qualities (whose concepts are clear since they serve to recognise them, but not distinct since we cannot find their content), the senses also enable us to know other qualities whose concepts are more distinct. These are ascribed by Aristotle to the common sense, which Leibniz interprets as an internal sense where the perceptions of the different external senses are united. Such are the ideas of numbers and figures. There are also objects of another nature, which are objects of the understanding alone. Such is the idea of self. Since we conceive other beings who perceive sensible objects, it is by the understanding that we conceive substances in general. Leibniz summarises his position in the statement that there is nothing in the understanding which has not come from the senses, except the understanding itself, or the one who understands. With the Platonists, we therefore have to agree that the existence of intelligible things, especially mind or soul, is incomparably more certain than the existence of sensible things. This lends plausibility to Leibniz's view that, speaking in mathematical rigour, there exist at bottom only intelligible substances, of which sensible things are only the appearances. The fact that we know necessary truths, which are independent of the senses, shows that there must be a natural light within us.

Turning to the second question, Leibniz argues that there must be some substance separate from matter, namely God, the ultimate cause of things including matter. Since active force and perception cannot be explained by any mechanism, he concludes that there is something immaterial everywhere in created things, particularly in us, where this force is accompanied by a fairly distinct perception. He was inclined to believe that all finite immaterial substances were joined to organs and accompany matter, and even that souls or active forms were to be found everywhere.

Sophie Charlotte showed Leibniz's exposition to Toland, who replied with mock modesty (quoting John the Baptist), though rejecting the author's conclusions without mentioning his name (*K* **10**, pp 167–77). Leibniz in turn communicated his reactions to Sophie Charlotte, scathingly dismissing his unnamed critic as one who had not taken the trouble to understand his position (*K* **10**, pp 181–8). Thus Leibniz accuses the critic of wishing to prove against him what he does not deny; that is, that we need organs of sense in order to have thoughts. By his principle of pre-established harmony, Leibniz had established an exact relation between mind and body, so that even the most abstract thoughts were represented by some traces in the brain.

The pre-established harmony, and in particular Leibniz's reply to the criticisms of Bayle, was another theme of the summer conversations. Sophie Charlotte had met Bayle in Holland and conversed with him after her visit to

Aachen (*GP* **6**, pp 8–9), so that the appearance of the second edition of his *Dictionary* in 1702 was a matter of great interest to her. Leibniz had endeavoured to explain to the Queen that Bayle's objections concerning the conformity of faith with reason were not as strong as some, hostile to religion, wished to make out, and she encouraged him to put his replies in writing so that they could be considered more carefully. This was the origin of the *Theodicy*, assembled from the pieces written at this time and augmented with others, which Leibniz published some years after the Queen's death (*GP* **6**, pp 9–10).

In August 1703 (*K* **10**, pp 212–13), when he was prevented by the Elector from accompanying Sophie to Lützenburg, Leibniz informed the Queen that he had defended his system of monads and the union of mind and body against the attacks of the French Benedictine François Lamy. He also told her of a rare philosophical conversation with the Elector, in the presence of Sophie, concerning the question whether goodness and justice were arbitrary or founded on eternal reasons. In April 1704, when the Queen was impatient for his visit to Berlin (*K* **10**, p 230), Leibniz told her that he had almost finished his comments on Locke. A fortnight later, still in Hanover, he mentioned to her for the first time his correspondence with Lady Masham (*K* **10**, pp 237–45) and then repeated for her the popular account of his metaphysics that he had conveyed to Locke's friend.

Leibniz arrived in Lützenburg at the end of August (*K* **9**, pp 92–5) and Sophie made a visit at the beginning of October. On the afternoon she arrived, having taken only a cup of chocolate for lunch, she walked for two hours in the park with the Queen, Leibniz and the princesses of Kassel and Ansbach. Leibniz spent most of December in Dresden trying to establish the Society of Sciences there. On his return to Berlin, he was prevented by his work for the Berlin Society from accompanying Sophie Charlotte to the Carnival in Hanover, so he made a visit to Lützenburg on 11 January 1705 in order to see her before her departure. Neither Leibniz nor the Queen could have suspected that this would be their last meeting.

Tragedy and the search for consolation

At the beginning of 1705 Leibniz (*K* **10**, pp 394–9) wrote a memorandum for the King recording his services to the Court, which included the organisation of the Berlin Society of Sciences, engagement in the negotiations between the theologians of Brandenburg and Brunswick on the reunion of the Lutheran and Reformed Churches besides various political and diplomatic assignments. During the last year he had incurred expenses in excess of 2000 taler, but this he regarded as a minor loss compared with that of his precious time. The King showed his appreciation by making him a present of 1000 taler. Writing to Sophie Charlotte on 31 January (*K* **10**, p 262), he promised that he would return to Hanover without delay as soon as he had received the payment. Alas, the Queen had become mortally ill, unexpectedly and in the prime of life. The Calvinist pastor Claude de la Bergerie had been called to her bedside at the Palace in the Leinstrasse and remained with her until she died in the early hours of 1 February.[9] Leibniz received the news on 2 February from Joachim Heinrich von Bülow, who had brought it to Berlin. Such was his affection for the Queen that the tragic news brought him to the verge of serious illness. Everyone at the Court acknowledged that, among individuals, he was one of those who had suffered the greatest loss and even the foreign ambassadors offered him their condolences. On the same day he wrote to Fräulein von Pöllnitz (*K* **10**, p 264), saying that he did not cry or moan, but he didn't know where he was. His letter, he added, was more philosophic than his heart. Asking her to convey his condolence to Sophie, he mentioned that he had not ventured to write himself, not knowing her state of mind. Despite his grief, he tried to persuade the Court officials to return packets of letters the dead Queen had in her possession to their senders (*K* **10**, pp 265–7), in order to save them from the fate of burning which, on the orders of the King, had already befallen many of her own letters (Guhrauer 1846 **2**, 261–2), as well as letters she had received. A month later Leibniz wrote to Johann Matthias von der Schulenburg (*K* **10**, p 270) confessing that, although his reason told him that regret was superfluous and that the Queen should be honoured rather than mourned, his imagination continually presented the princess with her perfections and reminded him that he had lost one of the greatest satisfactions of the world that he could expect to experience in the whole of his life.

Leibniz sought his own consolation in the writing of a long German poem in memory of the Queen (*K* **10**, pp 291–5). The poem begins (Hankins 1972) by praising the Queen's qualities of beauty, virtue and intellect while lamenting her death, concluding—in accordance with the mythological transformations common to such poems—that she must have been more than human to have manifested such excellence: 'Ein Engel muss es seyn, der Fleisch und Beine nimt'. Sadness is replaced by dissatisfaction concerning the limits of the divine power: 'Wo bleibt die Weissheit dann?' The answer to this question—'Die Weissheit lässet sich in allen Dingen spüren'—brings a ray of hope in which the tragedy can be seen in the wider perspective of the monadic philosophy. Then follows a sublime exposition of this philosophy, in which each soul or monad, like a mirror, reflects the totality of the universe, while the community of rational souls freely seek their perfection

[9]Sophie Charlotte died of pneumonia. She had already contracted a heavy cold on the journey from Berlin. With the death of Sophie Charlotte, the Electress Sophie, then 75 years old, had lost four of her children and her only daughter (Schnath 1978, pp 572–3).

under divine government. Consolation is achieved with the assurance of the Queen's eternal citizenship in this community:

> *Die Seelen die mit Gott in Innung können treten,*
> *Die fähig ihr Verstand gemacht Ihn anzubeten,*
> *Die kleine Götter seyn und ordnen was wie Er,*
> *Die bleiben seines Staats Mitglieder immermehr.*

9

Hanover, Wolfenbüttel and Berlin (1705–1710)

On leaving Berlin towards the end of February 1705, Leibniz visited Duke Anton Ulrich in Wolfenbüttel before returning to Hanover at the beginning of March. Here he found some consolation, as he explained to Caroline of Ansbach (*K* **9**, pp 116–19), on learning that Sophie Charlotte had herself said, 'Ich sterbe einen gemächlichen Tod, denn ich fühle nichts', and that she had died with a wonderful serenity of spirit and a tranquility of mind resigned to the orders of supreme Providence.[1] Leibniz did not rest long in Hanover. To Sophie (*K* **9**, pp 120–1) he explained that, as the Elector was in Celle, he had taken advantage of his absence in order to return to Wolfenbüttel, where Duke Anton Ulrich had invited him to discuss plans for the restoration of the Ducal Library to its former glory. From Wolfenbüttel he travelled on to Berlin in order to discuss the account of the life of Sophie Charlotte in Hanover before her marriage that he had been asked to contribute to the customary public oration. This was completed (*K* **10**, pp 273–84) soon after he returned to Hanover at the end of May to find himself subjected to further harassment from the Elector. As Georg Ludwig believed Leibniz to be strongly diverted from the history of the House of Brunswick by his frequent journeys and voluminous correspondence and feared that, in the case of his death, the history would remain uncompleted, he issued an order forbidding him to undertake any further journeys without

[1]Although this description of her last hours is well attested by eye-witnesses, who quote her as saying, 'Meine Seele ist schon bei Gott', it seems that her response to the words of comfort of the clergyman was rather reserved. This may have given rise to the legend, related by her grandson Friedrich the Great, that she died as a mocking freethinker. According to this story, she is supposed to have said, 'Do not torment me, for I now go to satisfy my curiosity on the principle of things that Leibniz has never been able to explain to me; on space, infinity, being and nothingness. And I prepare for my husband the King the spectacle of a funeral, where he will have a new opportunity to display his grandeur' (Schnath 1978, p 572).

the permission of the Elector himself. At the same time Duke Georg Wilhelm of Celle, who died later in the year, pressed him to accelerate the historical work.

In July Leibniz resumed his correspondence with Lady Masham, explaining that the death of the Queen had been the cause of the long interruption (*GP* **3**, pp 366–8). Offering condolences on the death of her own friend Locke, he informed her that he had completed his remarks on Locke's excellent work, adding however that he was inclined to the view of those who believed the source of necessary truths to be innate in the mind. Besides writing to Lady Masham concerning the philosophy of her father, Leibniz took up again the threads of his discourses in the Queen's garden at Lützenburg in letters to Sophie.

Before the end of the summer he composed a memorandum for the Elector on the differences between the Houses of Brunswick and Brandenburg, taking the opportunity to remind his unappreciative master of his services to the House of Brunswick over a period of thirty years. These included, in addition to the history, works and discourses relating to primogeniture, the ninth Electorate, the succession in Lauenburg and the Hanoverian succession in England. When in Berlin, he had tried to promote unity between the two Houses, which was highly desirable in view of the dangers from France. Shortly after composing this rejoinder to the Elector's order forbidding him to travel, he journeyed to Brunswick for the fair and then to Wolfenbüttel, where he was again a frequent visitor over the following years.

At the beginning of 1706, Leibniz was evidently visited by the Jesuit teacher of philosophy, mathematics and theology, Bartolomaeus Des Bosses (*GP* **2**, p 291), with whom he subsequently conducted a fruitful correspondence on philosophy. Although not in the first rank of scholars, Des Bosses was a man with whom Leibniz could discuss philosophical issues seriously both by letter, and for a time, face to face, for he taught at the Jesuit College in nearby Hildesheim until he moved to Cologne at the end of 1709. Among other scholars with whom Leibniz entered into correspondence at this time were the natural philosopher Nicolaus Hartsoeker, Johann Wilhelm Petersen, who had lived as a private teacher since the Rosamunde affair, Christian Maximilian Spener, the son of the renowned Pietist, and the philologist and naturalist Johann Leonhard Frisch, who gave him instruction in the Russian language. To his mathematical correspondents he added Jakob Hermann, Christian Wolff and Conrad Henfling.

Politics of the English succession

When Celle was united with Hanover on the death of Duke Georg Wilhelm on 28 August 1705, two of his ministers, Andreas Gottlieb von Bernstorff and Jean Robethan, entered into the service of the Elector Georg Ludwig. From this time, Robethan, who had been a French refugee in the service of King William before moving to Celle, was the Counsellor responsible for official communications on behalf of Hanover concerning the English succession. The influence of Leibniz was therefore exerted informally through his English correspondents. At the beginning of 1706 he composed for Sir Rowland Gwynne, then visiting Hanover, an open letter to the Earl of Stamford (*K* **9**, pp 188–200), criticising the politics of the Whigs, who had opposed a motion of the Tory Lord Haversham to invite the Electress Sophie to England. The Whigs had accused the Tories of only wanting to invite Sophie in order to oppose another Court to that of Queen Anne, and by thus sowing dissension, advance the affairs of France and the Jacobite pretender, the Prince of Wales. It was in fact those who opposed the invitation to Sophie, the author of the letter (supposedly Gwynne) declared, who were really advancing the Jacobite cause. There was no precedent, the writer argued, for keeping the heir away during the lifetime of a monarch. Indeed the history of England was full of examples of heirs excluded owing to their absence, and in this case the exclusion of the heir could lead to the end of the Protestant religion and liberty in England. Having praised the Electress Sophie's personal qualities, the author pointed out that her resolution to avoid meddling in the affairs of England had led some to accuse her of indifference, but this was wrong, for she had a great affection for the Queen. She had expressed her sentiments in a letter to the Archbishop of Canterbury (*K* **9**, pp 177–9), which she had given the author permission to communicate to his friends. In this she had declared that, although she was content in her present state, she would be willing to hazard her life in crossing the sea in order to affirm the Protestant succession, if this were deemed to be appropriate. However, she would leave it to the Queen and Parliament to decide what they considered to be most agreeable.

After Gwynne had translated the letter, Leibniz had it printed in Holland and circulated in England. It was not well received in Parliament, where it was described as a malicious libel, tending to create a misunderstanding between the Queen and the Princess Sophie (*K* **9**, p xl). The letter was condemned in the Lower House of Parliament by a vote of 141 to 97, which was followed by a petition to the Queen, to which the Upper House assented, to seek out the printer and publisher with a view to punishment. The Queen remarked that nothing could please her more than the zeal of the two Houses for the maintenance of the good understanding between herself and the Electress.

Leibniz's authorship of the letter was not discovered but Gwynne was refused permission to visit Hanover again in the following year. In a letter to Thomas Burnet of 26 May (*K* **9**, pp 215–20), Leibniz writes as if he had nothing to do with the affair. The letter, he adds, had not been approved by the Court and Gwynne had disclaimed all responsibility for the publication.

Yet Leibniz took pains to point out the merits of the letter. It contained many good things, he remarked to Burnet, for showing the need of assuring the succession by an effective establishment of the heir; in particular the presence of a representative of the House of Hanover in London. Moreover, he considered Stamford's reprimand of Gwynne for mentioning Sophie's letter to the Archbishop of Canterbury to be unfair, for Sophie had intended this letter to be communicated to others.

Towards the end of May Lord Halifax[2] presented in Hanover two Acts that had been passed in the English Parliament, confirming the Hanoverian succession. These were the 'Act of Regency', which provided for the setting up of a regency council under the Archbishop of Canterbury in the period between the demise of Queen Anne and the installation of her successor, and the 'Act of Naturalisation', which declared the Electress and her descendants to be English. On this occasion, the Electoral Prince Georg August was admitted to the Order of the Garter, as had been his father at the time of the 'Act of Settlement'.

Lord Halifax was accompanied on his visit by the English critic Joseph Addison and the herald van Bruck, who was also a poet and architect. Leibniz found their company very congenial (*K* **9**, p 226). Reporting to Thomas Burnet on his meeting with Lord Halifax (*K* **9**, pp 220–4), Leibniz expressed the view that the victories of the allies against France were a better guarantee of the Protestant succession than any Acts of Parliament. France had indeed lost the initiative after the decisive victory of Marlborough and Prince Eugene at Blenheim in 1704 and had been forced to retreat west of the Rhine. The victory at Ramilles in 1706 forced them out of the Netherlands at the same time as Prince Eugene was driving them out of Italy. The prospects for the pretender Prince of Wales were therefore not very promising. Writing to Davenant (*K* **9**, pp 224–7), the English ambassador in Frankfurt, on 26 July 1706, Leibniz repeated the views he had expressed to Burnet, adding that the Electoral Prince and Princess were making good progress in their study of English.

A wedding in Berlin

In June 1706, Sophie Charlotte's son, the Crown Prince Friedrich Wilhelm of Prussia, became engaged to the Princess Sophie Dorothea, daughter of the Elector of Hanover and the banished Princess of Ahlden. Owing to the political differences between Hanover and Berlin, the detailed arrangements

[2]Lord Halifax was a friend of Newton. As Chancellor of the Exchequer, he had appointed Newton to be Warden of the Mint and later he took Newton's niece Catherine Barton into his house as 'Super-intendant of his domestick Affairs', leaving her a handsome legacy in his will (Westfall 1980, pp 594–600).

for the wedding presented some problems (Schnath 1978, pp 581–4). The marriage took place in Hanover by proxy on 14 November, the Electoral Prince Georg August representing the bridegroom, and three days later the Princess travelled to Berlin. At the request of the Elector, Leibniz composed a memorandum demonstrating the validity of marriage by proxy, which had been questioned by the chief master of ceremonies in Berlin, Johann von Besser, who wished to please the King by claiming the ceremony that was to take place in Berlin as the true marriage.

As the Princess was unwilling to change her religion, Leibniz had suggested what seemed to him an acceptable solution that both parties could agree to with a clear conscience; namely, that they should marry according to the rites and liturgy of the Anglican Church. For they were both covered by the 'Act of Naturalisation', being descendants of the Electress Sophie, and moreover, the King was known to have an inclination for the rites and liturgy of the Anglican Church. This was strongly opposed by the Reformed Bishop Benjamin Ursinus von Bär and brought from the King a reprimand to his Court Chaplain Jablonski on account of his 'wholly improper correspondence' with Leibniz. In Hanover the Elector issued a rescript forbidding Leibniz to undertake any further collaboration in endeavours to unite the Protestant Churches. Religious freedom for his daughter having been assured by the King as a condition of the marriage, he evidently considered reunion to be henceforth superfluous as a political objective. Leibniz was in the embarrassing position of not being able to mention the rescript to the theologians with whom he had formerly collaborated. To Fabricius in Helmstedt he wrote: 'As things stand now, I expect nothing more from the reunion business. The thing will sometime accomplish itself.'

The wedding provided for Leibniz an opportunity to visit Berlin. On the way he stayed for a few days in Wolfenbüttel and Salzdahlum[3]—one of several visits he had made that year—where he met Pius Nicolas Garelli (later Imperial Counsellor and physician to the Emperor), who had been sent by the Empress for secret discussions with Duke Anton Ulrich concerning the arrangement of a marriage between his granddaughter Elizabeth Christine and her second son, the King of Spain (later Emperor Karl VI) (*K* **9**, pp 241–4). On the advice of Sophie (*K* **9**, pp 239–40), Leibniz recommended Ferdinand Orbanus, the Jesuit who had attempted to convert Caroline of Ansbach, as a suitable teacher for the young princess. She was received into the Catholic Church on 1 May 1707 in Bamberg Cathedral, at the age of fifteen. From there she travelled to Vienna and then to Spain for the marriage. The betrothal of the Princess to a Catholic brought an angry reaction from both town and state; the clergy, in particular, expressed their

[3]Duke Anton Ulrich's Palace near Wolfenbüttel. The building of this North-German 'Versailles', completed in 1694, ruined the economy of the Dukedom. It was demolished in 1813, the timber buildings being then beyond repair.

displeasure in strong words, even threatening the Duke with exclusion from the Eucharist (Bodemann 1888, p 91).

Having arrived in Berlin from Wolfenbüttel on 15 November 1706, Leibniz had an audience on the following morning with the King, who was impatient to see his new daughter-in-law. On the same day Leibniz wrote to Sophie (*K* **9**, pp 241–4) reporting that the King of Sweden had concluded a peace treaty with Poland, under the terms of which August (the Elector of Saxony) would renounce the Crown in favour of Stanislaus Lesczenski, though most Poles regarded the election of Stanislaus as illegal. The King in Prussia and the Elector of Hanover, he added, had been named as guarantors of the treaty.

On 26 November Leibniz visited the Crown Princess in Spandau and two days later attended the ceremony of ratification of the marriage in the Royal Chapel. At the end of the banquet in the evening, the King called him over to express his regret that the Queen was not there to witness the happy event (*K* **9**, pp 248–51). Leibniz took the liberty of suggesting to the King that the new Princess (Sophie Charlotte's niece) would remind him of the dead Queen, to which the King replied that her eyes resembled those of the Queen. A few days later there was a masquerade and a ballet at Charlottenburg (as Sophie Charlotte's Lützenburg was now called) and a firework display was planned to take place as soon as there was an improvement in the weather. Leibniz reported to Sophie that the Princess had allowed her hair to be cut. Although this pleased the Crown Prince, Leibniz described it as a terrible execution (*K* **9**, pp 252–4).

In a letter to her granddaughter, Sophie remarked that, if she wished to see this good man (Leibniz) often, the King would have to give him a regular income, as he received in Wolfenbüttel, where he also made many visits. It seems that the idea of enlisting the help of the Princess on his behalf had been suggested to Sophie by Leibniz himself (*K* **9**, p 254). He also looked to Sophie Dorothea for the kind of support he had received from Sophie Charlotte in the realisation of his plans for the Berlin Society of Sciences. On the occasion of this visit he spent six months in Berlin working for the Society.

From Berlin Leibniz sent Sophie on 4 March 1707 a poem entitled 'Le carneval des dieux', which was inspired by an amusing incident at the Court (*K* **9**, pp 273–4). It seems that the King had acted as a match-maker in arranging a marriage between his valet and a servant of Sophie Dorothea, which was very popular at the Court.

Late in May Leibniz left Berlin for Leipzig, where he persuaded Count Anton Günther von Arnstadt to publish the illustrations of the coin collection of Andreas Morell (*D* **5**, p 422) and no doubt viewed the manuscripts of Kepler that were then in the possession of Michael Gottlieb Hansch. While visiting his friend Jakob Heinrich von Fleming he also met the Counsellor of Legation, Ernst Christoph von Manteuffel (Guhrauer 1846 **1**, app, p 8). On a visit to the Swedish camp near Altranstädt, he had a sight of the three kings, Charles XII of Sweden, August (the deposed king of Poland) and Stanislaus, the new king of Poland. In a letter to Thomas Wentworth (Lord Raby) (Guhrauer 1846 **2**, app, p 27) he reported that he saw Charles XII eating at midday. The meal lasted half an hour during which the King said not a word. Leibniz waited over a week hoping to gain an audience when the King returned from reviewing his troops dispersed in the country. He was evidently not too disappointed at not meeting the King, for it would have been difficult to find a topic of conversation; the King was apparently only knowledgeable about military matters. Before leaving Leipzig for Halle, where he sought out Christian Wolff, Leibniz made an excursion to Probstheida to see his nephew Friedrich Simon Löffler (Bodemann 1895, p 153). He then travelled to Wolfenbüttel where he awaited the return of Duke Anton Ulrich for two days and finally arrived in Hanover on 16 June. Reporting his arrival to her granddaughter, Sophie mentioned that he had seen the three kings.

Progress on the history and visits to Wolfenbüttel

In the same month that Leibniz returned home, the first volume of the history was published in Hanover with the title *Scriptores rerum Brunsvicensium illustrationi*. This gave him the opportunity to press for support in the advancement of the work that the Elector had been evidently so anxious to see completed. In his memorandum concerning the continuation of the work, Leibniz claimed expenses for his assistants, whom he had up to then paid himself, a payment in advance for the printing costs of the second volume, and the purchase of 120 copies of the first volume for distribution to the scholars who had provided source material for it. Repeated requests for the settlement of these claims brought no response from the Court in Hanover, though the Duke of Wolfenbüttel agreed to pay a third of the printing costs of the first volume.

After three months in Hanover, Leibniz made another visit to Duke Anton Ulrich in Salzdahlum, this time in the company of the Electress Sophie, and then travelled further to Hessen. During his stay in Kassel he met the Landgrave Karl, who was interested in mathematics and physics, and also Denis Papin, who was about to leave for England. Although he had failed to secure permission for Papin to cover the first part of his journey by sailing his self-built ship downstream along the Weser through Münden, Leibniz was able to give him a letter of introduction to the Secretary of the Royal Society, Hans Sloane. While in Kassel Leibniz also sketched a design for raising loads with the help of a waterfall (Bodemann 1889, p 332). On the return journey he spent two days in Göttingen with Justus von Dransfeld studying historical documents (*D* **5**, p 486). In Hildesheim he had a conversation with the missionary Nicolas Agostino Cima, who had just returned from China, on the possibilities of disseminating knowledge of Chinese culture in Europe and gave him a recommendation to the Berlin orientalist Mathurin Veysiere La Croze (*D* **5**, p 485). Back in Hanover, he wrote to Duke Anton Ulrich on 11 October, telling him that he had to stay indoors owing to a foot injury that had been aggravated on his last journey.

About the middle of November Leibniz recorded a curious dream, in which he had sung a New Year carol remembered from his youth. He could only recall the last verse:

Der du von uns weg genommen
Alles Weh, alles Weh
Hilf, dass wir bald zu dir kommen
O Christe! O Christe!

Later in the month, he wrote a poem for King Friedrich I in Prussia, congratulating him on the birth of a son to Princess Sophie Dorothea (*K* **10**, pp 416–18). His good wishes were not fulfilled however, for this prince did not survive.

At the beginning of 1708, the Leipzig theologian Adam Rechenberg reminded Leibniz that it was forty years since their last meeting in Leipzig, when Leibniz had said goodbye to his 'ungrateful native town'. This brought from Leibniz the remark that he had never felt any resentment against Leipzig, which he loved as his homeland. As a young man he had been impatient. Yet he had not regretted leaving Leipzig, for the errors of men were guided through divine Providence, so that good often came out of bad decisions (Bodemann 1895, p 231).

Between visits to Wolfenbüttel in the spring of 1708, Leibniz had a meeting with John Toland, who made a short stay in Hanover on his way home from visits to Berlin and Vienna (*GP* **3**, p 317). Reporting this meeting in a letter to Thomas Burnet, Leibniz added that the young prince of Hanover (the son of Caroline) was growing wonderfully and cutting his teeth without pain. It seems probable that Leibniz at this time also made his first acquaintance with Prince Eugene of Savoy, who visited Hanover briefly in April.

In July, Leibniz proposed to the Elector that, following a health cure in Karlsbad, he should be allowed to make a journey to Munich in order to evaluate the source material in the archives there for his historical work. The Elector agreed but on condition that the journey was made at his own expense. This was too much for Leibniz, who was still trying to recover the expenses of printing the first volume of the history, and he abandoned the visit to Munich. Writing to the President of the Chamber Friedrich Wilhelm von Goertz on 30 July, he explained: 'I am not wealthy enough to follow the example of the Duke de la Feullade, who erected a memorial to the honour of the French kings at his own expense. I will leave the Bavarian Guelfs resting in their old documents until I am richer' (*MK*, p 210). At the same time, he declared his intention to make the journey to Karlsbad for his health cure following a short visit to Berlin and the Michaelmas fair in Leipzig, returning to Hanover within about two months. Meanwhile, he had made a short visit to Detmold, residence of the Count of Lippe, in order to see Queen Maria Anna of Portugal. Here he also met the Count's physician, Engelbert Kämpfer, who was famous for his travels to Persia, the East Indies and Japan.

A secret journey to Vienna

Towards the end of October 1708 Leibniz was in Brunswick planning with Duke Anton Ulrich a secret mission to Vienna with the object of restoring the bishopric of Hildesheim to Brunswick-Wolfenbüttel (Hohnstein 1908, p 380). This bishopric had been given to the Elector of Cologne at the peace of Westphalia in 1648. Duke Anton Ulrich gave him a recommendation to Emperor Joseph I explaining that, although he was making the journey incognito, the Emperor could have complete confidence in him. Leibniz left Brunswick in the direction of Vienna about the middle of November, travelling through Halberstadt and Erfurt, where he met his former student Philipp Wilhelm von Boineburg, then ambassador of Mainz in Erfurt. Continuing through Eger and Karlsbad, he arrived in Regensburg on 28 November. From here he sent an accurate report on his journey to Duke Anton Ulrich but in a letter to Sophie pretended that he was taking a long health cure in Karlsbad (*K* **9**, p 290). Continuing the journey by boat on the Danube, he arrived in Vienna at the beginning of December. He stayed with the Emperor's physician, Pius Nicolas Garelli, whom he had previously met in Wolfenbüttel. Through Garelli he gained admission to the Empress Amalia, daughter of the late Duke Johann Friedrich of Hanover. Concerning the bishopric of Hildesheim, he had many conversations with Prince Salm, who decided that a formal claim supported by legal arguments would have to be submitted, if the question were to be pursued further. During his visit, Leibniz composed a memorandum for the ambassador of the Duke of Modena on the claim of the House of Este on the town of Comachio (Ravier 1937, p 82). At the request of the Russian ambassador, Baron Johann Christoph von Urbich, he composed a plan for the development and advancement of science in Russia. On 28 December he left Vienna for Leipzig in the company of the Russian ambassador with whom he discussed the political relations of Hanover with Sweden and Russia, and also the possibility of a marriage of the Czarevitch Alexei with a Brunswick princess. Duke Anton Ulrich entertained the hope that his granddaughter Charlotte, sister of the Queen of Spain, would become the wife of the heir to the Russian throne (Bodemann 1888, p 97). Arriving in Leipzig early in January 1709, he made the acquaintance of the wife of the Polish First Lord of the Treasury, Johann Georg Prebendowsky, and had conversations with his Jesuit friend Carlo Mauritio Vota (*K* **9**, p 296) before continuing to Berlin. Here he visited the Crown Princess Sophie Dorothea and also met the new Prussian Queen Sophie Luise. Reporting these meetings to Sophie he mentioned also that there was in Berlin a lady astronomer of outstanding talents who often worked in the Observatory at night. To explain his long absence from Hanover, he invented the story that, after spending three weeks in Karlsbad, he had visited the universities of Saxony in search of an assistant for his historical work. This is what he told Sophie and also the Hanoverian Prime Minister von Bernstorff (*K* **9**, p 291).

Sophie replied (*K* **9**, p 294) that the Elector had spoken of offering a reward to anyone who knew his whereabouts when his presence in Berlin came to light. Unfortunately, news also reached Hanover concerning his journey to Vienna. The secret had been betrayed by someone who had dined with him and the Russian ambassador. On learning about the visit, the Hanoverian ambassador in Vienna, Daniel Erasmi von Huldeberg (whom Leibniz had carefully avoided meeting) immediately informed the Elector.

After staying in Berlin for a few weeks to work for the Society of Sciences, Leibniz returned to Hanover early in March 1709, when the Elector Georg Ludwig, during a meeting in Sophie's room, expressed his displeasure at the secret journey to Vienna. Unrepentant, Leibniz composed a justification (*K* **9**, pp 297–300) in which his polite language did not disguise his strong criticisms of the Elector regarding his attitude to Leibniz and the history of the Guelfs. In order to avoid disclosing the real purpose of his journey to Vienna, he invented another story. Having arrived in Karlsbad, he heard that the Empress wished to learn his opinion concerning the controversy between the Pope and the Duke of Modena. Being already half-way there, he decided to make the journey to Vienna, travelling incognito so as not to attract attention to his purpose. He had written to Sophie, asking her to inform the Elector of his intention to go to Vienna without waiting for his permission (in order to save time), but owing to the negligence of someone in Karlsbad, this letter was never delivered. He then went on to mention the Elector's coldness towards him and the apparent lack of appreciation of his historical work, contrasting the Elector's treatment of him with the support and appreciation he had received from Ernst August and Johann Friedrich, adding that his work had been generally applauded in the world, if not in Hanover. Comparing his case with the author of the history of the late Elector of Brandenburg, whose task had been incomparably easier but was paid 3000 taler per year, Leibniz complained that he could not even obtain the expenses that had been agreed. Even when he only requested the purchase of books he needed for the history, so many obstacles were placed in the way that he had to buy them himself. Lastly, he accused the Elector of not caring much for the glory reflected on him from his ancestors, because he was content with his own glory and grandeur. If he were really interested in the history, he would give it better support.

Despite his strained relations with the Elector, Leibniz continued to offer him political advice concerning the affairs of Hanover. For example, he wrote several memoranda on his conversations with the Russian ambassador von Urbich. Again, although the expenses involved in the publication of the first volume of the Brunswick history had still not been refunded, he worked with undiminished devotion to produce the next volume. Immediately on his return to Hanover, he had requested the Privy Counsellors to place at his disposal the documents that Friedrich August Hackmann had brought back from Italy and before the end of the year he was able to present to the Prime Minister Andreas Gottlieb von Bernstorff for his approval a proof of his composition of the annals of Guelf history for the years 768 to 785. Yet he was sufficiently dissatisfied with his life in Hanover to ask Duke Anton Ulrich to take him into his service (Guerrier 1873, pp 170–4). It would be better, he added, if the suggestion were seen to come from Anton Ulrich himself.

Correspondence with Lady Masham and Pierre Coste

In May 1705, there appeared in the *Histoire des ouvrages des savans*, in the form of a letter to the editor Basnage de Beauval, an essay of Leibniz entitled *Considérations sur les principes de vie et sur les natures plastiques* (*GP* **6**, pp 539–46). This was the outcome of a request from Jean Le Clerc, editor of the *Bibliothèque choisie*, for an opinion, based on the principle of pre-established harmony, of the controversy between Bayle and Le Clerc concerning the existence of vital principles and plastic natures, such as Lady Masham's father, Ralph Cudworth, had proposed. According to Bayle, Cudworth's system favoured the cause of atheism, since God was unnecessary if the plastic natures served his function. Cudworth had argued, for example, that his immaterial plastic natures were needed to form an animal, since this could not be produced by the laws of mechanism alone. In Leibniz's view, the celebrated authors More and Cudworth, who had occasioned the controversy with their vital principles and plastic natures, erred in supposing that souls have influence on bodies. Describing his own system, in which vital principles (that is, indivisible substances or unities) are spread throughout all nature, Leibniz explains that the course of motion of bodies is not at all changed within the order of nature, God having pre-established it as it should be. Indeed, his system furnishes a new proof of the existence of God, since the agreement of so many substances, none of which exerts an influence on another, can be brought about only by a general cause of infinite power and wisdom.

Writing to Lady Masham (*GP* **3**, pp 366–8) in July 1705, following the long gap after the death of Sophie Charlotte, Leibniz listed the points of agreement between her father's philosophy and his own. While he agreed with Cudworth, however, that animals have not been formed mechanically by something inorganic, he was of the opinion that the plastic force was itself mechanical and consisted in a preformation, that is, in organs already existing, which alone are capable of forming other organs. Leibniz thus identifies Cudworth's plastic nature with his own view of the organic nature of matter. To Lady Masham he remarks that he has simply added an explanation that her father had omitted.

In her reply, written on 20 October 1705 (*GP* **3**, pp 369–73), Lady Masham explains how the charge that her father's plastic natures favour atheism fails. Cudworth had in fact supposed the operation of the plastic natures to be essentially and necessarily dependent on the ideas of the divine intellect. Even if matter could be supposed to have of itself the same power that the plastic natures were said to have by the gift of God, this would not help the atheists. For the power given to the plastic natures was only the power to execute the ideas of a perfect mind, and this power would lie eternally dormant and unproductive, if there were no such mind in the universe.

On a personal note, Lady Masham mentioned the possibility of meeting Leibniz in Hanover in the following year, when she planned to undertake a journey in the company of her son, in order to broaden his mind by travel before he embarked on the study of law. The change of scene, she added, might also improve her poor health. Leibniz expressed his pleasure at the prospect of seeing her in Hanover (*GP* **3**, pp 373–5). Sophie, he added, would also be delighted to see her, for she loved to entertain people who spoke English, and nothing could be more agreeable than an English lady interested in philosophy. Concerning her father's philosophy, he agreed with her that it opened no door to atheism, though he himself did not have recourse to plastic natures.

Lady Masham composed a work on the *Divine Love*, in which she criticised the ideas of John Norris, a follower of Malebranche. This was translated into French by Pierre Coste, her son's tutor, who opened a correspondence with Leibniz by sending him a copy. Thanking Coste (*GP* **3**, pp 382–6) in a letter of 4 July 1706 for the presentation of this translation, Leibniz remarked that he had also profited from his translation of Locke's *Essay concerning human understanding*. In her work on the *Divine Love*, Leibniz remarked to Coste, Lady Masham had approached his own definition, given in the preface of his *Codex juris gentium diplomaticus* (1693), where he examined the sources of justice. With the letter he enclosed a copy of the relevant passage in this preface (*GP* **3**, pp 386–9). He was of the same sentiment as his English friend when she declared that Malebranche's doctrine of occasional causes had little application to the love of God and his creatures. Indeed, he considered the cause of the relation between mind and body, whether it be occasional causes or his own pre-established harmony, to be irrelevant to the theme. Referring to François Lamy as another author who had written on this subject in his book, *Connoissance de soy meme*, Leibniz remarked in passing that Lamy's objections to the pre-established harmony were easily countered. In fact he published a refutation in 1709 in a supplement to the *Journal des Sçavans* (*GP* **4**, pp 577–95). At the end of his letter Leibniz asked Coste to give his regards to Lady Masham.

Owing to a long delay in the delivery of the letter, Coste was not in a position to reply until 20 April 1707 (*GP* **3**, pp 389–91), when he reported that Lady Masham, who sent her regards, had been in London for several months, where she had been attacked by an illness from which she had not fully recovered. Having discovered that Leibniz was writing some criticisms of Locke, Coste was anxious to send him some corrections to the translation, for he did not wish him to refute things Locke had not said. Also he promised to send some additions to the chapter on liberty, which Locke had made shortly before his death in consequence of a dispute with a correspondent, Mr Limborch. Leibniz then explained to Coste (*GP* **3**, pp 391–2) that, following his correspondence with Lady Masham, he had hoped to confer with Locke himself. His aim had been to clarify things rather than refute the ideas of another. Although his work had been completed, he would be pleased to profit from the additions and corrections Coste had promised to send.

Coste sent the corrections to Leibniz on 25 August 1707 (*GP* **3**, pp 392–9), at the same time relating a strange story of events in London. Some refugees from the south of France had been making converts by their prophecies. One rich gentleman among the converts published a book of prophecies in languages unknown to him. These were evidently very severe on the clergy. The prophets later lost all credibility by predicting the resurrection of one of their number (*GP* **3**, pp 405–7).

After reading the clarifications Locke had written following his dispute with Limborch, Leibniz (*GP* **3**, pp 400–4) summarised for Coste his own position concerning the freedom of the will. Since God had chosen among an infinity of possibles that which he judged to be best, it must be acknowledged that all is comprised in his choice and nothing can be changed. But this hypothetical necessity, accepted by all the theologians and philosophers (except the Socinians), does not in Leibniz's view destroy the contingency of things. When a choice is proposed—for example, to go out or not—the question is whether the true proposition, 'I will choose to go out', is contingent or necessary. Leibniz replies that it is contingent, because neither he nor any other mind can demonstrate that the opposite of this truth implies contradiction. Our liberty, as well as that of God, is exempt from necessity, though it is not exempt from determination and certitude. Nothing happens without cause or determinate reason; consequently there is always that which inclines us without necessitating.

There followed a long interruption in the correspondence between Leibniz and Coste. Meanwhile Lady Masham died in her home at Oates on 20 April 1708.

Correspondence with Sophie on philosophy

An extract he had read in a review of a book on mathematics by the Duke of Bourgogne served Leibniz as a pretext for reminding Sophie (*K* **9**, pp 145–55) of the conversations in the Queen's garden in Charlottenburg, when he had explained to her and her daughter his theory of simple substances and true unities. For the extract in effect expressed Leibniz's theory of substance admirably but ended in a paradox which obviously called for clarification. Thus the Duke concluded: 'Geometry teaches us the divisibility of matter to infinity and we find at the same time that it is composed of indivisibles.' It is in the solution of this difficulty, Leibniz tells Sophie, that he believes he has rendered some service and established the true philosophy of incorporeal substances.

In his letter to Sophie Leibniz expounds his theory of simple substances in some detail. First, recalling the conversations in Charlottenburg, he reminds her that souls are true unities but bodies are only aggregates and, consequently, bodies perish by the dissolution of their constituent parts, whereas souls are immortal.

In order to understand better the actual division of matter to infinity (though all the parts are not in fact separated), it should be borne in mind, Leibniz suggests to Sophie, that God has already produced as much order and variety as it is possible to introduce up to now, so that nothing remains indeterminate but no two objects are exactly alike. A piece of stone, for example, is composed of certain grains which, in the microscope, appear as rocks exhibiting a thousand sports of nature. Sophie herself had recognised this order and variety; for he recalled an occasion when, in the park at Herrenhausen, she had challenged the late Carl August von Alvensleben to find two identical leaves. Matter only appeared to be a continuum in the same way that a toothed wheel, when rotated rapidly, appeared uniformly transparent. It could therefore be concluded that a mass of matter was not truly a substance—its unity being only ideal—but an aggregate of an infinity of true substances, a well-founded phenomenon. Likewise time and space were not substances but well-founded phenomena, or principles of relation—time the foundation of order in things conceived to exist successively and space the foundation of order in things conceived to exist simultaneously. From the ideal nature of time there followed a proof of the celebrated truth of the theologians and Christian philosophers of the conservation of things by a continual creation. Here, Leibniz declared, was the best application of the famous labyrinth of the composition of the continuum; the analysis of the actual duration of things in time demonstrated the existence of God, while the analysis of matter actually found in space demonstrated the existence of unities of substance and consequently of immortal souls.

Sophie sent Leibniz's letter to her niece Elizabeth Charlotte of Orleans, who replied on 27 December that she understood Leibniz's unities as little as if they had been described in Greek or Latin. When her son returned to Paris, however, she would ask him if he could understand what they were about. On 21 February, the Duke of Orleans wrote a piece for Leibniz expressing admiration for the way in which he had exposed the error of those who confused matter with extension but raising two objections (*K* **9**, pp 169–70): first, he asked how a unity could change, and secondly, he confessed that he could not understand the difference or relation between the spiritual unities and the material unities. The Duke also referred to a second letter of Leibniz without mentioning the contents. This was sent to the Duchess of Orleans for her son on 9 February 1706 (*K* **9**, pp 163–9) and concerned a report communicated to the Paris Academy of Sciences on the lack of development of religion in a young man of Chartres, a deaf-mute who had learnt to speak after his hearing was suddenly restored. Leibniz was not convinced that the mental development of the young man was so retarded as the report suggested. Some children of this kind, he pointed out, could achieve almost as much as other men, if they were given appropriate education. Words were not essential and communication could be through equivalent characters (like those of the Chinese) or pictures. For example, he knew of a German gentleman who had learnt to read and write, a girl born in Germany of French parents who assisted in the management of her father's household and a painter of merit employed by the last reigning Duke of Oldenburg, all deaf-mutes from birth. To the Duke of Orleans, he expressed the hope that the publicity given to the case in question would lead to greater vigilance on the part of magistrates and ecclesiastics to ensure that such children received suitable education. For the means of giving them all the necessary instruction existed; some deaf children had even been taught to speak.

Meanwhile, in a letter of 6 February 1706, Leibniz offered to Sophie further clarification of his system (*K* **9**, pp 155–63), explaining that God is a simple substance standing outside the series, who sees the universe clearly (that is, as it really is), and that each soul is a world apart, representing things outside it confusedly from its point of view. Also, he added, as there are two kinds of perceptions, the ones simple and the others accompanied by reflection, there are two kinds of souls, namely the common souls which have simple perception and rational souls which also have reflection; the first are only mirrors of the universe but the second are also images of the divinity.

In March 1706, Leibniz again wrote to Sophie with further elaboration of his philosophy and replies to the points raised by the Duke of Orleans (*K* **9**, pp 170–7). To elucidate what the Duke still found obscure in the nature of unities or simple substances, he would have found useful a logical calculus or universal characteristic. Although he had sometimes proposed this project, he confided to Sophie that he doubted whether he would ever be in a position to bring it to completion, for this was something that could not be done single-handed. Thus deprived of the instrument of which he had need for demonstration, he could perhaps only give occasion to the Duke to clarify his own thoughts. Following some polite flattery in which Leibniz declared that he had learnt something from the Duke's comments, he set out to resolve the difficulties on the basis of the principle of sufficient reason, or as he expressed it, the principle that there is always a why. First, he considered the question why there is something. There would not be a reason for the existence of anything, he explains, unless there were a last reason, which must have the reason of its existence in itself. Thus the last reason of things is no other than the necessary absolute substance, which is not subject to change. The Duke regarded this as the only true unity. Leibniz points out, however, that experience shows us that there are changes and substances subject to them. Moreover, he adds, this is confirmed by reason. For by the same reason that there is something rather than nothing, there is more rather than less, and if things were always the same there would be less, for all that follows would be excluded. Since all changes are modifications of simple substances, it follows that the nature of created substances consists in this liaison and each substance is carried by its nature from one state to another. Having thus established the existence of simple substances or unities subject to change, Leibniz goes on to explain the relation between the different unities and particularly between the mind and organic body in terms of the pre-established harmony.

At the end of the letter he expressed to Sophie, with his usual optimism, his belief that the young Duke of Orleans, who was known for his encouragement of research in Paris, would see great improvements in the world during his lifetime, to which he would himself have contributed notably.

New essays on human understanding

Although Leibniz had expended a great deal of time and effort on the composition of his *Nouveaux essais sur l'entendement humain*, his aversion to publishing refutations of dead authors (*GP* **3**, p 612) ensured that this work was not published during his lifetime. It seems that some extracts from his notes had found their way to Locke, who treated them with disdain. This did not surprise him, for he recognised that the principles he advanced were very different from those of Locke and must have appeared paradoxical to him. Despite the fundamental differences, however, Leibniz adopted a conciliatory approach, seeking to demonstrate whenever possible that Locke was only a short step from his own position.

The essays (*GP* **5**, pp 39–509) take the form of a dialogue between two friends, Philalèthe, a follower of Locke, and Théophile, who speaks for Leibniz. This literary form enables Leibniz to expound Locke's arguments before stating his own responses. As the division into books and chapters follows that of Locke's work, in which the same ideas recur in different settings, there is much repetition. Philalèthe (*GP* **5**, pp 62–3) is supposed to have had conversations with Locke in London and also at Oates, the home of Lady Masham. He begins by saying that Locke is against innate ideas and believes that we do not always think. Théophile (*GP* **5**, pp 63–6) begins by saying that he has been impressed by a new system, of which he had read something in the philosophical journals of Paris, Leipzig and Holland and also in Bayle's *Dictionary*. Besides an intelligible explanation of the union of mind and body, he finds in the unities of substance and their pre-established harmony, which this system introduces, the true principles of things. In this system, of course, all the thoughts and actions of the mind are innate. However, for the purposes of the debate, he will examine how in his opinion it must be said that, even in the common system (speaking of the action of bodies on the mind in the same way that Copernicans speak with other men of the motion of the sun), there are innate ideas and principles which do not come from the senses (*GP* **5**, pp 66–7).

At the beginning of the preface, Leibniz remarks that Locke has more relation to Aristotle, he himself to Plato, though both diverge in many ways from the doctrines of these ancients. The most fundamental question is whether the mind is entirely empty, like a blank tablet, as Locke believes with Aristotle, or whether the mind contains originally the principles of several notions which external objects merely awaken on occasions, as Leibniz believes with Plato and even St Paul, who writes that the law of God is written in the heart (Rom 2: 15). Mathematics, logic, metaphysics and ethics, Leibniz explains, are full of necessary truths, which are independent of the testimony of the senses and consequently can derive their proofs only from innate internal principles. Locke, says Leibniz, spends the first book denying innate ideas, then opens the second by claiming that ideas not having their source in sensation come from reflection. According to Leibniz, however, reflection is nothing else than attention to what is in us and the senses do not give us that which we already carry with us. That being so, he asks, can it be denied that there is much innate in our minds, since we are innate, so to speak, in ourselves, and since there is in ourselves being, unity, substance, duration, change, action, perception and a thousand other objects of our intellectual ideas? With this interpretation of reflection, Locke's view, he claims, does not really differ from his own. Leibniz supposes ideas and truths to be innate as inclinations, dispositions and natural capacities. The mind, however, is not only capable of understanding these truths but of finding them within itself (*GP* **5**, p 76).

In opposition to Locke, who believed that the mind may be without thought (in dreamless sleep, for example), Leibniz found a thousand indications of the existence of what we would call unconscious mental states (*GP* **5**, pp 46–7). He distinguishes between perception, which consists in being conscious of something, and apperception, which consists in being aware of a distinct perception (*GP* **5**, pp 121–2). The perception of light, for example, of which we are conscious, is composed of many minute perceptions, of which we are not conscious, and a sound, of which we have a perception but to which we do not attend, becomes perceptible by a small addition or increase. At every moment, Leibniz believes, there are numberless small perceptions in us, but without apperception and reflection; that is, changes in the mind itself of which we are not aware, because the impressions are either too slight or too great in number to distinguish. Even in deep sleep, some feeble and confused feelings are experienced and it would not be possible to be awakened by the greatest noise if there were not some perception of its small beginning, just as it would not be possible to break a rope by the greatest effort, if it were not stretched a little by smaller efforts. If some thoughts and perceptions were not unconscious, we would be unable to give attention to those that are important, since we would be constrained to think attentively of an infinity of things at the same time, not only including the innumerable impressions on our senses but also the traces of all our past thoughts that remain in the mind.

As a foundation for the doctrine that conscious perceptions are built up by degrees from minute insensible perceptions, Leibniz appeals to the law of continuity, according to which nature never makes leaps (*GP* **5**, pp 48–9). From the notion of insensible perceptions he then claims to be able to derive explanations of a number of important principles of his own philosophy. First, it is in consequence of these minute perceptions that 'the present is pregnant with the future and laden with the past', so that 'in the least of substances eyes as piercing as those of God could read the whole course of the things in the universe'. Secondly, they constitute 'the identity of the individual, who is characterised by the traces or expressions which they preserve of the preceding states of this individual, in making the connection with his present state'. Thirdly, they explain the pre-established harmony of the mind and the body, and indeed of all monads or simple substances. Fourthly, they prevent an indifference of equilibrium—the state of Buridan's ass when he starved to death because there was no sufficient reason why he should choose one of two bales of hay placed at equal distances from him. For it is these minute perceptions, Leibniz explains, which determine us in many a juncture without our thinking it. Fifthly, it is in virtue of insensible variations that no two individual things are perfectly alike. This in turn—the principle of identity of indiscernibles—destroys blank tablets of the mind, minds without thought, substances without action, voids in space, atoms, absolute rest and complete uniformity in any part of time, space or matter.

Having praised Locke for his rejection of occult qualities, Leibniz felt unable to conceal the fact that he had made a kind of retraction (*GP* **5**, pp 53–6) in letters to Edward Stillingfleet, Bishop of Worcester, where he had said that, after reading Newton's *Principia*, he had come to see that there was too much presumption in wishing to limit the power of God by our limited conceptions. The gravitation of matter towards matter, he held, was not only a demonstration that God can put into bodies powers that are inconceivable to us, but an incontestable instance that he has really done so. Here, in Leibniz's opinion, Locke goes from one extreme to the other. He had been squeamish concerning the operation of the mind when the question was merely to admit what was not sensible, and now he gives to bodies that which is not even intelligible, and he does so to maintain an opinion no less inexplicable; namely, the possibility that, in the order of nature, matter may think. This last idea of Locke was in fact just a speculative aside. Regarding substance as something purely indeterminate, he surmised that it was not possible for us to know whether God had not given to some systems of matter, fitly disposed, a power to perceive and think.

Leibniz expressed his disapproval of the appeal to miracles in the ordinary course of nature. This would give too much licence to bad philosophers and by admitting centripetal forces acting at a distance, the door would be opened to the occult qualities of the Scholastics. To explain himself distinctly (*GP* **5**, pp 58–9), he declared that it was necessary to consider the modifications which might belong naturally to a subject as arising from the limitations or variation of a constant and absolute original nature. Thus in the order of nature, it was not optional for God to give substances qualities other than those that could be derived from their nature as explicable modifications. By rejecting the distinction between what is natural and explicable and what is inexplicable and miraculous, Leibniz claims (*GP* **5**, p 59), we would renounce philosophy and reason, opening asylums of ignorance through a system that admits not only qualities which we do not understand (of which there are only too many) but even some that God does not understand; that is, qualities that would be miraculous or without rhyme and reason.

At a time when many people hardly respected revelation and miracles, those who wished to destroy natural religion, as if reason taught us nothing, were regarded by Leibniz as doing a disservice to the cause of religion and morals. Locke, however, he did not include in their number, for he maintained the demonstration of the existence of God and attributed to the immortality of the soul a probability in the highest degree (which could consequently pass for a moral certainty) (*GP* **5**, pp 60–1). By taking a small step, he could therefore accommodate himself to the doctrine of the natural immortality of immaterial souls, including those of animals, which Leibniz claimed to be fundamental in every rational philosophy.

Correspondence with Des Bosses

While Leibniz had played the role of a teacher in his correspondence with the Electress Sophie and Lady Masham, he could regard Des Bosses as a collaborator helping him to clarify and develop his own ideas. In his first letter to Des Bosses (*GP* **2**, pp 294–6), written on 2 February 1706, he explained that his views on the nature of bodies and on the basis of his philosophy had remained unchanged from the days of his youth, when he had been impressed by his agreement with the true Aristotle. Des Bosses readily agreed to Leibniz's request for help in clarifying the Aristotelian roots of his philosophy. To this end he would try to adapt as far as possible to the mode of expression of Aristotle the ideas advanced by Leibniz and to bring them into agreement with the dogmas of the Church. As a beginning (*GP* **2**, pp 296–9), he set up five propositions, taken from the metaphysics of Aristotle, which he thought to be in agreement with the principles of Leibniz. These propositions were:

1 Being and unity are interchangeable terms.
2 The continuum is divisible to infinity.
3 An actual infinity does not exist in nature.
4 Unity is the beginning of numbers.
5 Causes and principles of things do not proceed to infinity.

Although he had doubts about the third, for Leibniz supposed the existence of actual infinities, he thought Leibniz's intention could be satisfactorily expressed in terms of the Aristotelian potential infinite.

These five propositions led to discussion of some of Leibniz's most important ideas and form the subject of the correspondence up to the end of 1708. Concerning the first proposition, Leibniz (*GP* **2**, p 304) agrees that Being and unity are interchangeable terms, but he explains that, besides real entities and unities, there are also semi-real entities and unities. Thus, as there is Being by aggregation, so also there is a unit by aggregation. But in this case the whole is just a phenomenon, although the constituents are real. Leibniz (*GP* **2**, pp 300–1) could only give qualified approval to the fourth proposition. While he agreed that unity is the beginning of number, if ratios or priority of nature were being considered, he pointed out that the proposition was not true if magnitudes were in question, for these included fractions, which were less than unity, to infinity. The second proposition—that the continuum is divisible to infinity—he demonstrated easily from the mere fact that a part of a straight line is similar to the whole, so that, when the whole can be divided, so can the part, and similarly any part of the part. Points, he remarked, were not parts of the continuum, and there could be no more a smallest part of a line than a smallest fraction of unity. Confirming his belief in the existence of the actual infinite, Leibniz commented (*GP* **2**, pp 304–8) that the opponents assumed infinite number and the equality of all infinities to be implied. In his view, however, an infinite aggregate is neither one whole, nor possessed of magnitude; indeed, it was inconsistent with number. He thus defended his belief in the existence of the actual infinite while denying infinite number. It was of the essence of a number, he explained, to be terminated, so that, even if the world were infinite in magnitude, it would not be one whole in other than a verbal sense. The existence of the actual infinite followed from the fact that there was no part of matter, however small (and matter was supposed divisible to infinity) that did not contain monads (*GP* **2**, p 301).

At the beginning of 1709, Des Bosses raised the question whether the primary matter and the entelechy were inseparable and created at the same time (*GP* **2**, pp 367–8). He was evidently puzzled as to how Leibniz could suppose that all the mass of matter was created at the beginning and yet be able to maintain, following the account of Genesis, that the souls of animals and men were created after the third day (*GP* **2**, pp 368–9).

In response to this difficulty Leibniz (*GP* **2**, pp 369–72) pointed out that he preferred to leave the time of creation of the souls of animals an open question but explained how it was possible for new souls to be created without a new creation of matter. By matter, of course, Leibniz here meant secondary matter or mass, and it seems likely that the difficulty of Des Bosses was caused at least partly by a confusion in his mind with the primary matter of the monad. As Leibniz explains, God could create an infinity of new monads without increasing the mass, if he merely added old monads to the organic bodies of the new. The new monad, in this case, simply confers a new reality to an aggregate of existing monads, without changing the phenomena. For Leibniz was willing to admit some real metaphysical union between the soul and the organic body. But as he had already pointed out, in defending his principle of pre-established harmony against the criticism of the Paris Jesuit René Joseph de Tournemine (*GP* **6**, pp 595–8), since such a thing could not be explained by phenomena and changed nothing in them, he had not searched the formal reason for this union. It was enough, he remarked in his letter to Des Bosses, that it was tied up with the correspondence. There seems to be here a suggestion that the organic body is something more than a well-founded phenomenon, the union conferring on it a metaphysical reality or substantiality.

Having remarked that he had so far been speaking of the union of the soul or monad itself with the mass or aggregate of other monads, Leibniz turned his attention to the problem of the composition of the monad, which Des Bosses had raised in the first place. The primary matter and the entelechy, he explains, are created at the same time, for these together constitute the monad. When a monad is created, there is therefore an increase in primary matter, but since this is simply passive power internal to the monad, causing confused perception, it does not affect the phenomena.

Although Leibniz had shown that God could create new monads, he did not assert that new monads were created, but thought the opposite more probable. In the case of rational souls, he thought the transcreation of a non-rational soul into a rational soul (by the miraculous addition of an essential degree of perfection) to be more probable than an absolute creation.

Writing to Leibniz on 6 September 1709 (*GP* **2**, pp 385–8), Des Bosses remarked that, if God could create an infinity of new monads without increasing the mass, it would seem that he could also assemble an infinity of new monads and establish between them a connection such that no extension resulted. This, he surmised, could have some application to the problem of transubstantiation, which required the introduction of a new reality—the body of Christ—while leaving the phenomena unchanged, and he asked Leibniz for his opinion. Leibniz replied immediately (*GP* **2**, pp 389–91), stating that, according to the Lutherans, there was no place for transubstantiation of the bread but only for the view that when the bread is received, the body of Christ is perceived, so that only the presence of the body of Christ was to be explained. A presence, however, was something metaphysical, like a union, which could not be explained through phenomena, as he had already made clear in his response to Tournemine. At the beginning of 1710 (*GP* **2**, pp 398–400) Leibniz repeated this explanation, at the same time seeking to clarify the line of reasoning of Des Bosses. Since the bread is not a true substance but an aggregate, it must derive its reality or substantiality from a certain superadded union. The reality of the bread thus resides in the union and not in the monads themselves. From this, according to Leibniz, Des Bosses infers that God can simply destroy the union and replace it by another, which binds the same monads into a new reality, namely the body of Christ. Since the monads constituting the aggregate are the same, the phenomena are unchanged, so that the body of Christ appears as bread.

Besides discussion of the Eucharist and the problem of the substantiality of composite bodies or aggregates, which continued for many years, the correspondence at this time included a number of exchanges on China. For example, in August 1709 Leibniz sent Des Bosses a paper he had written on Chinese religion and the binary system (*GP* **2**, pp 380–4). Writing in August 1710 (*GP* **2**, pp 409–10), Leibniz laments the fact that Bouvet has evidently ceased the collaboration, and in a letter of 18 November (*GP* **2**, pp 413–14), he was very critical of the papal legate Charles de Tournon, whose lack of diplomacy in China had alienated the Emperor, so that he would only tolerate missionaries who promised to abide by the practices of Matteo Ricci, who had respected Chinese traditions.

Correspondence with Hartsoeker

Leibniz discussed other fundamental aspects of his philosophy with the Dutch scholar Nicolaus Hartsoeker, who lived in Düsseldorf in the service of the Elector of the Palatinate and in October 1706 sent him a copy of his *Principes de physique* through the Prussian Privy Counsellor Baron de Croseck. At the outset of the correspondence, Leibniz invited Hartsoeker to make a contribution to the *Miscellanea* of the Berlin Society of Sciences. When Hartsoeker visited Germany, Leibniz had conversations with him in Hanover.

Concerning Hartsoeker's book, Leibniz made a number of comments (*GP* **3**, pp 488–94), simply in order, as he explained, to give Hartsoeker the opportunity to develop and further clarify his thoughts, for he wished to avoid disputes on matters that were somewhat conjectural, preferring to leave such questions to posterity, when more experimental data would be available and physics would have attained a level of analytical rigour comparable with that to be found in geometry.

Leibniz was of course completely opposed to Hartsoeker's belief in the existence of atoms, and he explained to his correspondent that the infinite divisibility of matter was the effect of the wisdom and power of God. On other matters he was content to state his opinions. For example, he thought it unlikely that the planets floated in a heavy aether or that the atmosphere extended to the moon, for air was heavy and its density continually decreased with height. Again he conjectured that the ocean was a washing (*lessive*) of salts which remained from an ancient conflagration of the surface of the earth's globe. Besides explaining his views on geology, Newton's theory of colours and Halley's hypothesis concerning the magnetic variation, Leibniz expressed his scepticism regarding the claims of the alchemists—he was as yet a little sceptical chemist, he remarked (*GP* **3**, pp 494–6), as Boyle had been in his book—and he emphasised the need for the cultivation of practical medicine.

Following the publication of Hartsoeker's *Eclaircissemens sur les conjectures physiques* in 1710, the correspondence, which had hitherto been confined to questions of physics, took on a more general philosophical character. This came about in consequence of difficulties raised by Hartsoeker concerning Leibniz's explanation of cohesion by conspiring motions of the parts. Besides claiming that the conspiring motions would need a miracle (*GP* **3**, p 501), Hartsoeker mentioned that the theory seemed to resemble that of Malebranche (*GP* **3**, p 498). In reply, Leibniz (*GP* **3**, p 500) pointed out that his theory had appeared in his essay on motion of 1671—that is, before the publication of Malebranche's book—and he cited the experiments with magnets, where iron filings are bound together by the motion of the magnetic matter, as an analogy of his explanation of cohesion. Hartsoeker thought Leibniz's theory was more difficult to comprehend than the atoms in which he himself believed. At this point (*GP* **3**, p 502), he offered to Leibniz his own view concerning the constitution of the universe; namely, that it was composed of two entirely different substances, one active and the other passive. The active substance, which he called the first element (pure fire), was the soul of the universe, while the passive substance or matter consisted of an infinity of atoms. By these comments, Hartsoeker had in effect invited Leibniz to explain his own philosophical system. This he did in a letter of 30 October 1710 (*GP* **3**, pp 504–10), where he outlined his theory of substance and the principle of pre-established harmony. Atoms, he concluded, were the effect of the feebleness of the imagination, for while they satisfied the imagination, he believed he had shown that they shocked the higher reason.

Correspondence with mathematicians

Writing to Varignon in July 1705 with a request for news of the Paris Academy of Sciences (*GM* **4**, pp 127–31), Leibniz enclosed a paper he had composed on the dispute between Saurin and Rolle concerning the calculus. This paper, he explained, was not intended for publication but only for communication to the Academy through the Abbé Bignon, in the hope that the members would be satisfied by his response to the most frivolous objections imaginable, so that the unseemly dispute could be brought to an end. Failing this, he suggested to Varignon that they should abandon the attempt to seek justice for Saurin and the calculus within the Academy and instead solicit the judgments of others. Having heard that De la Hire had constructed an instrument for finding eclipses—Tschirnhaus had reported seeing it—Leibniz asked Varignon to let him know the cost of both the cardboard and brass versions which he wished to purchase. At the same time he informed Varignon that he had learnt something of importance from his demonstration of the formula for centrifugal force and on another occasion would explain a considerable consequence he had found (in fact the correction of his own calculation). While agreeing with Varignon in taking the 'tangent' at A (figure 9.1) as the prolongation of the chord EA, he believed Varignon to be in error in supposing the motion along FG to be uniformly accelerated. Supposing that the motion in the curve could be approximated by motion in a polygon with infinitesimal sides such as EA and AG, Leibniz in effect regarded the motion in the arc AG to be composed of the uniform inertial motion along AF and a uniform motion along FG caused by the centrifugal force acting instantaneously at A.

In his reply, written on 9 October 1705 (*GM* **4**, pp 131–9), Varignon informed Leibniz that the Abbé Bignon had appointed Cassini, De la Hire, Gallois and Fontenelle to consider the dispute between Saurin and Rolle on the calculus. After several meetings the judges had each given a written report to the Abbé Bignon but Varignon did not know when he would declare the result. Concerning the calculation of centrifugal force, Varignon defended his claim that the motion FG was uniformly accelerated, but he was evidently still worried by the lack of agreement between his result and that of Huygens. Besides asking Leibniz to show him the new calculation confirming Huygens' result to which he had referred in a previous letter, Varignon also enclosed for his judgment a demonstration of the commonly accepted result he had received from Johann Bernoulli together with his own critical response.

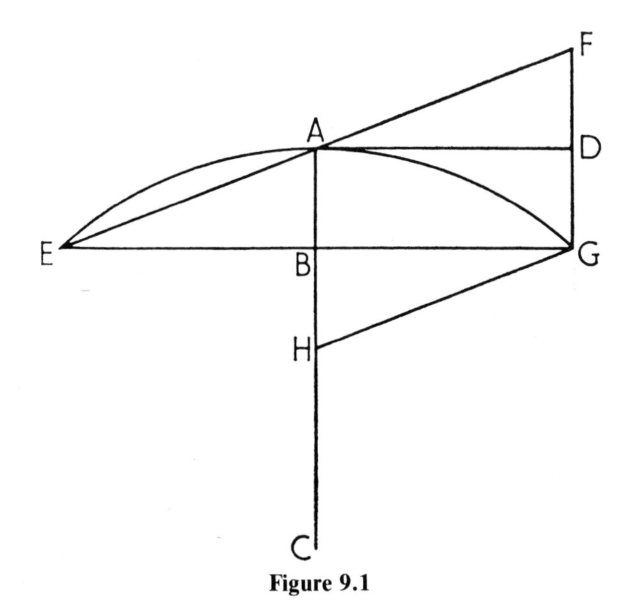

Figure 9.1

Less than a month after receiving this letter from Varignon, Leibniz sent an improved version of his essay concerning the causes of the planetary motions, entitled *Illustratio tentaminis de motuum coelestium causis* (*Elucidation of the essay concerning the causes of celestial motions*), to Otto Mencke for publication in the *Acta Eruditorum* (*GM* **6**, pp 254–76). The text, in two parts, contains not only a correction of the calculation of centrifugal force in accordance with Varignon's suggestion to take the 'tangent' (or line of inertial motion) at a vertex of the polygon as a prolongation of the side of the polygon terminating at that point but also a reply to the criticisms of David Gregory concerning the motion of the comets and the inconsistency of the harmonic circulation with Kepler's third law. In answer to the second objection he explained that the harmonic circulation was restricted to narrow bands within whose boundaries the planets moved, while he supposed the comets to move freely through the vortex, not having had sufficient time to accommodate themselves to the speed of the circulating fluid. Owing to its length, however, the paper was declined by Mencke, who suggested that Leibniz should have it printed as a separate publication. In consequence, only a short extract, including the replies to Gregory and the correction of the calculation of centrifugal force (with due acknowledgment to Varignon) appeared in the *Acta Eruditorum* in October 1706 (*GM* **6**, pp 276–80).

Meanwhile Varignon recognised his own mistake in the calculation of centrifugal force. As he reports to Leibniz in a letter of 26 November 1705 (*GM* **4**, pp 139–48), the solution came to him while walking. The element AG of the path must be regarded as a true arc and not as a side of a polygon. Then the motion in this path is composed of the uniform inertial motion in the tangent AD and a uniformly accelerated motion in DG. The instrument of De la Hire for finding eclipses, he added, would cost 8 or 10 pistoles in copper.[4] Following an illness of several months he wrote again to Leibniz on 29 April 1706 (*GM* **4**, pp 149–50), declaring with confidence that his hypothesis concerning motion in the arc was true while Leibniz's motion in the chord was only equivalent in the sense that it gave the correct result. Leibniz, however, was not convinced, for he maintained that it was simpler to keep the acceleration out of the elements. This was the method, he explained to Varignon, that he had used for over thirty years (*GM* **4**, pp 150–1).

Over the next few years Varignon communicated to Leibniz his researches on motion in a resisting medium and kept him informed of events in Paris. Although he was familiar with Leibniz's results on motion in a resisting medium, Varignon had to reconstruct the demonstrations himself, for these are lacking in the paper published in the *Acta Eruditorum*. By the end of 1709, he had reached the same impasse as Huygens and Leibniz before him in attempting to determine the trajectory of a projectile in a medium resisting as the square of the velocity (*GM* **4**, pp 167–8). Concerning events in Paris, Varignon mentioned that Madame la Marquise de L'Hospital was carefully preserving the papers of the late Marquis for her son, who was then only fourteen or fifteen years old (*GM* **4**, pp 165–7). Another item of interest he communicated was that, following the death of Abbé Gallois in 1707, Rolle, sensing the damage he had done to his reputation by criticising the calculus, pleaded that he had really been acting for Gallois, who had been rather cunning in using him in this way (*GM* **4**, p 158). Two years later, Rolle was in dispute with De la Hire concerning the solution of equations, which drew from Leibniz (*GM* **4**, pp 168–70) the remark that he seemed to have been born for making difficulties.

Following the death of Jakob Bernoulli on 16 August 1705, Johann left Groningen to succeed his brother in Basel, and on 14 November, Fontenelle

[4] 1 pistole = 5 taler.

read the customary éloge in the Paris Academy of Sciences. Reporting to Leibniz (*GM* **4**, p 148), Varignon mentioned that the éloge contained praise of the calculus. Evidently he had not listened attentively, for he failed to notice that Fontenelle had attributed the invention of the calculus to the Bernoulli brothers. In February 1706, this éloge was published in the *Nouvelles de la république des lettres*. Three months later Johann Bernoulli himself wrote to Leibniz expressing his disapproval (*GM* **3**, 2, p 792). Leibniz sent a refutation which was published in the journal in November (*GM* **5**, pp 389–92). In this he claimed to have invented the calculus in 1674. On his return to Germany he had neither the leisure nor the inclination to prepare it for publication, for he always preferred to press on with new ideas rather than to work on what was already in his power. When some of his old friends, and particularly Mencke, started the *Acta Eruditorum*, he had been pleased to contribute some illustrations of his geometrical meditations to that journal. At last, he contributed the elements of the calculus, then gave new illustrations, applying it to the motions of the planets, where he employed second-order infinitesimals. The Bernoullis had always attributed the invention of the calculus to him, though they had contributed applications. No-one, he added, had done more than the Bernoullis and L'Hospital to promote the calculus, and L'Hospital's book had been published with his approval.

When the same attribution of the calculus to the Bernoullis appeared in the *Mémoires de Trevoux*, Leibniz demanded a similar retraction (*GP* **6**, pp 595–7), this time indicating that Fontenelle considered his remarks to have been misinterpreted.

While Leibniz had easily countered the misunderstanding created by Fontenelle's éloge of Jakob Bernoulli, a more serious attack on his claim to the invention of the calculus was soon to emerge. For in a letter to Halley, published in the volume of the *Philosophical Transactions* for 1708 (which appeared in 1710), John Keill accused Leibniz of having stolen the calculus from Newton. According to Keill, Newton's arithmetic of fluxions was afterwards published by Leibniz in the *Acta Eruditorum* under a new name and symbolism (Keill 1708, p 185).

Leibniz continued to correspond with Johann Bernoulli on all the current mathematical literature. In January 1705 (*GM* **3**, 2, pp 760–1) he mentioned a report he had received from London, that Flamsteed was about to publish his lunar observations, covering a period of thirty years. This, he thought, would be of great value in arriving at a true theory of the moon. Among other subjects of discussion were Gregory's criticisms of Leibniz's planetary theory and Varignon's ideas concerning the calculation of centrifugal force (*GM* **3**, 2, pp 761, 770). At the end of 1705, Bernoulli irritated Leibniz by some tactless remarks in a paper which he published in the *Acta Eruditorum* (*GM* **3**, 1, pp 127–9) concerning the solution of a problem he had been given in 1703 by a mathematician in Groningen. The problem required the transformation of a given geometrical curve into one having the same length as the given curve. Although he mentioned that Leibniz also had a solution—this had been sent to him on 3 January 1704—he claimed that it was not general. Besides asking Bernoulli for an explanation, Leibniz composed a reply for the *Acta Eruditorum* (*GM* **3**, 2, pp 779–81), but did not send it for publication. Instead, he sent an original paper on the construction of curves through motion (the method used by Bernoulli in his solution), which was marred by errors probably caused by haste and his annoyance with Bernoulli. This was published in January 1706 (*GM* **7**, pp 339–44).

The solution which Leibniz communicated to Bernoulli was as follows (*GM* **3**, 2, pp 734–6). Let BB′ (figure 9.2) be the given curve and take any curve FF′ as a mirror, so that the rays such as BF (tangent to BB′) are reflected by the mirror into rays such as FL, where BF + FL is the same for all rays. The required curve is LL′, and since the mirror can be varied at will, an infinity of different curves will result. The solution seems quite general. However, Leibniz goes on to relate that in 1692 he had shown in the *Acta Eruditorum* (*GM* **5**, pp 279–85) how the curve LL′ (or envelope of the lines FL) could be found by means of the differential calculus. He pointed out that the line FF′ is the envelope of all the ellipses whose foci are corresponding points B and L. In January 1689, also in the *Acta Eruditorum* (*GM* **7**, pp 329–31), he had shown how these ideas provided a general method for solving problems in catoptrics and dioptrics. This, he claimed, marked the beginning of the theory of caustics that had been so well developed by L'Hospital.

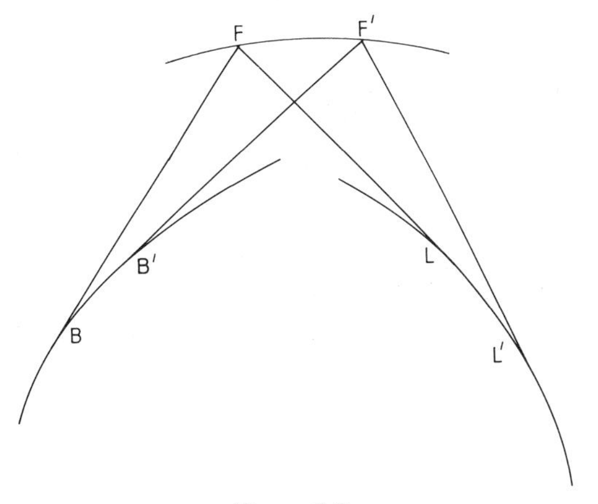

Figure 9.2

Bernoulli sought to justify his criticism on the grounds (*GM* **3**, 2, pp 785–7) that the mirror could be a circle, which Leibniz had excluded, and that the mirror-curve had to be a certain minimum distance from the given curve. In his defence, Leibniz claimed that he had illustrated his method by the special case in which the mirror-curve was an ellipse. At the beginning of 1707, in response to Leibniz's challenge, Bernoulli gave a construction of the transformed curve by his method, whereby he showed that there were circles between which all the boundaries of the transformed curve lay (*GM* **3**, 2, pp 803–10).

Jakob Hermann, one of the ablest students of Jakob Bernoulli, came to the attention of Leibniz as a result of his response to Nieuwentijt's criticisms concerning the principles of the differential calculus in 1700, which, on Leibniz's recommendation, earned him membership of the Berlin Society of Sciences. Leibniz corresponded with Hermann from 1704 onwards and in 1707 recommended him for the vacant professorship of mathematics in Padua. In Hermann Leibniz believed he had found a talented young mathematician who could be encouraged to develop further ideas which lack of time had prevented him from pursuing himself. In his first letter, written in reply to a communication of Hermann on the radius of curvature, Leibniz brought to his attention two such ideas; namely the concept of a differential and the binary system (*GM* **4**, pp 263–6). He then introduced Hermann to the method of using symbolic or fictitious numbers to represent general coefficients of equations, which he had developed in 1678. Thus, in his letter of 10 March 1705 (*GM* **4**, pp 268–70) he wrote a pair of simultaneous equations in the form

$$0 = 100 + 110x + 101y + 111xy + 120xx + 102yy$$

$$0 = 200 + 210x + 201y + 211xy + 220xx + 202yy$$

where the first digit refers to the equation and the other two indicate the term to which the coefficient belongs. For example, 102 refers to the term containing x^0y^2 in the first equation. From the expressions obtained by elimination using this symbolism had emerged the first ideas of the theory of determinants.

Another topic to which Leibniz encouraged Hermann to give his attention was the theory of polynomial equations. First, he sent his own attempt to prove 'Harriot's' theorem (Descartes' rule of signs) (*GM* **4**, pp 315–17). Then, following the appearance of Newton's *Arithmetica universalis* in 1707, he sent a copy of the method used by Newton for finding the divisors of an equation and asked Hermann to apply this method to equations of degree higher than the second (*GM* **4**, pp 324–5). At the same time he indicated a method of deriving from a given equation a new equation with a root equal to the difference of two roots of the first (*GM* **4**, pp 327–8). Finally, in a letter of 6 September 1708 (*GM* **4**, pp 335–9) he explained his own general method of divisors. After this, Leibniz seems to have abandoned any further attempt to incline Hermann towards the study of algebraic problems, for the subsequent correspondence is almost entirely concerned with dynamics.

One of the principal topics of discussion in 1709 was the calculation of centrifugal force. Hermann (*GM* **4**, pp 342–4) opened the debate by stating his objections to Antoine Parent's measure of centrifugal force. Parent's views are very curious—Varignon had already remarked to Leibniz that his book was of such terrible obscurity that he didn't have the patience to read it—for he quarrelled with Huygens while accepting his formula. Leibniz (*GM* **4**, pp 344–5) declared his agreement with Huygens and Parent, at the same time making the curious claim that the correction of his own calculation he had published in the *Acta Eruditorum* in 1706 was not so much the correction of an error as an improvement in expression. Also he rejected Hermann's idea that centrifugal force was essentially connected with circular motion. A body moving in any curve, he pointed out, had a tendency to move along the tangent; the tendency to recede from the curve to the tangent constituted its centrifugal force. Failure to recognise that, according to the theory of Huygens, the measure of centrifugal force was equivalent to a uniformly accelerated motion along DG (figure 9.1), and not a uniform motion, continued to be a source of confusion for Hermann (*GM* **4**, pp 346–55).

At the end of 1710, Hermann and Johann Bernoulli reached a landmark in the application of the differential calculus with their independent solutions of the inverse problem of central forces; that is, the demonstration of the proposition that a body moving under the action of a force directed towards a fixed point and varying inversely as the square of the distance from the point describes a conic section. Both solutions were presented to an assembly of the Paris Academy of Sciences on 13 December (Aiton 1964, pp 93–7).

Among other correspondents of Leibniz were Adolph Theobaldus Overbeck, a teacher at the school in Wolfenbüttel, with whom he discussed algebra, differential calculus and the binary system (Bodemann 1895, pp 212–13), and the Ansbach Counsellor Conrad Henfling, who addressed a letter to him on 25 March 1705 at the suggestion of Princess Caroline (Haase 1982, pp 44–7). The correspondence concerned mainly a writing of Henfling on the theory of music (Haase 1982, pp 59–81), which Leibniz later included in the first volume of the *Miscellanea* of the Berlin Society of Sciences. Another mathematician who contacted Leibniz about this time was Christian Wolff, who achieved fame as his faithful disciple. Educated in Jena under Erhard Weigel, Wolff first taught mathematics in Leipzig, where he composed a dissertation on the calculus. Following the advice of Otto Mencke, he dedicated this essay to Leibniz, to whom he sent a copy (*W*, pp 14–15). Recognising his ability, Leibniz immediately recommended him for a vacancy as Professor of Mathematics in Giessen (*W*, p 15). Instead of attempting original work, Wolff set himself the task of systematising the work of others, both in mathematics and philosophy, by assembling and ordering the material scattered in books and journals. When he explained his intentions to Leibniz in a letter of 4 April (*W*, pp 21–4), he received in return a summary of the theory of pre-established harmony, which was new to him (*W*, pp 24–8). At the end of 1706, again on the recommendation of Leibniz, he became Professor of Mathematics in Halle. Before taking up this appointment, he met Leibniz in Berlin (*D* **5**, p 160). On moving to Halle, he found the mathematics to be in a poor state and philosophy to be dominated by Christian Thomasius, whose teachings were not to his taste (*W*, p 10).[5] Thomasius, however, had a high opinion of Leibniz, to whom he referred as a living library (Bodemann 1895, p 336). Following the example of his master, whom he consulted concerning all difficulties, Wolff worked in several fields, including philosophy, mathematics, physics and natural history. About a year after Johann Burchard Mencke succeeded as editor of the *Acta Eruditorum* on the death of his father in 1707, Wolff took over the task of reviewing books on mathematics, though he seems to have submitted all his drafts to Leibniz for his approval. One of his first contributions, a review of Parent's book, *Recherches de mathématique et de physique*, called for a defence of the pre-established harmony and he asked Leibniz to indicate any changes or corrections he wished to make before publication. In a postscript to his letter, Wolff (*W*, pp 100–2) expressed his regret concerning the death on 11 October 1708 of Leibniz's old friend Tschirnhaus.

The Berlin Society of Sciences

Leibniz spent about six months in Berlin during the winter of 1706 and the spring of 1707 working for the Society of Sciences. Following a sitting under his presidency on 27 December 1706, he addressed a meeting of the mathematical members (*K* **10**, pp 405–7). Expressing the hope that the King would give orders for the building of the promised Observatory and pavilion, he suggested that they could help by encouraging the study of astronomy as much as the existing resources would allow and also by seeking applications of mathematics and mechanics that could be used to augment the funds of the Society. Up to that time, he reminded them, the only income had been the profit derived from the sale of calendars. He also asked them to give their minds to the plan to publish at least each year a volume of *Miscellanea* of the papers and reviews.

On 10 January 1707, Leibniz (*K* **10**, pp 407–9) set out detailed proposals for the promised patent concerning the planting of mulberry trees and silk-culture. Trees were to be planted in the Royal gardens, promenades, roads, ramparts and all suitable places, not only single trees but whole avenues. They would not only be useful for the silk-culture but would also serve as shade and ornament, like the limes. On 28 January, he set an example to the other mathematical members by presenting to the King a memorandum on

[5]On Leibniz and Christian Thomasius, see Utermöhlen (1979) and Heinekamp (1979).

the draining of swamps with the help of modern methods of surveying, part of the profits to be given to the Society. Before leaving Berlin in May, he presented to the King a detailed account of his work for the Society since 1700, and the King agreed to the purchase of a house opposite the Observatory for the Society (Bodemann 1889, p 223). However, the Exchequer refused to finance the purchase of the house until Leibniz, writing from Hanover, complained to the King in February 1708. The King responded by giving orders for three yearly payments of 700 taler, enabling the Society to secure a loan of 2100 taler for the purchase of the premises.

Also in 1708 the King referred to the Society for an opinion a proposal he had received from a Berlin resident, Caspar Rödecke, for a universal script. The report was composed on behalf of the Society by Leibniz, who evidently had conversations with Rödecke on the subject, and was presented to the Privy Counsellor von Ilgen in July 1709 (*GP* **7**, pp 33–7). Following a masterful analysis of the difficulties inherent in the perfection and carrying out of the project, Leibniz explained that Rödecke, like Wilkins and others who had tried before, could not hope to succeed single-handed. Unless he had the good fortune to find support in high places and the approbation of distinguished people for his project, Leibniz advised that he should apply himself to a more modest task that would nevertheless be a valuable contribution in the field; namely the translation of Wilkins' book into German or Latin.

At the beginning of 1709, when he stayed in Berlin on his return from his secret mission in Vienna, Leibniz once again held a conference on the publication of the *Miscellanea*, the first volume of which at last appeared in 1710 (Ravier 1937, pp 143–4). It contained a large number of contributions by Leibniz himself, including papers on the linguistic evidence for the origin of nations, the discovery of phosphorus and the Aurora Borealis, besides papers on mathematics and mechanics and a description of his calculating machine (figure 9.3).

Essais de Théodicée

Leibniz composed the *Essais de Théodicée* as a kind of literary memorial to his friend Sophie Charlotte, the late Queen in Prussia. While the work was in the press, he described its origin in a letter to Thomas Burnet (*GP* **3**, p 321). The greatest part, he explained, had been written piecemeal during the time of his conversations on philosophy with the Queen in the garden at Charlottenburg. On these occasions, he had attempted to show that the arguments of Pierre Bayle, whose writings were well known at the Court, were not so powerful as some people unsympathetic to religion had tried to make out. The Queen had often encouraged him to set out his answers to Bayle in writing so that they could be studied with more attention. After her

Figure 9.3 Leibniz's calculating machine. From C G Ludovici, *Ausführlicher Entwurf einer vollständigen Historie der Leibnitzischen Philosophie* (Leipzig 1737). Courtesy of the British Library.

death, he had collected these writings together, on the advice of friends, and arranged them into this large work on Divine Providence, the freedom of man and the origin of evil. The book was first published anonymously in Amsterdam in 1710—Leibniz was reluctant to appear as the author of a theological work (though his authorship was only thinly disguised)—but his name was added in the second edition of 1712. It is rather amusing that some readers of the first edition took the word *Theodicy* to be the name of the author and others as a pseudonym for Leibniz. It really signifies the justification of God's work, or as we might say, the trial of God for the creation of evil.

Written in French for the educated public, the work includes extensive discussions of current theological controversies and detailed criticism of the views of Pierre Bayle, whose great influence and the general interest of his theme made him a legitimate target even after his death in 1706. It contains also expositions and clarifications of the fundamental principles of Leibniz's philosophy. There are three appendices, in which Leibniz gives a summary of his theme and critical reviews of the works of Hobbes and William King, Bishop of Derry, on the origin of evil.

In the preface (*GP* **6**, pp 25–48), he refers to the two famous labyrinths wherein our reason is often led astray. While one, concerned with the continuum, exercises only the minds of philosophers, the other, which will be the subject of his work, is of interest to all. This concerns the question of freedom and necessity, especially in relation to the origin of evil, the liberty of man and the justice of God. Having described the origin of the work in conversations with an accomplished Princess (Sophie) and incomparable Queen (Sophie Charlotte), the anonymous author mentions that he has published a new system—the pre-established harmony—which provides a better explanation of the union of mind and body than the doctrine of Occasionalism favoured by Bayle and a more satisfying account of the formation of animals than the plastic natures supposed by Cudworth. Indeed the system of pre-established harmony, he claims, is confirmed by the observational evidence of preformation, which shows that new organisms are only a mechanical consequence of a preceding organic constitution. Concerning the connection between mind and body, while denying any influence of one on the other—that is to say, an influence whereby the one disturbs the laws of the other—he admits a real union (as he has already explained to Tournemine) but this, he remarks, is something metaphysical which does not change anything in the phenomena. According to his system, Leibniz adds, the laws of nature are neither necessary (as Spinoza believed) nor arbitrary (as held by Bayle) but depend on the principle of the best.

The work itself begins with a 'Preliminary discourse on the agreement of faith and reason' (*GP* **6**, pp 49–101), in which Leibniz sets out the history of the problem from the days of the Early Church and maintains, against Bayle, that the truths of divine revelation and those attained by the natural light of reason cannot be in contradiction. This, he points out, has an important bearing on his principal theme developed in the main work.

Leibniz classifies truths of reason into two kinds: eternal truths, which are absolutely necessary, so that the opposite implies contradiction—these are truths of logic, metaphysics and mathematics—and positive truths, which are laws that God has chosen to give to nature. We learn the latter by experience (*a posteriori*) or *a priori* by consideration of the fitness or harmony for which they have been chosen (elsewhere he describes this as the sufficient reason). Physical necessity is thus founded on moral necessity (*GP* **6**, p 50). The goodness and justice of God, as well as his wisdom, only differ from ours in being infinitely more perfect. So the necessary truths and their demonstrable consequences (being perfectly understood by God) cannot be contrary to revelation (*GP* **6**, p 51), from which Leibniz concludes that necessary and eternal truths should never be abandoned to sustain a religious dogma (*GP* **6**, p 64). Indeed, if the objections of reason against an article of faith are insurmountable, then it must be concluded that the supposed article of faith is in fact false; such, in his view, is the doctrine of the damnation of unbaptised infants. However, the mysteries of revealed religion, such as the Trinity, may be beyond reason while not being contrary to reason. For they are within the province of God's choice and hence possible, although the reasons or explanations are hidden from us.

To Leibniz, Bayle's demand that God should be justified in a manner similar to the plea of a man accused before a judge seemed to be one of the things that contributed most to his view that faith was incompatible with reason. For example, Bayle held that the goodness of God could not be maintained in view of his permission of sin, since the probability would be against a man who was found in a case which appeared to us similar to this permission (*GP* **6**, p 69). God foresaw that Eve would be tempted by the serpent, if he put her in the garden with the beast, and yet he put her there. In the case of a tutor placing his pupil in a dangerous predicament, the judge would not be impressed by the argument that he only permitted the evil without causing it or willing it. Leibniz responds by claiming that, although the universal moral law is the same for God as for man, the circumstances are entirely different in the case in question (*GP* **6**, p 70). For God had care of the whole universe, of which all the parts are bound together, and consequently had to take an infinity of factors into account, as a result of which he judged that it was not appropriate to prevent certain evils. The identification of general reasons for his choice (rather than the explanation of particular cases) would therefore suffice to eliminate the presumption of guilt that must be countered by reason if a contradiction of faith and reason is to be avoided.

The justification of God in permitting evil is one of the principal themes of the work. Leibniz explains that the existence of God is necessary, as the first cause of things, for all those we see are finite and contingent, having nothing in them that renders their existence necessary (*GP* **6**, p 106). God's understanding is the source of essences and his will is the origin of existences. Since the actual world is one of an infinity of possible worlds, God must have chosen this one. Moreover, in view of God's perfect wisdom and goodness, he could not fail to choose the best.

Evil could not have been avoided in creating this best of possible worlds, for in consequence of the binding together of all the parts with each other, the least change would alter the whole world (*GP* **6**, p 108). Leibniz goes on to explain that, although God had to permit evil in consequence of his decision to create the best of possible worlds, nevertheless he did not choose to create evil. First, Leibniz distinguishes metaphysical evil, which is simply imperfection, from physical and moral evil, consisting of suffering and sin respectively (*GP* **6**, p 115). Concerning metaphysical evil, he explains that God had no choice, for his creatures must have a natural imperfection, otherwise they would themselves be gods. According to Leibniz's interpretation, when we say that the creature depends on God, who conserves it by a continual creation, we mean that God is the source of the perfection in its nature and action (*GP* **6**, pp 120–1), but its imperfections, both in essence and behaviour, are a result of its necessary limitation. This limitation, he believes, is a kind of privation or passivity, which moderates by its receptivity the effects of God's grace. He finds a perfect image and illustration (*GP* **6**, pp 119–20) of the original limitation of creatures in the natural inertia of Kepler, which makes for a privation of speed, not by diminishing speed by itself when this has been acquired (for this would be to act), but by moderating by its receptivity the effect of the impressions received.

Physical and moral evil are not necessary but by virtue of the eternal truths they are possible. Although they exist, Leibniz argues that God did not choose them. He distinguishes between an antecedent volition, which relates to individual things considered in themselves apart from others, and a consequent volition, which results from the conflict and concourse of all antecedent volitions (*GP* **6**, pp 115–17). God has an antecedent volition to save all men, exclude sin and prevent damnation, and this would have effect if there were not some stronger cause that prevents it. Only the consequent volition, however, has effect. Physical and moral evil must therefore be permitted as a result of God's moral obligation to create a universe containing the greatest good. It follows that God wills *antecedently* the good and *consequently* the best. He does not at all will moral evil and he does not in an absolute sense will physical evil or suffering; there is no absolute predestination for damnation, and physical suffering often has a good outcome, leading to the greater perfection of those who suffer.

Having acquitted God of responsibility for the existence of evil, Leibniz turns to the other major problem, namely the freedom of the will, which in his view requires an exemption from both external constraint and necessity (*GP* **6**, p 122). Yet he believes that the individual is inclined by his preceding state, so that there is always a reason which carries the will to its choice. For the preservation of liberty, Leibniz maintains, it suffices that this reason inclines without necessitating (*GP* **6**, pp 127–8). There is a difficulty, however, in the apparent contradiction between this account of freedom and the view, generally accepted by the contemporary philosophers, that future contingents are determined (*GP* **6**, pp 123–4), since it would seem that what is foreseen by God cannot fail to exist. For example, it was true a century ago, Leibniz remarks, that he would write that day, as it will be true a century hence that he has written. While he agrees that what God has foreseen will happen, he describes this as a hypothetical necessity, whereas only an absolute necessity (whose opposite implies a contradiction) could demonstrate that his action in writing was neither contingent nor the effect of a free choice.

All is therefore certain and determined in man, as elsewhere, so that the human mind is a kind of spiritual automaton (*GP* **6**, p 131), although contingent actions in general and free actions in particular are not bound by an absolute necessity, which would be truly incompatible with contingency. If it is certain that we shall perform particular voluntary actions, it is no less certain that we shall choose to perform them. God foresees both the action and the choice we shall make. Foreseeing all that would happen freely, and even the prayers we should address to him, God has arranged other things accordingly (*GP* **6**, p 132). Thus neither the infallible foreknowledge of God nor the predetermination of causes can destroy contingency and liberty.

In answer to the question how man should act, knowing that all is determined, Leibniz recommends that, as we neither know how the future is determined nor what has been foreseen or resolved, we should endeavour to carry out the presumptive will of God, following the reason he has given us and the commandments he has prescribed. The rest we should be content to leave to God, knowing that he will not fail to bring about the best, both in general and in particular for those who have a true confidence in him (*GP* **6**, p 134).

Although voluntary action has its causes, Leibniz finds it good to make clear that this dependence does not prevent a wonderful spontaneity in us, which in a certain sense renders the resolutions of the mind independent of the physical influence of all other creatures (*GP* **6**, p 135). This spontaneity is a consequence of the pre-established harmony, of which he considers it appropriate to give some explanation. According to this system the mind, like all other simple substances distributed through nature, has within itself the principle of all its actions and passions (*GP* **6**, pp 137–8), though only minds or rational souls have liberty. In ordinary speech, however, we can say that the mind depends on the body in the same way that Copernicans refer to sunrise. The pre-established harmony is the cause of the perfect correlation between the actions of the mind and those of the body. The mind was originally created by God in such a fashion that it must develop of itself and represent in order all that happens in the body, and the body also in such a fashion that it must produce of itself that which the mind ordains. Each follows its own rules, the mind those of morals and final causes, the body those of mechanism and efficient causes, yet in perfect harmony, so that the body acts at precisely the time that the mind wills.

Clarifying further the relationship between mind and body, Leibniz (*GP* **6**, pp 138–9) explains that, in so far as the mind has a degree of perfection and distinct thoughts, God has accommodated the body to the mind, but to the extent that the mind is imperfect, he has accommodated the mind to the body, so that the mind allows itself to be inclined by the passions which are born of the corporeal representations. This has the same effect as if the one depended directly on the other by means of a physical influence.

Leibniz at home

On the afternoon of 10 January 1710, Leibniz received a visit from two travellers, Zacharias Konrad von Uffenbach and his brother, whose record of the meeting provides a rare glimpse of Leibniz in the informality of his home. Well over sixty years old, they reported, Leibniz presented a wonderful appearance, dressed in fur stockings and fur-lined night-gown, large socks of grey felt in place of slippers and a very long wig. He wore this comfortable attire on account of his gout (Eckhart 1779, p 225). A very kind man, he received them with the greatest politeness and entertained them with conversation on a medley of political and scholarly topics. Although he declined to show them the Electoral Library because it was still in such disorder, and his own library for the same reason, he brought out a few of the most valuable codices for them to see (Uffenbach 1753, pp 409–11). Two days later Leibniz paid them a return visit at the 'Red Lion' in the Neustadt.

A new commission from Anton Ulrich and an unexpected disappointment

When he visited Wolfenbüttel at the end of March 1710, Leibniz just missed Duke Anton Ulrich, who had departed for Bamberg, where, a fortnight later, he was officially received into the Catholic Church. The Duke had wished to meet Leibniz before his departure and had sent for him but received no reply. He promised that, when he returned, they would freely express their views on the conversion, though without entering into a dispute on the matter (Bodemann 1888, pp 189–90). King Friedrich I in Prussia expressed to Sophie his concern over the Duke's defection from the evangelical religion. There had been so many defections recently, he remarked, it seemed as if they were in the last days and the devil was at large (Bodemann 1888, pp 95–6). The Duke assured his subjects, however, that his personal conversion would entail no ecclesiastical or political changes in the country. The building of a Cathedral Church in Brunswick was in fact the only public consequence of the Duke's conversion.

In June, Leibniz was again in Wolfenbüttel, this time to discuss the purchase for the Ducal Library of the manuscript collection of the Danish Counsellor of State and historian Marquard Gude. For a long time Sebastian Kortholt, Professor of Rhetoric in Kiel, had corresponded with him about this valuable collection. To arrange the purchase for 2490 taler, he travelled to Holstein, through Schirensee to Kiel and Gottorp and finally to Hamburg. From there he reported to the Duke that the manuscripts were in two large boxes which could be transported as far as Lüneburg by water, and that, in addition to the manuscripts, he had secured thirteen portraits of scholars and seventeen Roman inscriptions.

Leibniz could feel pleased with his own literary output in 1710. Besides his major work, the *Essais de Théodicée*, the long awaited first volume of the memoirs of the Berlin Society of Sciences, *Miscellanea Berolinensia*, had also appeared. Another publication, which should have pleased the Elector, was the second volume of the Guelf history, *Scriptorum Brunsvicensia illustrantium*.[6] The Hanoverian contribution to the printing costs of the first volume had at last been paid, though he had still to press for the payment of the promised contribution from Wolfenbüttel. No doubt encouraged by

[6]This is a slightly altered form of the title of the first volume, *Scriptores rerum Brunsvicensium illustrationi inservientes* (*Writings serving to illustrate Brunswick matters*).

these achievements, Leibniz wrote a memorandum for the Empress Amalia, while on a visit to Brunswick, setting out his plans relating to Vienna, especially his long-standing ambition to be nominated as an Imperial Privy Counsellor. Then towards the end of the year came news from Berlin as unexpected as it was hurtful to dampen his spirits. For he learnt from a friend that, several months earlier, an assembly of the members of the Society of Sciences resident in Berlin had elected Baron von Printzen as Director.[7] Not knowing whether he was still President and being unable to understand why he should have been treated so badly, he appealed to the Crown Princess Sophie Dorothea to use her influence in the cause of justice, especially as he had spent the past year working on the *Miscellanea*, which had been applauded abroad if not in Berlin (*K* **10**, pp 418–21). Those who should have consulted him and kept him informed of the affairs of the Society, above all the Secretary Jablonski, had said nothing. Perhaps they were too ashamed of their conduct to write to him, he remarked to the Princess. Baron von Printzen, he suggested, had been used by a cabal within the Society.

On the same day that he wrote to the Princess, Leibniz also set out his complaints in a letter to von Printzen (*K* **10**, pp 421–2). In reply, the Baron (*K* **10**, pp 422–3) stated that the King had chosen him to be Director and he invited Leibniz to advise him, though he did not answer Leibniz's direct question as to whether he was still President. Leibniz learnt from another source that the King had confirmed him in this office (*K* **10**, pp 423–4).[8] Writing again to von Printzen on 30 December, Leibniz asked him to order the Secretary Jablonski to consult him in advance concerning all decisions and to keep him informed of everything that took place in the Society (*K* **10**, pp 424–6). Leibniz also sent to von Printzen a memorandum complaining about the lack of cooperation he had received from members of the Society. Only one volume of the *Miscellanea* had thus far been produced and to this he had contributed many of the articles himself.

[7]In fact, von Printzen had acquired the title of Honorary President (Harnack 1900 **2**, p 192).

[8]In the directory for 1712, von Printzen is described as President and Director (Praesident und Director) and Leibniz as Ordinary President (Praeses ordinarius) (Harnack 1900 **1**, 1, p 175).

10
Hanover and Vienna
(1711–1716)

On 19 January 1711 the official inauguration of the Society of Sciences took place in Berlin in the absence of Leibniz. It appears, however, that both the Crown Princess Sophie Dorothea and the Director of the Society, von Printzen, invited him to Berlin, and in a letter to the Hanoverian Prime Minister von Bernstorff, written from Wolfenbüttel on 17 February, Leibniz declared his intention to visit Berlin before returning to Hanover. It was more than two years since he had been there and in that time he had worked diligently on the history—indeed he would be staying for a while in Helmstedt on the way, where his assistants Eckhart and Hackmann were professors, in order to complete the combined index of the three volumes of the *Scriptores*—and he needed a little relaxation that would give him renewed strength to continue the writing of the history (Doebner 1881, p 255). To von Bernstorff he expressed his confidence that the Elector would have the kindness to grant him permission for this journey.

Leibniz arrived in Berlin on 25 February and on the same day signed the document admitting Christian Wolff as a member of the Society. Over the next three months he presided at several meetings, dealing among other things with the management of the *Miscellanea*, the promotion of the silk-culture to provide financial support and the admission of new members. On behalf of the Society he also petitioned the King to order the provision of new sources of revenue (Harnack 1900 **2**, p 222).

Having arrived in Berlin at a time of increasing political tension between the governments of Berlin and Hanover on account of the military intervention of Hanover in the affairs of the Bishopric of Hildesheim, Leibniz was suspected of being a spy. The Elector Georg Ludwig (Schnath 1978, p 598) justified his action in occupying the fortress town of Peine and compelling the town of Hildesheim to receive a Hanoverian garrison by the failure of his policy of sequestration of the revenues of the Bishopric in his Electorate, already in operation for several years, to prevent the oppression of the Lutheran population by the Catholic administration. At this time the town of Hildesheim had complained about the damage to its commercial interests caused by unauthorised brewing in the ecclesiastical establishments. Maintaining that he was only exercising the right of protection granted by the Treaty of Westphalia, the Elector indicated that, after removal of the abuses, his troops would be withdrawn. In Berlin, however, it was feared that he intended to seize the Bishopric, taking advantage of the fact that there had been an interregnum for ten years because the Archbishop of Cologne, who had the right of succession, had been unable to take possession as he was an enemy of the Empire by his alliance with France in the war of the Spanish succession (Schnath 1978, p 563). King Friedrich I in Prussia complained that the occupation of Peine and Hildesheim prevented him leaving his country without crossing through the Elector's territory. To this Sophie replied in a letter to Leibniz (*K* **9**, p 332) that the King had only to look at a map to see that he would always have to pass through Brunswick territory, whether Hildesheim was occupied or not. On 11 July, the Hildesheim Cathedral Chapter signed a treaty (declared invalid by Pope Clement XI in 1712) redressing the grievances of the Protestants, and on 16 November a settlement of the brewery dispute was concluded (Schnath 1978, p 601). Most of the Hanoverian troops were then withdrawn, leaving only a small protective force in Hildesheim.

During his stay in Berlin, Leibniz had a bad fall, which kept him in bed for a long time. This misfortune was regarded with some amusement by the Elector as a fit reward for preferring to be in Berlin rather than Hanover (*K* **9**, pp 324–5). Sophie was more sympathetic, writing to her granddaughter, the Crown Princess Sophie Dorothea on 7 March, that she was sorry poor Leibniz was still ill, adding that the Elector was not pleased that he travelled about so much, for he loved his conversation. To Leibniz himself (*K* **9**, pp 328–9) she expressed sympathy for the discomfort caused by his injury and remarked that his journey had been doubly unfortunate; while he was regarded as a spy in Berlin, his departure without permission had created a bad impression in Hanover, though no one there doubted his fidelity. Writing to Sophie on 21 March (*K* **9**, pp 326–8), Leibniz remarked that he had been visited by one of the Court physicians, ostensibly to offer advice on his injury, but whose real purpose, he suspected, was to make a report on whether he was a spy. Since he only saw scientists and did not speak of other things, it should be clear, Leibniz added, that he was not a spy and he hoped the King would soon be better informed. On the question of permission for the journey, he reminded Sophie that he had written to von Bernstorff and that was equivalent to writing to the Elector.

In a further attempt to counter criticism of his journey, Leibniz explained that he had been working on the history, having found several books in

Berlin that were neither in Hanover nor Wolfenbüttel besides receiving from Modena great packets of material on the Italian ancestors of the House of Brunswick, so that his work on the history should not be measured by the time he spent in Hanover. The historical material from Italy to which he referred came from the Ducal Librarian and Archivist in Modena, Ludovico Antonio Muratori, who wished to see his genealogy of the House of Este included in the Brunswick history that Leibniz was preparing, failing which he intended to publish it himself in Modena (Bodemann 1895, p 197). Leibniz was not satisfied with Muratori's work and on 27 January communicated to him the decision of the Hanoverian Court that the publication of a joint genealogy of the Houses of Brunswick and Este from Azo would not be appropriate. On 23 April he drafted for von Bernstorff a letter to the Duke of Modena, proposing that Muratori's genealogy should not be published until the research had been completed. A summary of Muratori's work, in the form of a letter to Leibniz, was however included in the third volume of the *Scriptorum Brunsvicensia illustrantium*, which appeared in 1711.

Following a visit to the Leipzig fair in May, where he hoped to find a new assistant for the historical work, Leibniz spent about a fortnight as the guest of Duke Moritz Wilhelm von Sachsen-Zeitz. This was the first of several visits to Zeitz in his later years. In the Duke's palace he admired the collection of books and made the acquaintance of the Court deacon Gottfried Teuber, who agreed to assist in the development of his second design for a calculating machine (Bodemann 1895, pp 330–3). Towards the end of the visit, Christian Thomasius was also a guest (Schmiedecke 1969). From Zeitz Leibniz travelled to Brunswick, where he stayed for a few days before returning to Hanover, where he arrived about the middle of June.

Writing to an English correspondent, Dr Hutton, Leibniz explained that, although relations between Berlin and Hanover were somewhat strained, he had worked for the reunion of the Protestants in the two countries, something that would be to the advantage of the Anglican Church (*K* **9**, pp 337–41). In Sophie's room, he added, he had spoken many times with the Elector on the dogmas of the Anglican Church, seeking to show that a man of the Confession of Augsburg could subscribe without difficulty of conscience to the thirty-nine articles.

Three months after his return to Hanover, Leibniz received from the Elector Georg Ludwig (on behalf of the Elector of Mainz) a commission to investigate the question of the authority of the Electors of Saxony and the Palatinate, the Vicegerents following the death of Emperor Joseph I on 17 April 1711, to continue the Imperial Diet in the interregnum (Schnath 1978, pp 434, 440). It was generally understood that the office of Emperor should remain with the Habsburgs and, before the end of April, King Friedrich I in Prussia had already decided in favour of Joseph's younger brother Karl, the

King of Spain. By secret diplomacy, Louis XIV sought to hinder the election of Karl. He hoped that the Protestant Electors could be persuaded to take advantage of Swedish and French support in order to make one of themselves Emperor. The plot was discovered when a message from Louis XIV to his ambassador in Poland was intercepted in The Hague and deciphered in Hanover. The Elector Georg Ludwig immediately informed Vienna and thus justified the confidence the Emperor's family had placed in him (Schnath 1978, pp 435–7). At the election in August, the Hanoverian delegates were the President of the Chamber, Baron von Goertz and the ambassador to the Imperial Diet, Christoph von Schrader. On returning from Spain, the new Emperor Karl VI was crowned in Frankfurt am Main on 22 December 1711.

Duke Anton Ulrich, who was in Frankfurt for the coronation, used his personal influence with the new Emperor, whose wife, the Duke's granddaughter Elizabeth Christine, was still in Barcelona, to extract from him a promise that his friend Leibniz would be granted the title of Imperial Privy Counsellor that he desired. Having missed Leibniz in Hanover, the Duke wrote to him from Brunswick on 3 February 1712 to tell him that he had succeeded in getting the promise from the Emperor (Bodemann 1888, p 212). When the Court Chancellor, Count Philipp Ludwig von Sinzendorf confirmed the promise, however, Leibniz was disappointed to learn that he was to be given an honorary title without any payment. In a letter for the Empress sent to Baron Jakob Wilhelm Imhof in Barcelona on 27 September 1712 (*K* **9**, pp 365–70) he pointed out that, in the course of his work on the history of Brunswick, he had already made considerable discoveries, especially in Italy, to establish certain rights of the Empire. Even if, as he desired, he were to be excused routine duties and allowed to continue the work on the history of Brunswick, he therefore believed that his services to the Empire would still be deserving of the emoluments of the office. Consequently, he requested the Empress to use her influence on his behalf and he enclosed a draft document for her to send to her husband in Vienna (*K* **9**, p 371). At the end of the letter to von Imhof he explained that he intended to visit Karlsbad for his health, and if he received a favourable reply, would then go on to Vienna.

The real purpose of Leibniz's visit to Karlsbad was to accept an invitation of the Czar, Peter the Great, to meet him there. Following his audience with the Czar in the previous year, arranged by Duke Anton Ulrich on the occasion of the wedding of his granddaughter Charlotte and the Czarevitch Alexei, Leibniz had been commissioned by the Czar with the promotion of mathematics and science in Russia, for which he was given an annual payment of 1000 taler and the title of Russian Privy Counsellor of Justice (Guerrier 1873, p 270). In Karlsbad, however, he took on a new role, for he had been commissioned by Duke Anton Ulrich to work in Karlsbad and Vienna for an alliance between Russia and Austria that would enable the new

Emperor to bring the war with France to a successful conclusion (Bodemann 1888, p 102).

A wedding in Torgau

Leibniz, it will be recalled, on the occasion of his secret visit to Vienna in 1708, had discussed with the Russian ambassador von Urbich the possibility of the marriage of the Czarevitch with a Brunswick-Wolfenbüttel princess. The Czar's choice fell on the Duke's granddaughter Charlotte, sister of the Empress Elizabeth Christine. The fifteen-year-old princess, who was brought up at the Court of Saxony and seldom saw her parents, had great fear of the proposed marriage; her forebodings were tragically borne out by subsequent events. According to Sophie, writing to Leibniz on 1 April 1711 (*K* **9**, pp 331–2), the princess and the Czarevitch had found 'un amour violent' for each other, following the visit of Alexei to Dresden in February when the betrothal took place. For Leibniz the marriage provided an opportunity to meet the Czar. The wedding ceremony took place on 25 October 1711 at the Court of Saxony in Torgau, all the expenses being paid by August, the Elector of Saxony (now restored to the throne of Poland), who was pleased by this means to be reconciled with the Czar, whom he had deserted in the Northern War.

Leibniz arrived in Torgau on the evening of 19 October, the same day as Duke Anton Ulrich, having travelled from Wolfenbüttel, and he stayed until the end of the month. To Sophie (*K* **9**, pp 349–50) he reported that Duke Anton Ulrich had wished him to go to Torgau and that the weather was very good. Sophie was both surprised and pleased by his visit to Torgau and remarked that she would like to have been there herself to meet the Czar (*K* **9**, p 350). Writing to Sophie from Leipzig on 31 October, Leibniz reported that, an hour and a half after the wedding ceremony, which was conducted in Russian according to the Greek Orthodox rites (though the Princess gave her consent in Latin), an earth tremor had been felt in Leipzig and some neighbouring places, though it had not been noticed in Torgau (Bodemann 1889, pp 258–9). During his visit, the Czar had spoken to him several times, always very courteously, and he had dined at the Czar's table. Two hours after the Czar's departure, he had also left the town, and having arrived in Leipzig, he assured Sophie, he would soon be in Hanover again. On the way, however, he spent ten days in Zeitz as the guest of his new friends the Duke and Duchess. In Zeitz he had the opportunity to confer with Teuber on the solution of technical problems relating to his calculating machine (Bodemann 1895, p 330). Then after a few days in Wolfenbüttel, he arrived again in Hanover on 28 November.

Sophie's youngest son, Prince Ernst August, made a disparaging remark about Leibniz, not for the first time, when he commented that the Czar no doubt held him to be the Duke of Wolfenbüttel's fool, for he looked very much like it. In fact the Czar had been receptive to Leibniz's suggestions concerning the promotion of science in Russia, and in particular had promised to have observations of the magnetic declination made throughout his vast Empire (*D* **5**, p 294). Suggesting also that the Czar should initiate communication with China for the purpose of learning the sciences and arts known in the east but not in Europe, he mentioned the relationship between the hexagrams of the *I ching* and his binary arithmetic, though here he claimed Bouvet's discovery as his own (*FC* **7**, pp 395–403). These ideas were followed up in correspondence, notably with the diplomat von Urbich and the Russian General James Bruce, to whom he sent detailed proposals for researches on languages and magnetic observations in the Russian Empire (Guerrier 1873, pp 239–49). He also sought to enlist the support in suitable form of the Society of Sciences in Berlin for the magnetic measurements in Russia (Guerrier 1873, pp 195–7). Reporting to the Russian Chancellor on 5 March 1712, von Urbich remarked that no one was more knowledgeable than Leibniz concerning history and antiquities, as also in mathematics, and that several years ago he had already recommended him to the Czar as one who could introduce science to Moscow (Guerrier 1873, p 209).

Leibniz spent most of September in Wolfenbüttel and at this time took on the painter Wilhelm Dinninger[1] as a secretary. In the same month he received the invitation to meet the Czar in Karlsbad that enabled him to begin work on Duke Anton Ulrich's commission to promote an alliance of the Czar and the Emperor (Guerrier 1873, p 252). On 9 November, Leibniz reported to Sophie (*K* **9**, pp 373–4) on his journey to Karlsbad and his meeting there with the Czar, who had commissioned him to prepare proposals for the reform of the law and administration of justice in the Russian Empire. He assured Sophie that his invitation to become a Russian Solon—Solon was the Athenian law-giver—would not take up much time, for in his opinion the shortest laws, as the ten commandments and the twelve tables of the Romans, were the best, and it was a subject he had cultivated from his youth. Although, at the Czar's request, he would follow him to Töplitz in the company of the Russian diplomat Count Nariskin, who had just returned from Vienna, he had no intention of travelling further than Berlin in the Czar's service. In fact he took leave of the Czar in Dresden on 24 November. The following day he wrote to Duke Anton Ulrich (Bodemann 1888, pp 221–2), who had been amused by his remark about becoming another Solon (presumably related to him by Sophie) and took the joke further by

[1]In 1730, the editor of the *Recueil de Littérature*, published in Amsterdam, alleged that Dinninger was Leibniz's natural son, quoting the astronomer Gottfried Kirch as having often remarked on the resemblance of Dinninger to Leibniz. Church records show that Dinninger was born on 15 December 1686 in Saarmund. It is difficult to see how Leibniz could have been in Saarmund in 1686 (Guhrauer 1846 **2**, p 365 and app, p 43).

hoping he would not become another Andrew. For according to legend, the Apostle Andrew came to preach the Gospel in Kiev and was crucified. To which Leibniz wittily replied that he would welcome the cross of Andrew if it was set with diamonds—Peter the Great had founded the order of St Andrew in 1698.

On 10 December, the Duchess of Orleans, having heard from her aunt of Leibniz's journey and meeting with the Czar, remarked that he was wise not to wish to go to Moscow, which must be a wild place, but she thought he would willingly follow the Czar to Berlin in order to visit the Crown Prince and Princess (*MK*, p 232). At this time he was already in Prague and arrived a few days later in Vienna, without having informed anyone in Hanover of his intention to go there.

Imperial Privy Counsellor in Vienna

From Prague, Leibniz wrote to Bishop von Buchhaim on 8 December 1712, advising him that he would soon arrive in Vienna, and sought through him an introduction to the Imperial Vice-Chancellor Friedrich Karl von Schönborn, nephew of the Elector of Mainz. Within a few days of his arrival he composed a memorandum for the Emperor Karl VI giving a detailed account of his activities and service to the Empire in support of his request for the promised appointment as Imperial Privy Counsellor. On 23 December, he wrote to the Hanoverian Prime Minister von Bernstorff explaining his journey to Vienna. Originally he had intended to return to Hanover when he took leave of the Czar in Dresden but had been forced to stay longer on account of a foot injury.[2] Then the opportunity had arisen to make a journey to Vienna with almost all expenses paid. On the same day, his amanuensis Johann Friedrich Hodann reported to him that the Elector was angry about his long absence, especially as he had given no detailed information of his plans before he left for Karlsbad, and von Bernstorff pressed for his speedy return.

Before the end of the year, Leibniz visited the dowager Empress Amalia and Sophie's son, Prince Maximilian Wilhelm, who was living in voluntary exile in Vienna. To Sophie he explained that, after his visit to the Czar in Karlsbad he had received news from Vienna that he would be welcome to pay a visit to the Emperor, especially as he was already half-way there (*K* **9**, p 378). In a further letter to von Bernstorff (*K* **9**, pp 376–7), he claimed that

[2]This was probably a recurrence of the condition that had delayed his journey to Berlin in the summer of 1704. Indeed, it seems that he was permanently afflicted to some extent by an open lesion on his right leg (Eckhart 1779, p 197). On 20 February 1712, the Duchess of Orleans remarked in a letter to her aunt that, while the wound remained open, Leibniz would live in good health, but if it closed, he would die. She based her opinion on her observation of similar cases.

his stay in Vienna was no obstacle to his work on the history and that, in any case, the work in Hanover had been held up at that time by the illness of his assistant Eckhart, which kept him in Helmstedt.

Following his first meeting with the Emperor about the middle of January 1713, Leibniz again wrote to von Bernstorff to tell him that the Emperor, recognising that the history of the House of Brunswick could not be written without regard to Imperial history, had shown great interest in his historical work and offered him the use of his jealously guarded library. With the remark to von Bernstorff that he had not yet been able to take advantage of this facility owing to the cold weather, he indicated at least by implication that he intended to stay in Vienna until the spring. Two months later, having presented to the Emperor a memorandum on the means of prosecuting the war (*FC* **4**, pp 239–47) and proposals for a cartographic survey of the country and the establishment of the Society of Sciences in Vienna (*FC* **7**, pp 328–31), he revealed to the Hanoverian Court the true purpose of his journey by asking von Bernstorff to procure for him the Elector's permission to accept the title of Imperial Privy Counsellor. To support his case in Hanover, he requested from the Emperor a letter in his hand that he could send to the Elector (*FC* **7**, p 331). This permission was granted, although the Elector had already secretly attempted to prevent the conferment of the title. Thus the Hanoverian ambassador in Vienna, Daniel Erasmi Huldeberg, in a conversation with the Empress Amalia on 25 February, reported that the Elector wished to warn the Emperor that Leibniz was not in the least a suitable person for the office; for his wish to do everything led him to find pleasure in interminable correspondence and travel, while he had either no talent or inclination to bring anything to a conclusion (Doebner 1881, p 217).

Despite this lack of support from Hanover, Leibniz received the document confirming his appointment as Imperial Privy Counsellor in April 1713. This was dated 2 January 1712,[3] the day on which the Emperor had given his formal promise, and a salary was paid after all from that date. Also, permanent accommodation was found for him in Sweckat, near Vienna, but because the bed there was uncomfortable, he was advised to bring one with him.

Duke Anton Ulrich, in a letter to Leibniz (Bodemann 1888, pp 228–9), expressed his pleasure that the Emperor recognised his merit so well, and if

[3]Eberhard (1795, p 174) in his biography states that Leibniz was made a Baron of the Empire in 1711 at the same time as he was promised the office of Imperial Privy Counsellor. Comparison of the draft prepared by Leibniz of the document naming him as Imperial Privy Counsellor with the wording of the decree shows, however, that the 'von' before his name had in every case been removed. Evidently the officials in Vienna, where all documents relating to the nobility were registered, refused Leibniz's claim to the title. Moreover, the archives contain no trace of a patent of nobility in preparation. It now seems beyond doubt that Leibniz was never created a Baron (Müller 1966, pp 9–10).

he might say so, better than his own countrymen; no prophet, he added, found honour in his own country. The Emperor in fact granted to Leibniz as a special privilege the same right of audience as one of his ministers (*K* **9**, pp 412–14).

In Vienna, Leibniz was received as a friend by Prince Eugene of Savoy (*FC* **4**, p lii), the General commanding the armies of the Empire and the greatest statesman of Germany. The Prince could speak better on theology than Leibniz on military matters, because he had studied theology in his youth whereas Leibniz had never been in a war. Among the topics of conversation was the accommodationist theory of the Jesuit missionaries in China, which retained ancient authentic Confucianism as a social element in Chinese Christianity. Prince Eugene argued against the position of the Jesuits, which Leibniz defended. On the question of the continuation of the war with France, Prince Eugene and Leibniz were in complete agreement. When the envoys of Louis XIV brought the peace proposals to Vienna, Prince Eugene took part in the deliberations, which resulted in the rejection by the Emperor of the Treaty of Utrecht, which his allies signed on 11 April 1713. It was almost certainly for Prince Eugene that Leibniz composed his essay, *Paix d'Utrecht inexcusable (FC* **4**, pp 1–140). In this he explains that Prince Eugene and Marlborough were on the point of entering France and restoring Europe to the equilibrium of the Treaty of Westphalia, when England and Holland, who had undertaken to support the Austrian claim to the Spanish succession, had chosen to detach themselves from the alliance and make their separate peace. Germany, he maintained, could not be secure without the return of Strasbourg and Alsace.

Soon after his arrival in Vienna, Leibniz received from Wolfenbüttel a letter (Bodemann 1888, pp 223–4), in which his old friend Duke Anton Ulrich confided disturbing news concerning his granddaughter Charlotte. On 1 December 1712, the seventeen-year-old princess, brought up in the splendid Court in Dresden, followed her husband to Torun in Poland, where she found herself isolated among hostile courtiers. From her husband she obtained no support, for besides having to leave her alone for long periods on account of military duties, he was given to a drunken and vulgar life. On 13 December she announced to her father that, as the distance was not yet too great and the Czarevitch was away, she would regard it as sinful to miss the opportunity to see him once more in this life. A few days later she returned to Wolfenbüttel without having sought permission from her husband or the Czar. Duke Anton Ulrich was worried about the Czar's reaction, and when he reported to Leibniz a month later that the princess was still at home, he remarked that when children tried to rule themselves, it seldom turned out well. Leibniz (Bodemann 1888, pp 224–5) sought to pacify his friend and excuse the princess, who evidently feared that she would never see her family again.

At the beginning of March 1713, the Czar visited Hanover to discuss personally with the Elector the military operations against Sweden. From Hanover he travelled to Wolfenbüttel, where he was entertained by Duke Anton Ulrich, who was pleased to inform Leibniz that all was well, the Czar having been quite friendly to his daughter-in-law and her relations (Bodemann 1888, pp 231–2). The princess travelled to Russia in April 1713, but her position became increasingly desperate. To her mother she wrote that, although she had always tried to hide the character of her husband, the mask had fallen without her wish. She was more unhappy than one could believe and words express, and she prayed that heaven would be moved to pity and release her from this world. After two and a half more years of misery, the young princess died at the end of October 1715, ten days after giving birth to a son.

In the early summer of 1713, Duke Anton Ulrich had a happier experience in a joyful meeting with his granddaughter Elizabeth Christine, the Empress. On 3 April, he asked Leibniz to enquire in Vienna (Bodemann 1888, p 232) by what route she would travel from Barcelona, for she had written to him to say she would like to speak to him on the way. The Duke, then 79 years old, met his granddaughter at Innsbruck before she entered Vienna on 2 June. Back in Wolfenbüttel, he wrote to Leibniz on 19 June, expressing his delight that the Empress, as Leibniz would himself discover, was esteemed by everyone and adored like a goddess (Bodemann 1888, pp 234–5). That day, another pleasure awaited him, for he expected the arrival in Salzdahlum of Sophie and Molanus.

Reporting to Leibniz on the visit of the Czar to Hanover in March 1713, Sophie mentioned that the Czar spoke well of him (*K* **9**, pp 389–90). She also referred to Friedrich I in Prussia, who died on 25 February 1713, as a very Christian king (*K* **9**, pp 388–9). This brought from Leibniz the comment that the title had been well applied, to judge by the exterior, but only God knew the interior. He added, however, that the action of the young king in rehabilitating Danckelmann (against whom no offence had ever been proved) was more Christian than that of his father in banishing him from the Court and confiscating his property; it remained to be seen whether this would be returned. All the world was in Vienna for the treaty,[4] he told her, but the Emperor would be the last to sign and with reason. Writing to Leibniz at the end of April (*K* **9**, pp 393–5), Sophie remarked that she had hoped to see him in Hanover sooner than his last letter. She expressed her thoughts on the loss of her children, especially the one who lived (Maximilian) but no longer wished to see her—Leibniz had promised to try to persuade Maximilian to visit his mother in Hanover. Concerning the treatment of Danckelmann, she thought he had judged well, but added, not without a touch of irony, that his Observatory in Berlin would not be as well cared for under the new king as his tax on the calendars. On 16 May she was

[4]The Treaty of Utrecht, which the Emperor did not sign.

able to report the birth of a small princess in Berlin to augment the number of her grandchildren. That day, she told him, they departed for Herrenhausen, where the air of May, the song of birds and the frogs would give her pleasure, but not the consideration that the trees last longer than us. Remarking on a recent crime, in which a member of the Elector's guard had led a gang in beating to death a poor pastor and his family for their money, she had been saddened that the soil of Brunswick could produce such monsters. Sophie's melancholy thoughts were echoed in Leibniz's reply (*K* **9**, pp 400–2). Having remarked that the Empress would arrive in Vienna by water in a few days, he added that the many bridges and the fast flowing stream demanded caution to avoid striking against pillars. He remembered that one day a Count von Sternberg, travelling by the same route, was drowned with all his family, but he had no doubt the Empress would be taken good care of. In his next letter (*K* **9**, pp 408–10), he expressed some romantic notions concerning the affairs of England. A triple alliance between Sophie, the Emperor and the Czar, he declared, would be most agreeable. She would furnish the successor, the Emperor the troops and the Czar the fleet. This would be an action worthy of the ancient heroes of the Round Table. For King Arthur, he reminded Sophie, was of Great Britain.

The visit of Peter the Great to Hanover and Wolfenbüttel gave new impetus to the plan of an alliance between Russia and the Empire. Writing to Leibniz on 3 April 1713, Duke Anton Ulrich (Bodemann 1888, pp 232–3) reported that the Czar was willing to provide 20 000 troops to accomplish something on the Rhine and had asked him to communicate this to Vienna through his 'Solon'. These words of the Duke encouraged Leibniz to make renewed efforts to prevent the disastrous peace with France and also provided him with a patriotic motive for prolonging his stay in Vienna. Over the next year, he presented to the Emperor several memoranda with proposals for the continuation of the war in order to force the restoration of Strasbourg and Alsace to the Empire. For example, in May 1713, he proposed measures that could be taken to win back the support of England and Holland, who had signed the Treaty of Utrecht out of despair of achieving more (*K* **9**, pp 403–7). France, he believed, would move heaven and earth to prevent the Protestant succession in England, and as England would disarm in peace time (having no standing army) the way would be open for an invasion by the Jacobite pretender. In his view, the possibility of the Emperor and the Czar moving against France would bring the support of the Elector and also the King in Prussia, who had an interest in the English succession as the closest heir after the line of the Electoral Prince (*FC* **4**, pp 189–206). By 15 July 1713, when he presented a further memorandum to the Emperor (*FC* **4**, pp 214–17), Leibniz's reading of the political situation suggested that an open alliance with the Czar would then be inadvisable, as this would be used by the French as propaganda to alarm the Ottoman Court. He therefore proposed that the Czar should be urged to act secretly through the two kings who were his allies, persuading them, in their own interests, to increase their contingents to the Emperor's forces. Just before the signing of the Treaty of Rastatt on 7 March 1714, which gave Spain to the Duke of Anjou and the Spanish Netherlands (together with the Italian territories, apart from Sicily) to Austria, but left Strasbourg and Alsace in the possession of France, Leibniz presented to the Emperor an analysis of the political, economic and military steps that should be taken on the assumption that the war would still continue for several years (*FC* **4**, pp 148–53). Of particular interest is his suggestion that the cost could not be met simply by raising taxes. The people must be given the means to pay either by the revival of industry or by providing opportunities for making profit out of the use of natural resources. A commission, he proposed, should be established to determine ways of making the country more prosperous.

At the beginning of 1714, a Scotsman, John Ker of Kersland, arrived in Vienna with a bizarre proposal for the Emperor concerning operations against France. He sought out Leibniz, who offered assistance to enable him to put his plan to the Court. The idea was that Ker, in commission of some English Protestants, should lead a privately financed pirate war (though with the Emperor's approval) against France and Spain in the West Indies. Such attacks would not only deprive France of her vast resource of riches but also divert her naval force so as to prevent her making any attempt on England in support of the pretender. In his memorandum to the Court, Ker (*FC* **4**, pp 292–308) warned that the Scots, being by nature ready to adventure upon any enterprise that affects their inclination without having regard to the consequences, were easily manipulated; the Highlanders, being men of no thought, would upon any encouragement appear in arms for the pretender. Although the Emperor was offered a share of the prizes and the supremacy over all conquests, he had reasons for not accepting Ker's proposals, as Leibniz reported to von Bernstorff on 30 June 1714 (Doebner 1881, pp 288–9). Leibniz himself favoured the plan and offered advice to the Emperor on the legal implications and on the best way of exploiting the plan for the establishment of German colonies in the West Indies. Evidently he did not regard the Treaty of Rastatt as an obstacle to this kind of harassment of France and her ally Spain.

When Ker was about to return to England in the summer, Leibniz informed von Bernstorff that he would pass through Hanover incognito and advised that he should enter into communication with him on account of his knowledge of affairs in Great Britain and especially in Scotland. At the same time he advised Sophie (*K* **9**, pp 438–41) of Ker's journey through Hanover and suggested that a correspondence should be maintained with him. As a suitable correspondent, he proposed Sophie's niece, the Raugräfin Louise.

Only weeks after the signing of the Treaty of Rastatt, Leibniz lost one of his most loyal and gracious friends by the death of Duke Anton Ulrich on 27 March 1714. In his last letter to Leibniz, written on 6 March (Bodemann

1888, pp 237–8), he expressed the feeling that he would soon be transferred to a better Salzdahlum, where he would find a complete knowledge of all the sciences that he knew only imperfectly on earth. According to Sophie (*K* **9**, pp 433–4), he was attended in his last hours by a Lutheran pastor, to whom he confessed that he did not believe in Catholic superstitions, and a Catholic priest from whom he received the last sacrament. Deeply moved himself by the Duke's death, which, even at the age of 81 years he felt was too soon, he told Sophie that the Empress, who had received a letter from her grandfather only a few days before he died when he was too weak to hold a pen, was almost inconsolable (Bodemann 1888, pp 109–10).

Soon after he was given permission to accept the title of Imperial Privy Counsellor, Leibniz received a commission from the Elector to put to the Emperor the claims of Brunswick-Lüneburg on the Duchy of Lauenburg on the northern frontier. Without delay, he composed a memorandum for the Emperor and discussed it with him at an audience on 14 May 1713 (*FC* **7**, pp 332–6). A decision was reached in June 1714, when Leibniz was able to inform von Bernstorff that the claim had succeeded. Meanwhile Leibniz was subjected to increasing pressure from the Elector to return to Hanover. In the summer of 1713, von Bernstorff admonished him to return. Then, in October, after repeated warnings by his secretary Hodann of the growing anger at the Court caused by his absence, Leibniz received news that the payment of his salary had been stopped. Yet he used every pretext to delay his return. First, there was plague in Vienna, so that anyone moving out of the city would be in mortal danger from the hostility of those in unaffected areas. Then, as he recounted to Sophie in a letter of 29 November (*K* **9**, pp 412–14), he had a bad attack of arthritis, which prevented him from walking and made him take more rest in bed. In reply, Sophie quipped that the pestilential air of Vienna seemed more agreeable to him than that of Hanover (*K* **9**, pp 415–16). No doubt his good reception by the Emperor and the two Empresses retained him, but several English who had been in Hanover, she added, regretted his absence. At this time, one of the topics of discussion in the correspondence with Sophie was the authenticity of the head of St Gregory of Nazianze that had been sent to Vienna from Hanover. Amalia had asked him to consult Molanus about it, because it was rumoured that the real head was elsewhere (*K* **9**, pp 412–14). Telling Sophie about it, Leibniz mentioned that a celebrated French author had written a very learned book on the multiplicity of heads of John the Baptist. Molanus, she informed him (*K* **9**, pp 419–21), had nothing more concerning the relic than the testimony of Henry the Lion. She was disappointed that, in his last letter, he had said nothing about his return. Princess Caroline also pleaded for his return (*K* **9**, pp 417–19), assuring him of her continued friendship and that of her husband.

At the end of January 1714, Leibniz informed Sophie that, on the advice of friends who feared for his health, he would postpone his departure from

Vienna until the beginning of warmer weather (*K* **9**, pp 425–8). Sophie, he imagined, would be diverted by the annual Carnival in Hanover. An opera was in preparation in Vienna, but he was not eager to see it, because the auditorium was small and the cumbersome dress of an 'infinity of ladies' would make it very uncomfortable.

At the end of March 1714, von Bernstorff informed him of the Elector's impatience and advised him, as a friend, to return. His response was to ask permission to take a health cure for his arthritis at a nearby spa. Writing to Sophie on 9 May (*K* **9**, pp 438–41), he reported that after five days at the spa he had felt some benefit, but he had been advised to interrupt the treatment at this stage and continue with it after an interval. Nothing, he added, would prevent him from returning to Hanover that summer. On 20 May, Sophie addressed her last letter to Leibniz (*K* **9**, pp 446–8), in which she expressed the wish that he were in Hanover, for she found greater pleasure in conversation than in writing letters. Concerning the health of Queen Anne, she quoted the Flemish proverb, 'Krakende Wagens gân lang' (creaking carriages go far). For herself, she counted her age more dangerous, having passed 83 years, although she was wonderfully well in the circumstances. Just over a fortnight later, on 8 June 1714, Sophie died in the park at Herrenhausen.

Leibniz was informed of Sophie's death by Johann Mathias von der Schulenburg in a letter of 13 June (*K* **9**, pp 481–3). Besides explaining the circumstances, which are described more fully in a letter of the Countess von Bückeburg to the Raugräfin Louise (*K* **9**, pp 457–62), he attributed her death to chagrin over an unfriendly letter she had received from Queen Anne (*K* **9**, pp 454–5). Owing to a misunderstanding, the Queen believed that Hanover wished to send the Electoral Prince to London without advising her Court, and she made clear her objection to the presence of the Prince in London during her lifetime. On 9 June (not knowing that Sophie had died the day before), Leibniz sent a letter to von der Schulenburg (*K* **9**, pp 478–9) defending the action of the ambassador Ludwig Justus von Schütz in addressing his request concerning the Prince to the Chancellor rather than the Queen, for that was the recognised diplomatic procedure. Sophie was certainly upset by the Queen's letter (and similar letters to the Elector and the Electoral Prince), for two days before her death, she remarked to the Countess of Bückeburg that the affair would make her ill and that she would die of it. On the fatal Friday, Sophie dined in public and in the evening she walked in the park discussing the affairs of England with Caroline. Turning to the Countess von Bückeburg, who was walking at a few steps distance out of respect, she took her hand and walked between the Countess and the Princess, talking about all sorts of things, among them the beauty of Madame Bousch, who was walking with her sister at the other end of the avenue. When they were in the middle of the park and had traversed the avenue up to the first fountain, Sophie began to falter. Caroline asked if she was ill. She sighed and indicated her stomach. It started to rain heavily and the park emptied. With difficulty she tried to make the ten steps to the nearest shelter. Then she said, 'I am very ill; give me your hand'. They let her down to the ground, where they took her on their knees. A few moments later she was dead.

To Caroline, Leibniz expressed his great sense of loss (*K* **9**, p 462–5), adding in consolation that it was the death Sophie herself had wished (in her beloved park without doctor or priest). He composed a short commemorative verse (*K* **9**, p 465) and urged that the Duchess of Orleans should be charged with the conservation of her correspondence.

Leibniz was not yet ready to leave Vienna, for an important part of his mission remained incomplete; namely, the establishment of a Society of Sciences. Having obtained the Emperor's agreement for the project at the beginning of his stay in Vienna, he had already composed several memoranda containing detailed proposals and suggestions for the raising of revenue. He envisaged that the Society should be not just for Austria but for the whole of Germany and he proposed that one of the principal prelates of the hereditary states should have the overall direction as President (*FC* **7**, pp 343–7). Besides pensioners with specific duties and adjoints or assistants, there should be honorary members, including noblemen, ecclesiastics and gentlemen to give weight and eminence. The proposed organisation was modelled on that of the Society in Berlin and the resources were to include an Observatory, laboratories, botanic gardens and reserves of rare animals. Revenue could be raised by some of the means adopted for the Berlin Society, such as a calendar monopoly, and also by a tax on stamped paper, though Leibniz pleaded that the poor should have exemption from such a tax (*FC* **7**, pp 302–11).

In August 1713 the Emperor nominated Leibniz himself as President. While this was an honour he had not sought, presumably because he thought the office inappropriate for a Lutheran, he could not establish the Society single-handed, so that by the summer of 1714, the project still existed only as an idea. In order to promote the Court's interest in its realisation, he composed a memorandum for Prince Eugene, pointing out the great advances in knowledge that had taken place in recent times, such as the discovery of almost half the globe by means of the magnetic compass, a better knowledge of the heavens by means of the telescope and the revelation by the microscope of a small world in the parts of the large. The invention of printing had provided the means of conserving and communicating this expanding knowledge, to which a Society of Sciences could contribute notably. It was true, he remarked, that the late King in Prussia had established 'une espèce de Société des Sciences', but the difficult times had rendered it very limited. With the prospect of peace over the best part of Europe and a long reign for a young Emperor, it seemed that the circumstances conspired to raise their hopes concerning the progress of useful knowledge, especially under the auspices of the new Society of Sciences. On 23 June 1714, having returned from one of his visits to the neighbouring spa, he followed up his memorandum to Prince Eugene with a request to the Emperor for the setting up of a commission for the establishment of the Society (*FC* **7**, pp 337–8).

Just at the time he was most concerned with the establishment of the Society of Sciences in Vienna, Leibniz received through Eckhart a request from von Bernstorff for an early decision on whether he wished to return to Hanover or not. In reply he composed a detailed justification of his stay in Vienna, pointing out that he had faithfully served the Hanoverian Court for almost forty years. About a month later, on 12 August 1714, Queen Anne died and the Elector Georg Ludwig became King George I of England. As Leibniz, on his own initiative, had taken a decisive first step in prompting the Duchess of Celle to seek from William III the naming of Sophie and her descendants in the line of succession and had faithfully promoted the Hanoverian cause ever since, he had reason to expect some show of appreciation for his services. He therefore resolved to return to Hanover in order to see the new king before he departed for England. At the end of the month, he gave a farewell address to the scholars of Vienna and left for Hanover on 3 September. On the following day the Imperial Chancellor von Sinzendorf, at Leibniz's request, wrote to the Hanoverian ministers, Hans Caspar von Bothmer and Friedrich Wilhelm von Goertz, to tell them that the Emperor had seen Leibniz depart with regret and hoped he would soon return (*K* **11**, p xx).

Isolation in Hanover

After passing through Dresden and Leipzig, Leibniz visited his friend Duke Moritz Wilhelm in Zeitz, made a short stay in Wolfenbüttel and arrived in Hanover on the evening of 14 September to find that King George I and the new Prince of Wales had left in the direction of England three days earlier (*K* **11**, p 12). On the following day, he paid a visit to Caroline, now Princess of Wales, who invited him to stay with her in Herrenhausen until her departure for England (*K* **11**, pp 14–15). He had been surprised to learn from Caroline that the King had said of him, 'he comes only when I have become king' and he expressed to her husband Georg August the hope that it was said only in jest (*K* **11**, pp 9–11). In a letter to von Bernstorff again setting out the justification of his stay in Vienna, he remarked that the King's lack of appreciation of his work hurt him more than the withholding of his salary. It seemed that while Europe rendered him justice, he did not receive it where he had most right to expect it.

Leibniz reported to the Emperor through the President of the Imperial Privy Council, Ernst Friedrich von Windischgrätz (Bodemann 1895, pp 389–90), that Princess Caroline wished him to accompany her to England. After his visit to England, he wished to assure the Emperor, it was his intention to return to Vienna.

Plate 8 Princess Caroline of Ansbach. Painted in 1704 by Johann Karl Zierle. Courtesy of the Verwaltung der Kunstsammlung, Georg-August University, Göttingen.

Caroline departed for England on 12 October 1714 without Leibniz. To a correspondent (*GP* **3**, pp 634–40), he gave the uncertainty of his health as the reason that hindered his journey. Caroline took her children with her, apart from the eldest, the seven-year-old Friedrich Ludwig, who remained as regent, destined not to see his parents again for fourteen years. Leibniz sought relief from his depression by a visit to Zeitz, where he stayed for a month as a guest of Duke Moritz Wilhelm. Here the Duke arranged for him to see a speaking dog, which could say about thirty words, such as tea, coffee, chocolate, and recite the alphabet. This made a great impression on him (Schmiedecke 1969, pp 139–40).

When he returned to Hanover on 3 December 1714, Leibniz received a letter from the Prime Minister von Bernstorff, advising him that he should not come to London, where letters addressed to him had already arrived (*K* **11**, p 22). Instead, he would be well advised to stay at home, so that he could make up for his long absence in Vienna by presenting to the King on his next visit to Hanover a good part of the history for which he had waited so long. In reply to von Bernstorff (*K* **11**, pp 22–5), Leibniz promised that, during the winter, he would complete a volume, taking the history as far as the beginning of the present House. To complete the work, a further volume would be needed. These volumes would constitute the *Annales Imperii Occidentis Brunsvicenses*, based on the sources contained in the *Scriptores rerum Brunsvicensium illustrationi.*[5] Even if he did not live to produce the second volume, he remarked to von Bernstorff, the first could stand alone as a work on the antiquities of Brunswick. Yet, at the same time, he petitioned the King, through the Prime Minister and Caroline, to make him historiographer of England, on the grounds that in his work on the history of Brunswick he was often obliged to touch on the history of England.

At the beginning of 1715, the King issued an order to Leibniz forbidding him to undertake any long journeys until the history was complete (Doebner 1881, p 308). Both von Bernstorff and his friend Caroline explained that they would only be in a position to help him if he produced the tangible results that the King desired. The chief obstacle to the granting of any favour, the Prime Minister told him, was the King's belief that he would never see the history completed. Caroline persevered on his behalf. In February 1715 (*K* **11**, pp 34–5) she expressed her belief that, if he applied himself to the history during the winter, the King would listen to her pleas on his behalf. Then, in September (*K* **11**, pp 46–7), she told him she had spoken again to the King about his desire to be historiographer of England, but had received the reply, 'he must first show me that he can write history; I hear that he is diligent'. She took this to be a good sign and pleaded with Leibniz to do all he could to satisfy the King. At the beginning of 1716 (*K* **11**, pp 71–3), she expressed her belief that her representations for the restoration of his salary had been successful. This was in fact paid again from 29 May 1716 after an interval of over two and a half years.

On 5 March 1715, Eckhart wrote to a friend that if Leibniz worked as diligently on the history as himself, it would soon be completed. Showing little sympathy for Leibniz's state of health, he implied that the gout was just

[5]This is the title of the first volume, which was altered slightly for the other two.

an excuse. Then, besides the delays caused by his journeys, he had a tendency to extend the work to infinity. Certainly he was not willing to compromise his own scholarly standards for the sake of speed, though, as he remarked to von Bernstorff, he could easily deceive most readers (Doebner 1881, p 355). Again, he was determined to include as an introduction two chapters on the geological history of the country and the origins of the people. By the end of 1715, however, he was able to inform the President of the Chamber, von Goertz, that one volume was ready for printing. As, however, he would have to supervise this work, and illustrations were still needed, he preferred to defer the printing until the completion of the second volume, which would happen in 1716, if God gave him sufficient health. The King expressed his satisfaction with the reports he had received and indicated that, on his next visit to Hanover, he would reward Leibniz in such a way that he should be completely satisfied. But on 21 February, having heard rumours that Leibniz wished to go to Vienna, the King commissioned Eckhart to continue the history from the year 1024, though nothing was to be said to Leibniz in advance (Doebner 1881, p 363).

In April 1716, Leibniz wrote to the Duke of Modena pointing out that Muratori was in too great a hurry to publish his work, for more research was needed (Bodemann 1895, pp 200–1). His primary concern seems to have been to prevent the publication of Muratori's work before his own, on the grounds that Muratori hardly mentioned his researches, which gave the first indications to the Italians on the common origin of the Houses of Este and Brunswick, and that there were many errors of omission and commission. Justice, according to Leibniz, required that if his work did not appear first, then the two works should appear at the same time.

Soon after his return to Hanover from Vienna, Leibniz had written to Queen Sophie Dorothea in Prussia (*K* **10**, pp 455–7), telling her that, if the great changes in the Court at Hanover had not caused him to hurry, he would have passed through Berlin to spend a little time there on work for the Society of Sciences, which he wished was more active. He had himself been busy on behalf of the Society, preparing several things for the second volume of the *Miscellanea*, which should appear soon. He was therefore disappointed to learn from the Secretary on 6 April 1715 that his salary as President was to be halved (Harnack 1900 **1**, 1, p 198). While he was content to let this reduction pass without comment, when he was informed by the Court Chaplain Jablonski on 3 September 1715 (Harnack 1900 **1**, 1, p 205) that payment was to be discontinued altogether, he wrote a letter of protest to von Printzen (*K* **10**, pp 458–9), pointing out that his absence from Berlin had not prevented him from working for the Society and he had, for example, commissioned articles for a new volume of the *Miscellanea*. He was surprised, he told von Printzen, that the King himself had cancelled the payment. In reply, von Printzen (*K* **10**, pp 459–60) explained that payment had been stopped by officials and not the King himself. Attempting to justify this action, he pointed out to Leibniz that he had neither written any letter nor visited the Society for three or four years and seemed to have abandoned all connection with it; the payment, he added, was promised not as salary but in recompense for the expenses of travel and correspondence. With dignity and reasoned argument, Leibniz attempted to demonstrate, in his last official communication to the Society, dated 19 November 1715 (Harnack 1900 **1**, 1, pp 207–9), the injustice of the treatment he had received from its members. He had in fact written to Jablonski and corresponded with several scholars, including Varignon and Bernoulli, concerning articles for the *Miscellanea*. On his last visit to Berlin, he had commissioned material for the *Miscellanea* from Naudé and d'Angicourt besides initiating other studies. On the other hand, the Secretary had given him no information, although he had asked for it. New members had been admitted without consulting or even informing him. Some of these, he observed, served only to augment the list and contributed nothing. As to his presence, it was not necessary, provided the officers of the Society were prepared to act on his advice. Concerning his payment, he would not ask the impossible, nor did he wish to bring a complaint against anyone to the King, for such actions were hardly fitting for a man of his kind. He only asked that von Printzen should render him justice before the King, who would then decide what was reasonable regarding him and whether his advice should be sought concerning the re-establishment of the tottering reputation of the Society. The Court Chaplain Jablonski, who was Vice-President of the Society, reported to von Printzen on 11 December 1715 (Harnock 1900 **1**, 1, pp 209–10) putting the blame on Leibniz for the interruption of the correspondence and the delay in the production of the second volume of the *Miscellanea*, which he attributed to his failure to visit Berlin. It is not known what answer von Printzen gave to Leibniz, but it seems probable that Jablonski regretted the outspoken tone of his report (which fortunately Leibniz did not see), for he was soon in friendly correspondence with Leibniz again, keeping him fully informed concerning the affairs of the Society and seeking his advice. In a letter of 7 July 1716, he held out the hope of the restoration of the annual payment if Leibniz would have patience (Harnack 1900 **1**, 1, p212).

In the second half of June 1716, Leibniz visited the Czar, Peter the Great, in Bad Pyrmont, where he had been staying for relaxation since 26 May. The Czar spoke to him on many occasions and he presented a number of memoranda concerning the magnetic observations, the improvement of the arts and sciences in Russia, and the establishment of government ministries (Guerrier 1873, pp 346–69). During his stay, Leibniz developed a friendship with the physician Johann Philipp Seip, who was the author of a book on Bad Pyrmont (Bodemann 1895, p 279).

On leaving the spa, Leibniz made a short stay in Brunswick and Wolfenbüttel, then continued on to Zeitz, in order to visit his friends the Duke and Duchess and to inspect the progress on the construction of his new calculating machine. He returned to Hanover on 26 July, just before the arrival of King George I, with whom he dined the following day. To Caroline (*K* **11**, pp 128–30) he reported that the King seemed cheerful; the King, however, told Leibniz that he seemed less so than before. In August he followed the King to Bad Pyrmont, where they had further conversations. From the reports to Caroline, it seems that there was no longer any friction on account of his long absence in Vienna.

Despite the burden of the historical work and the distractions of his life in Vienna combined with the incidence of ill-health and the painful loss of his patrons Duke Anton Ulrich and the Electress Sophie, Leibniz continued to develop and clarify his ideas, chiefly in correspondence with friends and scholars, so that, in his last years, he produced some of his best writings. Among these was his *Discourse on the natural theology of the Chinese*, in which he gave a systematic exposition of his own interpretation of Chinese science and philosophy. His correspondence with Des Bosses led him to place a greater emphasis on the reality of the physical world, for he had to recognise that composite physical bodies had a real unity and were therefore something more than just well-founded phenomena. This important development in his view of the natural world is reflected in two works written in his later years, *The principles of nature and grace, based on reason*, consisting of a popular introduction to his philosophy of nature and metaphysics, and the *Monadology*, in which he offered a more rigorous clarification of his principles to his followers. At the time of his visit to Vienna, the priority dispute on the invention of the calculus had entered its most unpleasant stage, for the Royal Society, of which he was a distinguished member, had evidently judged against him without giving him a hearing. It was probably this event more than any other that led him to send to Princess Caroline the letter on the decline of natural religion in England, which initiated the correspondence with Clarke. This controversy provided for him the means of expressing his most penetrating criticism of Newtonian philosophy.

Nicolas Remond and the natural theology of the Chinese

Leibniz wrote his most substantial philosophical work on China, the *Discours sur la théologie naturelle des Chinois*,[6] in response to a request from the French Platonist Nicolas Remond for his opinion on the writings of two China missionaries, the Jesuit Nichola Longobardi and the Franciscan Antoine de Sainte-Marie (Antonio Caballero a Santa Maria). Remond, who was chief counsellor of the Duke of Orleans, entered into correspondence

[6]The best edition of the French text, with German translation, is that of Loosen and Vonessen (1968). There is an English translation by Rosemont and Cook (1977).

with Leibniz in 1713, in order to express his admiration of the *Theodicy* (*GP* **3**, pp 603–4).

When Remond's letter reached him in Vienna, after a delay of several months, Leibniz replied at the beginning of 1714 (*GP* **3**, pp 605–8), giving an account of the development of his philosophical thought, beginning with his youthful reading of Aristotle and Plato, then the moderns and his deliberations in the Rosenthal. At this time mechanism prevailed and led him to mathematics, which he studied intensively after meeting Huygens in Paris. But when he looked for the ultimate reasons of things, his search led him to metaphysics, to entelechies, and after many steps, to monads or simple substances as the only real entities. Material things were therefore only phenomena, though well-founded. Plato, he added, had caught some glimpses of this. The Platonists and Aristotelians were right in seeking the source of things in final and formal causes, but wrong in neglecting efficient and material causes and inferring from this, as Henry More for example, that there were phenomena that could not be explained mechanically. On the other hand, materialists, or those who accepted only a mechanical philosophy, were wrong in rejecting metaphysical considerations. His own distinctive contribution, he explained to Remond, was to have penetrated into the harmony of these different realms, so that by means of the principle of pre-established harmony he could demonstrate that everything in nature happens mechanically and at the same time metaphysically, but that the source of mechanics is in metaphysics.

Leibniz also gave Remond a description of his project of a universal characteristic, explaining that, if he had been less distracted, or were younger, or had the assistance of young men of talent, he would still hope to produce a kind of universal symbolism (*spécieuse générale*) in which all truths of reason would be reduced to a kind of calculus. This could be at the same time a kind of language or universal writing, but infinitely different from all those projected previously, for the characters and words would direct the reason and the errors (except those of fact) would only be errors of calculation. He had not succeeded, however, in generating much interest in the project. When he had spoken to L'Hospital and others about it, they had not given more attention than if he had related a dream to them (*GP* **3**, pp 611–13).

It seems that Leibniz was pleased to cultivate the friendship of Remond, whose influential position in the service of the Duke of Orleans—a patron of science and destined to become regent on the death of Louis XIV in September 1715—would enable him to promote his ideas among the leading circles in Paris. In a letter of 2 September 1714 (*GP* **3**, pp 626–9), after referring to the distress of Madame (the Duchess of Orleans) at the death of her aunt Sophie, Remond remarked that he had himself conceived a high estimate of the merits of the Electress on account of the great confidence she had placed in him. Also he suggested that, following the establishment of peace in Europe, Leibniz might take the opportunity to make another visit to France, a country where his merit was recognised. The Duchess of Orleans, with whom Sophie had shared her correspondence with Leibniz, let him know, through the Abbé de Saint Pierre, that she too would be pleased to see him in France. Leibniz entered into direct correspondence with the Duchess (known as Liselotte) in September 1715 (Bodemann 1884).

Remond first mentioned the work of Longobardi in a letter of 12 October 1714 (*GP* **3**, pp 629–30), where he also referred to a short dialogue between a Christian philosopher and a Chinese philosopher written by Malebranche. It was only a year later, however, that Leibniz received from Remond copies of the works of Longobardi and Sainte-Marie. Two months later, in January 1716, he informed Remond that he had written a whole discourse on the Chinese, but in March he said he needed more time to finish it. In fact the work was never completed and breaks off abruptly in the middle of a paragraph.

Matteo Ricci (Mungello 1977, pp 26–32), the first missionary in China, had adopted the accommodationist position with respect to the conversion of the Chinese, having decided that their customs and rituals—well-entrenched elements of their civilisation which they could not lightly abandon—were compatible with the beliefs and practices of Christianity. Longobardi, who succeeded Ricci briefly as head of the mission, was almost alone among the Jesuits in opposing this view, though he was supported by most of the missionaries of other orders, including the Franciscan Sainte-Marie. These opponents of Ricci took the view that the ancient Chinese were materialists and the modern Chinese atheists, so that conversion to Christianity required renunciation of the traditional Confucian beliefs. In Rome, the 'Rites controversy', as it was called, raged for nearly 150 years until Pope Benedict XIV in 1742 decided against the accommodationist position. In 1710, Leibniz had pronounced himself sympathetic to the Jesuit position in a letter to Louis Bourguet (*GP* **3**, pp 549–54) and, two years later, again defended this position in conversation with Prince Eugene. It was therefore to be expected that, when Remond sent him the works of Longobardi and Sainte-Marie, which were revived in French translation in the early years of the eighteenth century (though the authors were long dead), Leibniz would take the opportunity to defend what he and his missionary friend Bouvet considered to be the enlightened position of Ricci, for it must have seemed obvious that only the accommodationist position held out any prospect of success for the mission in bringing about the conversion of the Chinese.

While Leibniz accepted the view of Longobardi and Sainte-Marie that the modern Chinese were atheists, he sought to demonstrate in the *Discourse* that the ancient Chinese produced a natural religion (compatible with Christianity) from which the later practitioners had deviated. For his Chinese sources he was content to rely on the quotations given by

Longobardi and Sainte-Marie, although more extensive extracts were available in the collection published by the Jesuits in Paris in 1687 with the title *Confucius Sinarum philosophus*. The text cited most often was the *Compendium*, a collection compiled in 1415 and regarded by Longobardi as a summary of the classical texts, but which was in fact a product of the much later Neo-Confucian school of interpretation. As in the case of the interpretation of the hexagrams of the *I ching*, Leibniz was therefore led astray, in his attribution of a natural theology to the ancient Chinese, by a false assumption concerning the antiquity of the sources on which he had to rely.

A favourite device adopted by Leibniz in the *Discourse* was to show that some aspect of ancient Chinese thought was compatible with his own philosophy. Then it followed that, as his own philosophy was compatible with Christianity, so was that particular aspect of Chinese natural theology.

In the first part of the *Discourse* Leibniz sets out to show that the ancient Chinese had concepts of God and spiritual substances. They did not perhaps recognise these substances as separated and existing apart from matter, but in so far as created substances were concerned, Leibniz did not see any great harm in that, for he was himself inclined to believe that angels had bodies, a view shared also by several early Church Fathers. Moreover, in Leibniz's system, neither the souls of animals nor the rational souls of humans were ever entirely separated from a body, however subtle or ethereal this might be. With regard to God, it might be that some Chinese gave him a body and regarded him as the soul of the world, thus making an error similar to that of the Greeks. Leibniz argued, however, that when the most ancient authors attributed to the first principle or *li*—in fact a Neo-Confucian term—the production of matter or *ki*, they demonstrated their belief in a purely spiritual substance that could be identified with the Christian God. In further support of this conclusion, Leibniz draws parallels between Chinese statements concerning *li* and Christian images of God. For example, the Chinese speak of *li* as a globe or circle, which he compares with the saying of Hermes Trismegistus that God is an infinite sphere whose centre is everywhere and circumference nowhere.

The term *li*, as used by the Chinese, Leibniz judged to be ambiguous, sometimes meaning God and sometimes meaning spiritual substances in general, equivalent, in his own system, to monads. Not having knowledge of the pre-established harmony, however, they had erred, like some western philosophers, in seeking the diversity of spirits in their bodies. Nevertheless he believed he could claim for the ancient Chinese a belief in spirits such as those of humans or angels which were substances different from *li*, though emanating from it. There was another name for God, *Xangti*, meaning Lord-on-high (in fact dating from the classical period), which was more comprehensible to the common people than the more abstract *li*.

The second and longest part of the *Discourse* treats of the Chinese doctrines concerning the creations of God: matter and spirits. According to the interpretations of Leibniz, the most ancient Chinese philosophers, and Confucius after them, had knowledge of God and the celestial spirits who served him, under the names of *Xangti* and *Kuei-Xin*. For these philosophers ascribed to the *Kuei-Xin* the defence and protection of men, cities, provinces and kingdoms, not as if they were souls or substantial forms of these things but as if they were pilots of vessels, like the assisting intelligences of western philosophers. There was a great likelihood, he believed, that the Chinese came to their knowledge of these things through the tradition of the Patriarchs. Evidently, the ancient sages either believed that certain angels, as ministers of *Xangti*, presided over earthly things, or they wished God to be worshipped through the qualities of individual things, which the popular imagination could more easily conceive under the names of the spirits of these things. While the first alternative was not intolerable to Christianity, the second was preferable; for, according to this, the spirit of the seasons, the mountains and the rivers, was the same *Xangti* that governed the heavens. To explain the silence of Confucius himself on the spirits of natural things, Leibniz speculated that he thought what should be revered in inanimate things was only the supreme spirit, *Xangti* or *li*, but did not believe the people capable of detaching this supreme spirit from the objects which fell under their senses. On this question, Leibniz could even find satisfaction with the interpretations of the modern Chinese, who preferred explanation in terms of natural causes to those involving miracles. The new discoveries in Europe, which gave virtually mathematical reasons for some of the great wonders of nature, could be used for their enlightenment, but it would also be necessary, Leibniz added, to acquaint them with the demand of reason, that these natural causes could not be brought about without mechanisms created by the wisdom and power of the supreme substance, which with them could be called *li*.

To complete his account of the natural theology of the ancient Chinese, Leibniz offered, in the third part of the *Discourse*, an interpretation of their statements on human souls. Although these were subsumed in some fashion under spirits, he felt that they merited a separate discussion, in order to determine the Chinese doctrines concerning their nature and their state after this life. From the fact that the ancient Chinese spoke of the spirits of the ancestors whom they worshipped as if they were capable of obtaining good and evil for their descendants, Leibniz claimed to be justified in concluding that they conceived the human soul to continue its existence after this life. It was true, he added, that the Chinese mentioned neither hell nor purgatory, but it was possible that some among them believed the souls wandering here and there in mountains and forests were in a kind of purgatory. Here again he was able to point to a western parallel in the life of St Conrad, a bishop of Constance whose biography was included in the second volume of his *Scriptores rerum Brunsvicensium illustrationi*, who discovered souls in the form of birds condemned to the waterfalls of the Rhine and saved them by his prayers.

In the fourth and final part of the *Discourse* Leibniz gave an account of his binary system and the discovery, for which he now claimed the credit jointly with Bouvet, of its relationship with the hexagrams of the *I ching*. The supposed invention of the binary system by Fu-Hsi, the mythical founder of Chinese culture, thousands of years before it was rediscovered by Leibniz, was an essential element in his advocacy of the adoption by the missionaries of the accommodationist position. For there could be no better way of establishing among the modern Chinese a respect for their ancient texts and the natural theology contained in them, which should predispose them to Christianity on account of the striking similarities, than by demonstrating that they also contained mathematics of a level achieved in Europe only within the last few decades.

Correspondence with Des Bosses

In a letter to Des Bosses of 5 February 1712 (*GP* **2**, pp 433–8), Leibniz introduced a new term into his discussion of the nature of composite bodies. For a long time he had left open the question whether these aggregates of monads were simply well-founded phenomena or corporeal substances having real metaphysical existence. Now he explains to Des Bosses that, if corporeal substance is something real, it must consist of a real unifier or substantial chain (*vinculum substantiale*) superadded to the monads by God. Although the discussion with Des Bosses takes place in the context of a search for an explanation of the Eucharist satisfactory to the Catholics —Leibniz himself, as a Lutheran, did not believe in transubstantiation—his interest in the problem of the reality of bodies was evidently philosophical rather than merely theological. For there is no mention of the Eucharist in the notes he made in preparation for his letter to Des Bosses (*GP* **2**, pp 438–9), while the metaphysical reality of bodies is implied in the philosophical writings of his last years. Yet in his correspondence with Des Bosses, he frequently emphasises the alternative view that bodies are just well-founded phenomena, while his failure to use the term *vinculum substantiale* elsewhere than in his letters to Des Bosses perhaps indicates that he was not entirely satisfied with the new theory (Blondel 1972, pp 281–7).

Two different unions are envisaged by Leibniz. First, there is a real union of the soul and the body effected by the dominant monad, and secondly, there is a binding together of all the subordinate monads into a real corporeal substance brought about by the substantial chain. In fact Leibniz identifies the corporeal substance or body with the substantial chain. The monads themselves are not parts of this chain. While the monads—which are simple indivisible substances—remain intact, their passive and active elements (primary matter and entelechy) are combined separately to form the chain, which thus consists of an extended mass, arising from the primary matter, and a substantial form composed of the sum of the entelechies. Whatever can arise in this way, he explains, can also be extinguished, and is indeed destroyed by cessation of the union, unless miraculously conserved by God. It follows that the substantial chain or corporeal substance differs fundamentally from a soul or monad. While the latter is a simple indivisible substance, the substantial chain is in a perpetual flux, just like matter.

At this point in the correspondence, Leibniz indicated his preference for the view that only monads are real and that the union in the phenomena is supplied by the action of the perceiving soul. Then he adds that, if faith urges us to assent to corporeal substances (in order to explain the Eucharist), these must consist in that unifying reality which adds something absolute and hence substantial, even though fluid, to the things to be united.

Writing to Des Bosses from Wolfenbüttel on 20 September 1712 (*GP* **2**, pp 456–61), Leibniz expressed his approval of his friend's translation of the *Theodicy* into Latin, which he remarked was beautiful and would throw light on the work. He also explained how the Catholic teaching concerning the Eucharist could be accommodated in his philosophical system by means of the *vinculum substantiale*. At the consecration, God would simply need to replace the substantial chain of the bread and wine by another uniting the body of Christ to the monads of the bread and wine. As the monads are not parts of the chain and are unaffected by it, the phenomena of the monads of bread and wine would remain unchanged. Even if the bread and wine were not substances but mere aggregates of monads, they would still be aggregates of organic bodies (for every monad has a body) constituting real unities, whose substantial chains could be destroyed and replaced by the substantial chain of the body of Christ.

Des Bosses raised objections concerning the nature of the substantial chain, claiming that it could not be a substance, since it had no principle of action. Moreover, he could not be convinced that the view he wished to hold—that the chain was an accident rather than a substance—was contradictory, and in consequence the discussion dragged on without resolution. After his return from Vienna and a winter suffering from gout, Leibniz wrote to Des Bosses on 15 March 1715 (*GP* **2**, pp 492–3), emphasising the real nature of bodies. Dismissing Bishop Berkeley, who attacked the reality of bodies, as a man who wanted to be known for his paradoxes, Leibniz added that bodies were rightly regarded as things, for phenomena too were real. In his system, of course, this reality stemmed from the well-founding resulting from the pre-established harmony between the monads. But anyone who wished to consider bodies as substances (that is, metaphysically real like the monads), he explained, would need some new principle of real union such as his substantial chain. In letters written during the spring and summer of 1715, he replied to the objection of Des Bosses that the substantial chain lacked a principle of action. While the monads influence the chain, the chain itself can change nothing in the laws of the monads, for whatever modification it receives from them it will have as an echo (*GP* **2**, pp 495–6). A body which returns an echo, he maintains, is acting (*GP* **2**, pp 502–5). Also Leibniz made clear his view that only organic bodies—that is, those possessing a dominant monad—need be assumed to have a chain, and that the chain will always adhere to the dominant monad.

Popular works on philosophy

In 1714, while he was in Vienna, Leibniz composed two popular accounts of his philosophy, which contain eminently readable summaries of the chief principles. These works were the *Principes de la nature et de la grace, fondés en raison*, written for his friend Prince Eugene, and an essay given the title *La Monadologie* by a later editor, which he wrote for his new friend in Paris, Nicolas Remond. Neither work was published in his lifetime but the *Principles* appeared at The Hague in 1718 and the *Monadology* was first published in a German version in 1720.

Writing to Remond in July 1714 (*GP* **3**, pp 618–21), Leibniz explained that the account of the monads, which he had requested, was in progress, but owing to distractions he had not been able to complete it. To understand the whole of his system, he added, it was necessary to supplement the reading of the *Theodicy* with various articles published in the journals. A month later (*GP* **3**, pp 624–5), having completed the *Principles*, he sent Remond a copy of this, pending the completion of the *Monadology*. Remond (*GP* **3**, pp 629–30) found the clarifications very useful and suggested that Leibniz should add a second volume to the *Theodicy*. This should begin with the dynamics (which seemed to be the foundation of his system), include the journal articles and end with the *Principles* (*GP* **3**, pp 640–4). Leibniz pointed out, in a letter of 22 June 1715, that the dynamics, which he had not yet completed, would require a whole volume (*GP* **3**, pp 644–7).

Neither work contains any mention of well-founded phenomena and both refer to compound substances. This may indicate a shift in Leibniz's position away from phenomenalism and towards a belief in the reality of the natural world. There is, of course, no mention of the substantial chain, for even if he were completely satisfied with this concept, the kind of detailed explanations given in the correspondence with Des Bosses would be out of place in a popular account.

The *Principles of nature and grace* (*GP* **6**, pp 598–606) begins with definitions of substance as a being capable of action, simple substance as that which has no parts, and compound substance as a collection of simple substances or monads. The Greek word *monas*, it is noted, means unity. Since there could be no compound substances or bodies without simple substances, the conclusion is drawn that simple substances or monads must exist everywhere, so that the whole of nature is full of life. As Leibniz remarks in the *Monadology* (*GP* **6**, pp 607–23), monads are the true atoms of nature; that is, the elements of things. Continuing his account in the *Principles*, Leibniz explains that, as monads have no parts, they can neither be made nor destroyed. Their internal qualities and activities consist of (i) perceptions; the representation of that which is without, and (ii) appetitions; their tendencies from one perception to another. He makes a distinction between perception—which is the inner state of the monad, representing external things—and apperception, which he describes as consciousness (that is, self-consciousness) or the reflective knowledge of this inner state itself, something that is not given to all souls or to any soul all the time. For lack of this distinction, the Cartesians, he remarks, made the mistake of supposing that only self-conscious beings had perceptions, which led them to believe that animals were without souls. In his view, however, the perception of animals is accompanied by memory—so that an echo of the perception remains for a long time, making itself heard on occasions—and this, he supposes, means that the animals have souls. When these souls are raised to the level of reason, as in humans and angels, he calls them spirits.

Since nature is a plenum, Leibniz infers that there are monads everywhere, separated from each other by their own actions and continually changing their relations. Together with a particular body, consisting of the surrounding monads which it dominates, each monad constitutes a living substance, so that not only is there life everywhere but also infinite degrees of it, as the monads are organised in increasingly complex hierarchies. After explaining the pre-established harmony between the perceptions of the monad and the motions of its body, Leibniz turns his attention to the problem of the generation of animals, giving more detail than in his previous accounts. First he notes that there is a connection between the perceptions of animals which has some resemblance to reason. This, however, is grounded only in memory of facts or effects and not on a knowledge of causes. Reasoning in the true sense depends on necessary or eternal truths, as are those of logic, number and geometry, and is to be found only in rational souls. These souls are capable of performing acts of reflection and it is this, he notes, that enables us to acquire knowledge of the sciences and demonstrative truths. Plants and animals, Leibniz supposes, come from preformed seeds and therefore from transformation of living beings existing prior to them; he cites the metamorphosis of caterpillars into butterflies as an obvious example of transformation of the same animal. There are little animals, he explains, in the seeds of large animals, which assume in conception a new vesture providing them with a method of nourishment and growth so that they may emerge into a greater stage and propagate the large animals. Spermatic animals he defines as those which are raised by conception to the level of greater animals. Only a small number of chosen ones pass into the greater theatre. The spermatic animals themselves are enlargements of other smaller spermatic animals and so on, in accordance with the hierarchical organisation of the monads, to infinity. Not only souls, therefore, but animals as well, are neither generated nor destroyed—they do not completely perish in what we call death—but are transformed. Souls never leave the whole of their bodies and do not pass from one body to another entirely new to them. Animals change but take on and put off only parts; in nutrition this takes place through imperceptible changes but continually, while in conception and death, the changes occur suddenly and noticeably. In the case of human beings, the souls of the spermatic animals are not rational but become so only when conception determines them for human nature.

At this point Leibniz takes his exposition from the plane of natural sciences to that of metaphysics, where he uses the principle of sufficient reason to demonstrate the existence of God and the creation of the world in accordance with the principle of the best. In particular, he points to the fact that the laws of nature could not be explained by consideration of efficient causes or matter alone, but required an appeal to final causes in the form of a choice based on perfect wisdom, as one of the most useful evidences for the existence of God. To this *a posteriori* proof from the existence of contingents, Leibniz added in the *Monadology* an *a priori* proof from the reality of the eternal truths. In this version of the ontological argument, Leibniz in effect uses the concept of understanding rather than that of perfection to establish the possibility of a necessary Being. For the reality of the eternal truths (that is, their meaningfulness as true statements) must depend on the existence of a mind for which they are objects of understanding. The same is true of essences or possibilities, so that, if a necessary Being whose essence includes existence (that is, whose possibility implies existence) does not exist, then nothing is possible and there can be no eternal truths. Since there are eternal truths, the possibility and hence existence of God (a Being who understands them) is established.

Both the *Principles of nature and grace* and the *Monadology* end with an account of the special role of man in the divine scheme. To indicate the essential difference between God and the monads, Leibniz again appeals to the image of God as a sphere whose centre is everywhere and circumference nowhere. While each monad represents the universe confusedly from its point of view, God sees everything distinctly as it really is. Rational souls or spirits, he believes, are something more than monads or even simple souls (as those of animals), for they are not only mirrors of the universe of creatures but also of God himself. Thus, in discovering the sciences according to which God has regulated things, the soul imitates (in its own realm) what God performs in the great world. For this reason, Leibniz believes, spirits (the souls of humans and angels) enter into a kind of society with God and are members of the City of God. This takes place, not by a dislocation of nature, but by virtue of a harmony pre-established between the realms of nature and grace, so that God's plans for souls are in agreement with the laws of bodies. Finally he says of the beatific vision or knowledge of God that it can never be complete, since God, being infinite, cannot be known perfectly. Thus we shall never be in a position that leaves nothing to be desired and which would stupefy our spirit.

Correspondence with mathematicians

Guido Grandi, Professor of Philosophy in Pisa, learnt the calculus by his own efforts and entered into correspondence with Leibniz in 1703 by sending him a book he had written on the subject. In 1710 he published a paper on the infinitely great and small, in which he claimed to demonstrate that the sum to infinity of the series $1/(1+x)=1-x+x^2-x^3+\ldots$ obtained by putting $x=1$, namely $1-1+1-1+\ldots$, was equal to ½. This result, which he compared to the mystery of the creation out of nothing (for $1-1+1-1+\ldots$ may be written $0+0+0+\ldots$, so that $0+0+0+\ldots=½$), had been the subject of a controversy with the Professor of Mathematics in Pisa, whom he succeeded in 1714. Grandi's result became a theme of discussion between Leibniz and several of the mathematicians with whom he corresponded, especially after he published a letter to Wolff containing what he believed to be a resolution of the paradox. In this letter, which appeared in the *Acta Eruditorum* (*GM* **5**, pp 382–7) in 1713, Leibniz accepted Grandi's result. For, despite the appearance of absurdity, he believed that it could be justified. In the case of an even number of terms, the sum of the series is 0 and in the case of an odd number of terms, the sum is 1. When the number of terms is infinite, however, Leibniz holds that the distinction between even and odd disappears and it is therefore reasonable to take the sum to be ½. Although he seems to express some disquiet in his letter to Grandi of 6 September 1713 (*GM* **4**, pp 217–20), Leibniz defends his resolution of the paradox in letters to Varignon and Bourguet.

Writing to Leibniz on 19 November 1712 (*GM* **4**, pp 187–91), Varignon ridiculed Grandi's comparison with the creation and suggested that the paradox could be avoided by writing $1/(1+1)=1/(3-1)=1/3+1/9+1/27+\ldots=1/2$. He observed also that an infinity of other geometric series could be produced with the same sum. Louis Bourguet, who resided in Venice from 1711 to 1715 and was especially interested in ancient languages, corresponded with Leibniz on a variety of themes, including China, philosophy and mathematics. At the beginning of 1714 (*GP* **3**, pp 561–4) Leibniz suggested to him that, if he had any influence, he should recommend the invitation of Johann Bernoulli to succeed Hermann as Professor of Mathematics at Padua. If Johann Bernoulli, whom he described as 'a luminary of our century', could not be attracted by a suitable salary, then his nephew Nikolaus, he added, would be the next best choice. A few months later he expressed to Bourguet (*GP* **3**, pp 576–8) the hope that Count Jacopo Riccati and Bernardo Zendrini would continue to introduce the new sciences into Italy and asked if they had seen his resolution of Grandi's paradox in the *Acta Eruditorum*. Again writing to Bourguet on 3 April 1716 (*GP* **3**, pp 591–3), this time in a philosophical context, he made the interesting remark that, while the sums of the series $1+1+1+\ldots$ and $1+\frac{1}{2}+\frac{1}{3}+\frac{1}{4}+\ldots$ were both infinite, the sum of the first was infinitely greater than that of the second.

With Johann Bernoulli Leibniz discussed the question of the existence of logarithms of negative numbers. According to Bernoulli (*GM* **3**, 2, pp 885–8), $\log x = \log(-x)$, a result he inferred from the relation $d(\log x) = dx/x = d(-x)/(-x) = d(\log(-x))$. Leibniz (*GM* **3**, 2, pp 895–6) showed Bernoulli's result to be contradictory and argued that negative numbers do not have logarithms.[7]

From Varignon, Leibniz continued to receive news concerning the Academy of Sciences and his friends in Paris. When Hermann and Bernoulli presented to the Academy their solutions of the inverse problem of central forces—namely the proof that, in the case of a centripetal force varying inversely as the square of the distance, the orbit is a conic section—Leibniz took the opportunity to encourage his friend to persevere with his own researches. The inverse problem (to which, he remarked, Newton had not given enough attention) having been solved, he recommended to Varignon an investigation of the orbit in the case of a body attracted to two or more moving centres. This, he added, might provide a means of determining more accurately the motion of the moon (*GM* **4**, pp 174–6). In a letter of 19 November 1712 thanking Leibniz for his admission as a member of the Berlin Society of Sciences, Varignon reported that, although he had investigated orbits of bodies attracted to multiple centres, even in a resisting medium, he had not yet found anything relating to moving centres, which he had considered only in passing. In the same letter, he mentioned the death of Cassini at the age of 88 years, 'without any illness and simply by the need to die', and also sent the compliments of Leibniz's old friend Des Billettes, then 80 years old. Varignon also told Leibniz that he had spent two months reading a large work on magic squares for the Academy. Several years later, on 27 February 1716, Leibniz sent him a magic cube with 27 cells. This was presented to the Academy and examined by De la Hire, who found it true but could not discover the method of construction.

A close friend of Varignon, the Abbé Carl Irenaeus Castel de Saint Pierre, advocated the establishment of a kind of International Court that would settle political disputes. Through Varignon, the Abbé sent Leibniz a copy of his *Projet d'une paix perpétuelle* with a request for his opinion of it (*GM* **4**,

[7]On this controversy, see Marchi (1974).

pp 195–6). He was pleased to receive a detailed reply from Leibniz early in 1715 (*FC* **4**, pp 328–36). Leibniz pointed out that the Landgrave Ernst von Hessen-Rheinfels, who had commanded armies in the great war before the Treaty of Westphalia, had proposed a similar project. Expressing approval of Saint Pierre's view of the Empire as a model of Christian society, Leibniz agreed that things should as far as possible be left as they were. For this reason, he preferred the proposal of having the Emperor speak for the Empire to that of disbanding the Empire and giving the Emperor one vote (just like the Electors) as an hereditary ruler, which Saint Pierre had offered as an alternative. Leibniz was concerned that nothing should be done to weaken the influence of the Emperor in the Christian world. Again, he preferred the existing Imperial Chamber, where the judges had freedom of conscience, to the proposal of Saint Pierre, according to which the judges would follow the instructions of their princes.

The problems of planetary motion and gravity continued to be a theme of Leibniz's correspondence with Hartsoeker, who believed that the planets floated in equilibrium with the surrounding fluid without need of gravity or centrifugal force. Not being a mathematician, Hartsoeker wished to have from Leibniz an explanation of the elliptical orbit 'in three words only' (*GP* **3**, pp 530–2). While agreeing with Hartsoeker that the attraction could not be an essential quality of bodies (for this would be a return to the occult qualities that philosophy had abandoned), Leibniz maintained that there was such a force, having its source in the impulsions of fluids. He had shown mathematically that, when a planet is carried about the sun by a circulation with velocity inversely as the distance, this circulation combined with the gravity produces perfectly the planetary laws of Kepler. The same circulation, he added, arises when, following Newton, the same gravity is combined simply with the force of projection. In order to justify these results, however, long mathematical demonstrations were needed, for which he referred Hartsoeker to the *Acta Eruditorum* and Newton's *Principia*.

In his last years, the major theme of Leibniz's correspondence with mathematicians was the priority dispute on the invention of the calculus. Writing from Berlin on 4 March 1711 (*NC* **5**, pp 96–8), he thanked Hans Sloane, the Secretary of the Royal Society, for sending him the latest issue of the *Philosophical Transactions*, in which Keill made his accusation of plagiarism against Leibniz in the invention of the calculus. Seeking a remedy from the Royal Society in the form of a retraction by Keill, Leibniz remarked to Sloane that no-one knew better than Newton how false the accusation was, adding that he had never heard the term calculus of fluxions spoken of nor seen the symbolism employed by Newton before it appeared in the works of Wallis. How then, Leibniz asked, could he have published the arithmetic of fluxions invented by Newton, after altering the name and style of notation, as Keill asserted? The Royal Society asked Keill to express in a letter his account of the affair and this was sent to Leibniz by Sloane in May

1711 (*NC* **5**, pp 132–52). While declaring that he had not intended to suggest that Leibniz knew the name or symbolism used by Newton, he maintained that, in two letters transmitted to Leibniz by Oldenburg (the *Epistola prior* and *Epistola posterior*), Newton had given 'pretty plain indications . . . whence Leibniz derived the principles of that calculus or at least could have derived them'. In a letter of 29 December 1711 (*NC* **5**, pp 207–9), Leibniz again appealed to the Royal Society for justice. In response, the Royal Society set up a committee to consider the matter. The report, written by Newton himself and condemning Leibniz, was evidently accepted by the Society and also the recommendation that it should be published together with the relevant documents. The *Commercium epistolicum* appeared at the beginning of 1713 but Leibniz did not see a copy until he returned from Vienna in the autumn of 1714.

Leibniz was first informed of the contents of the *Commercium epistolicum* in a letter of 7 June 1713 (*NC* **6**, pp 1–6) from Johann Bernoulli, whose nephew Nikolaus had brought to Basel a single copy given to him in Paris by the Abbé Bignon, one of several sent from London for distribution to the learned. Expressing his displeasure at 'this hardly civilised way of doing things', Bernoulli told Leibniz that he was accused before a tribunal consisting of the participants and witnesses themselves, documents against him were produced and sentence was passed; the case was found against him and he was condemned. In the same letter, Bernoulli pointed to a mistake his nephew had found in proposition 10 of Book II of Newton's *Principia*, which, he claimed, served to demonstrate that Newton did not understand second and higher differentials. There is indeed an error, which Newton later corrected, but its source does not lie in the supposed mistake identified by Bernoulli (Whiteside 1970, pp 128–9). If, as Bernoulli asserted, Newton identified the terms of the binomial expansion of $(x+o)^n$ with the differentials, then this would reveal a confusion concerning higher differentials, but Newton always claimed that he had only said the terms were proportional to the differentials, which of course was true. Having thus provided Leibniz with some rather dubious evidence that he could use against Newton, Bernoulli asked Leibniz not to involve him in the dispute.

Before the end of July, Leibniz had composed a statement of his own position, which his friend Wolff arranged to have printed and circulated. In this document, generally known as the *Charta volans* (flying sheet) (*NC* **6**, pp 15–21), Leibniz, under the cloak of anonymity, expressed the view that he had probably been too generous to Newton in believing that he possessed something similar to the differential calculus, for the subsequent unfair dealing of the English had made him suspect that the calculus of fluxions had been developed in imitation of the differential calculus. Then, quoting Bernoulli's letter, he added that Newton was confused about higher differentials even when he wrote the *Principia*, as a certain eminent mathematician (the writer of the letter) had shown. A French translation of the *Charta volans*, together with some anonymous remarks of Leibniz on the dispute (*NC* **6**, pp 30–2) was sent by Wolff to the *Journal litéraire* in The Hague and published at the end of 1713. A Latin version also appeared in the *Deutsche Acta Eruditorum oder Geschichte der Gelehrten*.

Varignon expressed his support for Leibniz in a letter of 9 August 1713, and after his return from Vienna, Leibniz informed Varignon (*GM* **4**, pp 198–9) that he thought some day to give his own *Commercium epistolicum*, but first he would need to seek out his old papers. At that time he was very busy with the history and inconvenienced by his gout.

Just before Leibniz returned to Hanover in the autumn of 1714, Keill published in the *Journal litéraire* a response to the *Charta volans* and Leibniz's remarks which had appeared in this journal. Using material he had received from Newton, Keill attempted to show that Leibniz had made errors involving second differentials in his paper on planetary motion. Accepting the interpretation of Leibniz's calculations given earlier, which is justified by the documentary evidence, this criticism is seen to be without foundation (Aiton 1962). Wolff, who feared that Leibniz's silence might be taken as an indication that the English had a good case, pressed him to answer this criticism of Keill (*W*, pp 164–7). Writing to Wolff on 18 May 1715, Leibniz explained, however, that he wished to have no dealings with a man like Keill. In his paper on celestial motions, he declared, there was no error, only a phrase concerning centrifugal force which he later rendered more suitable. For what he had demonstrated—that the Keplerian ellipse results from a combination of the harmonic circulation with gravity—was certainly true and would remain so. Apart from the curious euphemism for a mistake later corrected, this was a fair description. By the autumn of 1715, Wolff had adopted Leibniz's attitude to Keill, saying that he could not approve his Excellency's replying to an idiot. However, he advised Leibniz to present his own *Commercium epistolicum* to the public (*W*, pp 174–6).

Towards the end of 1715, Leibniz learnt that he had an advocate for his cause in London. For he received a letter written several months earlier from the Abbé Antonio-Schinella Conti in Paris (*NC* **6**, pp 215–16) telling him that he was about to leave for London and would support his cause there as he had in Paris. Born in Padua, the Abbé had travelled to France in 1713 in search of information for a general history of philosophy. On his journey to England, he was accompanied by another correspondent of Leibniz, the mathematician Pierre Remond de Monmort, elder brother of Nicolas Remond, the chief minister of the Duke of Orleans. Writing to Conti on 6 December 1715 (*NC* **6**, pp 250–5)—the letter was intended to be shown to Newton—Leibniz defended his position and appended a challenge problem on orthogonal curves (devised by Johann Bernoulli) for the English mathematicians. According to an account of Conti written much later (*NC* **7**, pp 137–40), Newton asked him to invite the foreign ambassadors to examine the original documents in the archives of the Royal Society in order to check the accuracy of the *Commercium epistolicum*. When this had been done, Baron von Kielmansegg, speaking on behalf of the ambassadors, declared that this was not sufficient and recommended that, in order to settle the dispute, Newton should himself write a letter to Leibniz setting out his case and asking for a direct reply. The King was consulted and gave his approval. By this time Newton had already written a letter for Leibniz (*NC* **6**, pp 285–90); in effect, a shorter version of the anonymous review of the *Commercium epistolicum* he had recently published in the *Philosophical Transactions*. The Countess of Kielmansegg had this translated into French by Pierre Coste and it was transmitted to Leibniz with a covering letter by Conti in March 1716 (*NC* **6**, pp 295–6). The tone of Conti's letter is that of an impartial observer rather than a partisan of Leibniz, for it seems that he had acquired a taste for the Newtonian philosophy.

In his reply to Conti for Newton, written on 9 April 1716 (*NC* **6**, pp 304–14), Leibniz explained that he had not wished to enter the arena with the 'lost children' Newton had sent against him, including the author of the preface to the new edition of the *Principia*, which was full of sour remarks concerning those who do not accept gravity (as an essential quality of bodies) and the void. But since Newton now wished to appear himself, he would be pleased to satisfy him. First, he pointed out that he had no knowledge of the 'Committee of Gentlemen of several nations' who had, according to Newton, collected and published the documents forming the *Commercium epistolicum*, for no-one had consulted him. Even then he did not know the names of all the members of the committee, especially those who were not English, and he did not believe that they approved all that had been written against him in the work. Secondly, he claimed that the *Charta volans* had been written by a zealous friend at a time when he himself was in Vienna and had not seen the *Commercium epistolicum*. Although Newton had described it as a defamatory letter, it was, he added, no stronger than what had been published against himself. Thirdly, Leibniz explained that when at last he saw the *Commercium epistolicum* on his return to Hanover, he found that the letters were all about series with glosses voicing baseless suspicions but irrelevant to the real issue. There was not a word, he believed, that could cast doubt on his invention of the differential calculus. To reply point by point would require a work at least as long and involve a laborious search through a mass of old papers, some of which had been lost, for which his other occupations (especially the writing of the history) did not leave him the time. In the circumstances, he was content to treat with contempt the judgment of those who wished to pronounce against him on the basis of such a work, especially as the Royal Society itself had not done so, as he had seen from an extract of its record. Leibniz went on to give a detailed account of his discovery of the differential calculus. A few days afterwards, he sent a similar account to his friend the Countess of Kielmansegg (*NC* **6**, pp 324–30).

The correspondence with Clarke

Leibniz first informed Princess Caroline of his dispute with Newton in a letter of 10 May 1715 (*K* **11**, pp 37–40), in which he referred to the opinion of a French journalist writing in Holland, who thought that the quarrel concerned Germany and England rather than the individuals themselves. To restore the honour of Hanover and Germany, he suggested to Caroline, the King should appoint him historiographer of England, thus demonstrating that he was esteemed no less than the Master of the Mint. He had not the leisure to reply to Newton or his supporters, he told her, though able men in France and Switzerland had defended him. However, his friends had pressed him to examine himself the new philosophy of Newton, which was somewhat extraordinary. They believed, for example, that a grain of sand exercised an attractive force as far as the sun, without any medium or means. Yet they regarded the presence of the body and blood of Christ in the Lutheran Eucharist (without any obstacle of distance) as absurd. Leibniz's criticism was two-fold. First, the Newtonians deserved censure for denigrating the Hanoverian religion and, second, they were inconsistent in denying the miracle of the Lutheran Eucharist while proposing an attraction that was equally miraculous. His own view, Leibniz added, was that miracles should be reserved for the divine mysteries and not used in the explanation of natural things. He was pleased that Caroline had herself seen the weaknesses of Locke's philosophy, according to which all was material and matter was capable of thought, things which would lead to the ruin of religion. Locke, he remarked, was less of a philosopher than he had once thought, a change of mind possibly occasioned by the fact that, in a letter to Molineux, Locke had treated his objections with contempt. Modesty did not prevent him from telling Caroline that, when this letter had been published in a collection of Locke's posthumous works, a knowledgeable critic had remarked that his objections were the best thing in the collection.

Caroline (*K* **11**, pp 52–3) wished to see an English translation of the *Theodicy* and when she consulted her friend the Bishop of Lincoln, he recommended for the task Samuel Clarke, who was a chaplain to the King. This, she thought, was not a good choice, for Clarke was a friend of Newton and he was too opposed to Leibniz's views to maintain the required impartiality. Having already sent copies of Clarke's books, she wrote to Leibniz on 26 November 1715 telling him that she was herself in dispute with Clarke and asking for his help. In response to Caroline's request, Leibniz (*K* **11**, pp 54–5) sent a brief statement of his view that natural religion seemed to be in decline in England. After repeating the criticisms of Locke he had mentioned to Caroline in the spring, he described what he considered to be two fundamental errors in the philosophy of Newton. First, Newton held space to be God's sensorium, an organ that he used in order to perceive things. If God had need of such an organ, then it would follow, Leibniz explained, that the objects perceived were not entirely dependent on him and consequently could not have been created by him. Second, Newton and his followers held that God needed to intervene in the natural world from time to time in order to make adjustments to the working, in the same kind of way that a clockmaker needs to mend his handiwork and set it right. According to Leibniz, the machine of nature was created perfect in the beginning. In consequence of the conservation of *vis viva*, which he had established, the same force always remains in the world and only passes from one part of matter to another in accordance with the laws of nature. When God works miracles, Leibniz concluded, it is not to supply the needs of nature but those of grace.

Caroline showed this writing to Clarke and asked him to reply. Thus began the correspondence between Leibniz and Clarke. Altogether Leibniz wrote five letters in French. Clarke's replies were written in English, the last arriving in Hanover when Leibniz had only a few more days to live. Caroline carefully preserved all the letters, which were published in English and French, on facing pages, in 1717.[8]

In reply to the criticisms contained in Leibniz's first letter, Clarke claimed that Newton had not described space as God's sensorium in a literal sense but only figuratively and attacked Leibniz's idea that the world was like a perfect clock as the notion of a materialist, which excludes God's providence and government. After pointing out to Clarke that, in the *Opticks*,[9] Newton had expressly described space as God's sensorium, Leibniz explained that, far from excluding God's influence, he held that God continually creates the universe. True providence, however, required that God must have foreseen everything and provided for it. The kind of intervention envisaged by Clarke would be either supernatural or natural. If supernatural, then natural things could not be explained without recourse to miracles. If natural, then God would be comprehended under the nature of things; that is, God would be the soul of the world. He hoped Clarke and his friends did not hold this heretical doctrine and advised them to take care that they did not fall into it unawares. Against the charge of materialism, Leibniz pointed out that, in the *Theodicy*, he had demonstrated metaphysical principles which were opposed to such a view. While the foundation of mathematics is the principle of contradiction, he explains, to proceed from there to natural philosophy another principle is needed, namely the principle of sufficient reason. By means of this principle, which states that nothing happens without a reason

[8]The letters have been published in the original languages in *GP* 7, pp 347–440. There is an English translation by Alexander (1956).

[9]More precisely, in the Latin translation of the *Opticks*. It seems that Leibniz read a copy of this translation, *Optice* (1706), in which space is expressly described as God's sensorium. In most copies, however, the relevant page had been replaced (before binding) by a substitute, in which space is described as being 'as it were God's sensorium'. See Koyré and Cohen (1961).

why it should be so and not otherwise, he had demonstrated the existence of God and all the other parts of metaphysics or natural theology. For example, the principle could be used to demonstrate the error of the Newtonians in supposing the existence of a void. For the more matter there is, the more God has occasion to exercise his wisdom and power, so that God's wisdom and power is the sufficient reason for a plenum.

When Caroline sent Clarke's second letter to Leibniz on 10 January 1716 (*K* **11**, pp 71–3), she remarked that Clarke's replies were not written without the advice of Newton. With the help of Conti, she added, she would like to effect a reconciliation between himself and Newton. On 25 February, Leibniz (*K* **11**, pp 78–9) sent his third letter for Clarke, at the same time explaining to Caroline the obstacles in the way of a reconciliation with Newton. Newton, he told her, had connived at the efforts of others to attack him and had even employed the name of the Royal Society. He believed that the Royal Society owed it to him to make a declaration in his favour.

The chief complaint of Clarke in his second letter was that Leibniz, by supposing that God cannot exercise an arbitrary act of will (that is, act without a predetermined cause) has in fact deprived God of all power of choosing. In his reply, Leibniz pointed out to Clarke that he had granted the principle of sufficient reason in words but denied it in reality, for when he claimed that this sufficient reason was often the mere will of God, he was asserting that God sometimes acts without a sufficient reason for his choice. Leibniz, on the other hand, maintains that God has a power of choice which he exercises in accordance with his wisdom, so that his choice is always for the best. Since Clarke had taken his concept of real space as an example to illustrate his point, Leibniz took the opportunity to demonstrate how the principle of sufficient reason supported his own relational view of space. For if space were something absolute and real, there could be no sufficient reason why God should have placed bodies in it in one manner rather than in another which preserved the relative positions. The same was true of time, for if time were something absolute and real, there could be no sufficient reason why God should create the universe at one time rather than another.

In his third letter, Clarke repeated his belief in the reality of space and the mere will of God as the sufficient reason for many things. In addition, he claimed that the natural diminution of the active forces (that is, motion) in the universe, so that, in the words of Newton's *Opticks*, 'there is a necessity of conserving and recruiting it by active principles', implied no imperfection in God's workmanship. The analogy with a man-made machine, he asserted, did not apply, because such a machine continued to work independently of the artificer, whereas the universe was continually dependent on its creator.

Leibniz sent his fourth letter for Clarke to Caroline on 12 May (*K* **11**, pp 100–3), at the same time complaining that Conti was like a chameleon, for he seemed to have been converted to some of the Newtonian doctrines. Caroline (*K* **11**, pp 90–1) confessed that she had herself been almost converted to a belief in the void by the clear reasoning of Clarke, who spent many hours in conversation with her, often in the company of Conti and sometimes of Newton as well. For her own enlightenment, Leibniz sent her a paper on the void (separate from the letter for Clarke), which she showed to the King, in order to keep Leibniz in his thoughts (*K* **11**, pp 112–13).

In support of his view that, on some occasions at least, God had to make an arbitrary choice, Clarke had argued that, even if space were ideal as Leibniz supposed, there could be no reason why three equal particles were placed in a particular order rather than the reverse. This gave Leibniz the opportunity to explain his principle of the identity of indiscernibles. Having described the incident in the park at Herrenhausen concerning the identical leaves, he inferred from the principle that the situation envisaged by Clarke was an impossible fiction. Following some scathing words about the idea of space as God's sensorium and a lengthy account of his explanation of the relation of mind and body, Leibniz turned to the other main point of Clarke's third letter, which he dismissed with the remark that those who fancied active force (*vis viva*) to diminish of itself in the world did not well understand the principal laws of nature.

On 26 June, Caroline (*K* **11**, pp 114–16) sent Leibniz Clarke's fourth letter together with a renewed plea for a reconciliation with Newton. What did it matter, she asked, whether he or Newton invented the calculus? They were the great men of the century. She also described a visit she had made to Greenwich, where Flamsteed received her in the Observatory. Leibniz, he had told her, was an honest man but Newton a great knave, since he had stolen two stars from him. She had not been able to prevent herself from laughing. Both his house and his appearance, she observed, had the air of Merlin.

Writing to Caroline at the end of July (*K* **11**, pp 128–30), Leibniz sought to justify his continuing animosity towards Newton, suggesting that if she knew how Newton's followers had attacked him, she would approve his moderation. On 18 August, he sent her the first half of his fifth letter for Clarke, promising the remainder by the next post. Having heard that Pierre Des Maizeaux, a member of the Royal Society, had found a translator for the *Theodicy*, he asked Caroline for permission to dedicate the translation to her and to state in the preface that she herself had wished the translation to be made. To this she readily agreed.

Leibniz had become increasingly frustrated by Clarke's failure to understand his philosophy—he wondered whether Clarke had read the *Theodicy* —and he hoped that the explanations in his fifth letter would bring Clarke to see reason. If not, then he wished to bring the correspondence to an end. In his fourth letter, Clarke had dismissed the pre-established harmony as a mere word that in no way explained the cause of so miraculous an effect. Leibniz therefore explained in some detail the relation between soul and body by means of the principle of pre-established harmony. This harmony, he declared, was not a perpetual miracle, as Clarke supposed, but an original miracle, worked at the creation, as all natural things were. Since a simple substance, soul or monad was such that all its following states were consequences of the preceding state, God needed only at the beginning to make it a representation of the universe from its point of view, for it would then remain so perpetually. Souls perceived what passed without by what passed within, answering to the things without in virtue of the pre-established harmony, whereby every simple substance was by its nature a concentration and living mirror of the whole universe, according to its point of view. Again he explained to Clarke that soul and body did not disturb each other's laws; the soul acted freely, according to the rules of final causes, while the body acted mechanically, according to the laws of efficient causes. This was possible because God, foreseeing what free choices the soul would make, regulated the machine from the beginning so that it could not fail to agree.

In the course of his letter Leibniz clarified a number of points relating to the running down of the universe, the nature of space, the existence of a void and the freedom of the will. Clarke had objected that two inelastic bodies meeting together lose some of their force. Leibniz replies that this is not so. Their wholes lose it but their parts receive it. In other words, force (*vis viva*) is not lost but scattered among the small parts. Consequently the total force is not diminished and no winding up of the universe is needed. Clarke claimed space to be a property and objected that the resemblance of the parts of space contradicted Leibniz's principle of the identity of indiscernibles. In reply, Leibniz first dismissed the idea that space was a property—for there was nothing that it could be a property of—and then explained that, when he said two drops of water were not identical, he did not mean that it was impossible to conceive them so (as Clarke supposed) but that such a thing was contrary to divine wisdom. There was no difficulty in conceiving parts of space and time to be identical because they were ideal, as he had demonstrated by the principle of sufficient reason. Nevertheless, he distinguished between an absolute true motion of a body and a mere relative change. For when the immediate cause of the change was in the body, that body was truly in motion. Since no body was completely inert, it followed that no body was entirely at rest.

Clarke had referred to the experiments of Guericke as evidence for the existence of a void. Leibniz declared himself to be of the opinion of the Aristotelians and Cartesians that there was not a true void. Referring to the experiments of Torricelli, he remarked that glass had small pores which beams of light, the effluvia of the lode-stone and other very thin fluids could pass through. The fluid causing gravity, he believed, was of this kind; attraction, he repeated, was an occult quality.

On the question of the freedom of the will, Leibniz accused Clarke of unreasonable obstinacy in refusing to understand the arguments set out in the *Theodicy*. There he had carefully distinguished between an absolute necessity (where the opposite implies a contradiction) and a moral necessity, whereby a wise being chooses the best and every soul follows the strongest inclination. Moral necessity, he again explained at some length, in no way derogates from liberty. If Clarke's reply to his final letter had arrived in time for him to give attention to it, Leibniz would have been disappointed to find that his last efforts to bring his antagonist to see reason had evidently failed.

The last months in Hanover

In the spring of 1716, Leibniz and Jablonski took up again the question of the reunion of the two evangelical Churches. This they did at the request of King Friedrich Wilhelm I in Prussia, who wished to enter into negotiations also with the Anglican Church (Bodemann 1895, p 101). Recounting to Caroline the earlier negotiations, whose failure he attributed to the lack of resolve of the late king in Prussia (who allowed himself to be influenced by the unreasonable demands of the Pietists), Leibniz declared that the possibilities for reunion had never been so favourable. On the one hand, the Elector of Hanover had become King of Great Britain and entered the Anglican Church without changing his religion, from which it could be inferred that he saw no essential difference in the religion of the Anglican and Lutheran Churches. On the other hand, the Anglicans believed they had the same religion as the Reformed Church. From these statements, taken as premises, it followed that the Lutheran and Reformed Churches had the same religion. Thus they were in fact already united and it only remained to bring this to the understanding of the people by a declaration of the kings supported by the theologians. Leibniz suggested to Caroline that she should speak secretly to someone who could put it to the Archbishop of Canterbury. Having written his letter, Leibniz heard that the Archbishop had died and that the King had chosen Caroline's friend the Bishop of Lincoln as his successor. He therefore added a postscript suggesting that she should speak to the new Archbishop herself, stressing that it would be better if the initiative appeared to come entirely from the English. On no account, he added, should his name be involved (*K* **11**, pp 85–90). In her letter of 26 May 1716 (*K* **11**, pp 112–13) Caroline reported her belief that the King regarded reunion to be without practical value and almost impossible for the bishops, who would raise difficulties concerning the validity of orders. Owing to this lack of interest of the king, the official negotiations between the Anglican and Lutheran Churches, for which Leibniz had hoped, did not come to pass. In the first half of October, however, he invited Jablonski to Wolfenbüttel for private discussion of the proposals Jablonski had submitted to the King in Prussia for the reunion of the Lutheran and Reformed Churches. Reporting on these conversations in a letter to von Printzen on 3 November,

Leibniz expressed his hopes for the progress of the Society of Sciences in Berlin under the protection of the new king.

Although he suffered from gout and arthritis, these ills were tolerable and even without pain when he rested, as he remarked to Remond in a letter of 27 March 1716 (*GP* **3**, pp 673–5). They had not prevented him from making a visit to Brunswick to see the Duchess before she visited Vienna for the confinement of her daughter Elizabeth, the reigning Empress. Later in the year, when he visited the Brunswick fair, he spent a few days with Duke Ferdinand Albrecht II of Brunswick-Bevern and his wife, who was a sister of the Empress. The Princess of Bevern, he reported to Caroline (*K* **11**, pp 182–6), was as radiant as her sister.

Soon after returning from the Brunswick fair, Leibniz heard rumours from Vienna that his salary as Imperial Privy Counsellor was to be suspended. On 20 September, he wrote separately to the dowager Empress Amalia (*K* **11**, pp 192–5), her lady-in-waiting Fräulein von Klenck (*K* **11**, pp 191–2) and his friend Theobald Schöttel, who managed his financial affairs in Vienna, asking for their help, for if the rumours were true, he would be unable to return to Vienna as he had hoped. The writing of the history of the House of Brunswick, he explained, was almost finished, and as this was the only thing keeping him in Hanover, he was planning to return to Vienna as soon as the King had crossed the sea again to England. He felt degraded to be regarded as holding only an honorary title, despite the decree of the Emperor, his actual services, the very large double-tax he had to pay and the considerable expenses, which had absorbed all the pension he had received so far, besides the loss of two years he had employed in Vienna, which at his age was a great price. Almost from his youth he had rendered service to the Empire by his historical researches and other works, while he had begun to choose people of merit for membership of the Society of Sciences as soon as it would be effectively founded. Convinced that the rumour was absolutely false, Fräulein von Klenck (*K* **11**, pp 195–6) resolved with Schöttel to prevent the Empress from taking any action until they would see whether the next instalment, due in a few days, would be paid. Karl Gustav Heraeus (*K* **11**, pp 233–5), the Emperor's Inspector of Antiquities, explained the situation in a letter written on 18 November, too late for Leibniz to receive. It was true that all the pensions of the old Counsellors had been abolished, and the case of Leibniz needed to be clarified. He owed thanks to Amalia and Fräulein von Klenck for their support, but all the learned ministers recognised his merit and the Vice-Chancellor did not imagine that the Emperor ever intended that he should be included among those whose pensions were to be deleted.

Princess Caroline still expected to see Leibniz in England and, in a letter to him at the end of August (*K* **11**, pp 181–2), she expressed the hope that the King would bring him to London when he returned from his visit to Hanover. In his reply (*K* **11**, pp 186–90) Leibniz held out little hope of seeing her again. The history, he explained, would occupy a good part of the following year and the state of his health made it uncertain whether he would be able to go to England later. In the same letter, he told her that her young son, whom she had asked him to visit as often as he could, chased the rats and charmed the visitors from England, though he was as yet unable to converse in English. Writing to Leibniz on 15 September (*K* **11**, p 190), Caroline explained that she was not to blame for the fact that her son had not started to learn English. She feared that those responsible for his education would leave it too late for him to acquire a good accent. She was sorry to hear that Leibniz would not soon be coming to London. There was no reason, she suggested, why he should not work at the history in London rather than Hanover, so that his friends could have the pleasure of his company and conversation.

At the beginning of October, Leibniz reported to Hans Caspar von Bothmer in London on the progress of the history. In reply, Bothmer (*K* **11**, p 198) expressed his belief that, in recognition of his industry and devotion to his work, the Prime Minister von Bernstorff would use his influence with the King, on his return, to secure for Leibniz the appointment as historiographer of England that he desired. In fact Leibniz did not complete the history of the House of Brunswick[10] and he did not have to decide whether to go to London or Vienna, for the end was closer than he had anticipated. In the first few days of November, gout came to his hands and he had to stop writing. After lying in bed for eight days, attended by his amanuensis Johann Hermann Vogler and his coachman Heinrich, he agreed on the evening of Friday 13 November to consult a doctor.[11] As Dr Seip, who had earlier in the year befriended him in Pyrmont, was in Hanover, he was sent for. He informed Leibniz that his condition was serious and prescribed medicine which brought him a little rest during the night. At noon on the following day, Dr Seip confirmed that there was no chance of recovery and Vogler asked to be allowed to call a lawyer and Lutheran pastor. Leibniz gently declined, saying there would be time the next day. In the evening, Vogler was in the ante-room for a moment when he heard a noise and hurried back to find Leibniz attempting to burn a sheet of paper in the candle

[10]Leibniz had originally planned to cover the history up to Ernst August, but the difficulties encountered in the early part led him to confine his attention to the *Annales* from 768 to 1024 and leave to his successors the writing of the later history, for which the source material was to be found in the House Archives. After his death, efforts were made by a succession of historiographers to bring the work to completion, until in 1746, George II decided that the genealogical treatises prepared by Eckhart should be published first. Then nothing was done for seventy years until G H Pertz was appointed historiographer in 1832 and published the *Annales Imperii Occidentis Brunsvicenses* in three volumes in 1843–6.

[11]An eye-witness account of Leibniz's death and funeral is given by Vogler in two letters, which have been published by Ritter (1916).

flame. Meanwhile the coachman had also returned. Dr Seip called again at 9 PM and an hour later, at 10 PM on Saturday 14 November 1716, Leibniz died peacefully at the age of 70 years. Vogler immediately informed Eckhart, Leibniz's secretary, who undertook all the formalities.

On Sunday evening the coffin was taken to the Neustädter Church. After the arrival of his only heir, his sister's son Friedrich Simon Löffler, on 26 November, an official Commission was entrusted with the examination of the legacy. Besides his library, manuscripts and calculating machine,[12] which were placed in the care of the Electoral Library, he left a black box containing money and securities in excess of 12 000 taler.

The funeral and burial took place on 14 December in the Neustädter Church. Eckhart had covered the coffin in black velvet and placed Leibniz's coat-of-arms at the head. The Chief Court Chaplain, H Erythropel, conducted the service, which was accompanied by a school choir, in the presence of Leibniz's nearest relations and acquaintances. Eckhart relates, in his biography, that although the whole Court was invited, no-one (apart from himself) appeared (Eckhart 1779, p 192). At the time, the King and his retinue were at the hunting lodge in Göhrde, near Lüneburg, within easy reach of Hanover. Eckhart speculates that this complete neglect by the King and his ministers may have been a result of the widely held opinion that Leibniz was an unbeliever. During the nineteen years of their association, Eckhart had never known him to take communion. On many occasions, he recalled, Leibniz had described himself as a priest of natural justice and remarked that he found nothing other than this in the New Testament. Eckhart testifies further that Leibniz always spoke well of everyone and made the best of everything. A more likely reason for the official indifference would seem to be the fact that he had incurred the King's displeasure.

John Ker of Kersland (*K* **11**, pp xxxv–vi), later recalling an act of Christian charity, when Leibniz without his knowledge had settled his debts in Vienna where he had been in financial difficulties, described how he had arrived in Hanover on the day his friend died and had been surprised to see how little regard was paid to his ashes by the Hanoverians, for he was buried more like a robber than what he really was, the ornament of his country. Neither the Court he had served for over forty years, nor his nephew, who had inherited a handsome sum of money, commissioned any monument, and for over fifty years following his death, his grave was unmarked. Today it is identified by a copper plate with the bare inscription 'Ossa Leibnitii'.

Neither the Royal Society nor the Berlin Academy of Sciences marked the

[12]His nephew Johann Freiesleben also made a claim on the estate, but this was refused. There followed a long dispute between the Government of Hanover and the Löffler family concerning the payment of a fair sum for the books and manuscripts retained in the Electoral Library (Lackmann 1969).

death of Leibniz with a memorial address or publication. On 13 November 1717, however, Fontenelle read an Éloge to the Paris Academy of Sciences. This was based on a short biography which the Duchess of Orleans had obtained from Eckhart for the purpose (Eckhart 1779, p 125). When she heard the details of Leibniz's death, the Duchess wrote to Friedrich von Harling in Hanover, that Leibniz did not need to have priests about him; they could teach him nothing, for he knew more than they knew, and she had no doubts about his salvation (Bodemann 1895a, pp 102–4).[13] Thus it was Sophie's niece, Liselotte (whom he had never met), who remained his most loyal friend among the great and secured for him in a foreign land the ceremonial recognition that his own countrymen had been either too indifferent or too bigoted to give.

[13]Friedrich von Harling was the husband of the late Anna Katharina, who had been governess of Sophie's children and of the Duchess herself. In her last letter to Leibniz, written a few days before his death, the Duchess confided to him her feelings of sorrow when she reflected that the old Duchess of Celle (who had just been visited by her granddaughter the Queen in Prussia) still lived, while her beloved aunt Sophie was no more (Bodemann 1884, pp 53–4).

Epilogue

Almost the whole of Leibniz's legacy of manuscripts and personal library is still preserved today in the Niedersächsische Landesbibliothek, the successor of the Electoral (and later Royal) Library in Hanover. At the end of his biography, Eckhart, who succeeded Leibniz as Librarian, declared his intention to publish the writings he had left behind. A complete edition of Leibniz's works, he estimated, would require three volumes. In the first he intended to bring together all the writings already published. The second volume would include all the completed unpublished writings, while the third volume would consist of Leibnitiana, including Leibniz's thoughts, discourses and short reflections on all manner of themes and also his poetry. Although Eckhart published a selection of Leibniz's linguistic writings under the title *Collectanea Etymologica*, his plan for an edition of Leibniz's collected works came to nothing.

In 1718 J F Feller, who had been Leibniz's secretary from 1696 to 1698, edited and published a volume of Leibniz's correspondence over these years. Three further volumes of letters, collected by Sebastian Kortholt, were published by his son, Christian, between 1734 and 1738. Yet another three important selections of letters appeared in 1745. These were J M Bousquet's edition of the correspondence with Johann Bernoulli, J E Kapp's edition of German writings including part of the correspondence with Jablonski, and the first two volumes of J D Gruber's projected edition of Leibniz's complete correspondence. Only three letters to or from Leibniz were included in Gruber's volumes, which were devoted to the correspondence of Boineburg and Conring from 1651, an unnecessarily lengthy introduction since Leibniz first met Boineburg in 1667. Gruber's health prevented him from continuing the project. In 1749, C L Scheidt, then Librarian in Hanover, published the *Protogaea* in Latin and German. This was followed by four volumes on the origin of the Guelfs, which had been prepared by Eckhart as the first stage in his plan for the history of the House of Brunswick. In 1765, R E Raspe published a volume of Leibniz's philosophical writings, which included the *Nouveaux Essais* as the principal item.

The first collected edition of Leibniz's works was that of L Dutens, published in six volumes in Geneva in 1768. Although he was refused access to the manuscripts in Hanover, Dutens included unpublished letters of Leibniz which he located in the libraries of France and Italy besides works and letters scattered in journals and the publications of other authors. However, he missed the publications of Bousquet, Kapp and Raspe.

At the beginning of the nineteenth century, J G H Feder, who was then Librarian in Hanover, planned to continue the project of Gruber, treating Leibniz's correspondents systematically in alphabetical order. He achieved only one unsatisfactory volume, having failed to find the correspondence with Arnauld. In 1837 G E Guhrauer published a second edition of *De principio individui*, which had become so rare even in the eighteenth century that Dutens had been unable to find a copy. This was followed by two volumes of the German writings (intended to show the importance of Leibniz for the history of German literature) and a volume on the Electorate of Mainz in 1672, which included the documents relating to Leibniz's political mission in Paris. In 1842, he published the first edition of his biography of Leibniz, citing many French and Latin letters in rather poor German translations. A second, improved edition was published in 1846, on the occasion of the bicentenary of Leibniz's birth. J E Erdmann published three volumes of Leibniz's philosophical works in 1840, attracting attention to his universal characteristic, logic and scientific method.

Another nineteenth century Librarian in Hanover, G H Pertz, embarked on the gigantic enterprise of bringing out a complete edition of Leibniz's works. He undertook the historical work himself, at last completing the history of the House of Brunswick in three volumes, which he followed with a fourth volume of writings to mark various political occasions, arranged chronologically. The philosophical works were assigned to C L Grotefend, who published the correspondence with Arnauld and Ernst von Hessen-Rheinfels. The mathematical works, assigned to C I Gerhardt, were published between 1849 and 1860 in seven volumes. This edition included for the first time mathematical manuscripts in the Hanover Library. L A Foucher de Careil proposed a collected edition of Leibniz's works to the Academy of Sciences in Vienna. Between 1859 and 1875 he published seven volumes, including (in some disorder) documents on the reunion of the Catholics and Protestants, history and politics. O Klopp published between 1864 and 1884 eleven volumes of historical and political writings arranged in chronological order. These included the correspondence with Sophie, Sophie Charlotte and Caroline, besides documents relating to the foundation of the

Society (later Academy) of Sciences in Berlin. C I Gerhardt at last produced (between 1875 and 1890) a reasonably complete edition of the philosophical works in seven volumes.

At the beginning of the twentieth century, two important collections of manuscripts were published by L Couturat, relating to logic, and E Gerland, concerning questions of physics, mechanics and technology. A definitive edition of Leibniz's works, in seven series, by the Prussian Academy of Sciences (today the Akademie der Wissenschaften der DDR), has been in progress since 1923. Eleven volumes have been published in the first series, giving the general, political and historical correspondence to October 1695. Only a few volumes in the other series have so far appeared. These include four volumes of philosophical writings and one volume of mathematical, scientific and technical correspondence, though the second volume is in the final stages of preparation. At the present time, however, only a small fraction (perhaps as little as 10 per cent) of the mass of manuscript material on mathematics and science in the possession of the Niedersächsische Landesbibliothek exists in any printed edition.

Leibniz was not well served by his disciple Christian Wolff, under whose influence a simplified and materialised version of his philosophy came to be taught in the universities. Through the *Nouveaux Essais*, however, his theory of knowledge, with its emphasis on the activity of the mind (which for him included the unconscious) had a profound effect on Kant, who completed the distinction between necessary and contingent propositions by the introduction of the synthetic *a priori*. Although the *Theodicy* retained its popularity in the eighteenth century, despite the ridicule of Voltaire, the doctrine of monads and their pre-established harmony ultimately suffered the common fate of speculative metaphysical theories. Yet this theory was not without influence, for Boscovitch, one of the acknowledged founders of field theory, pointed to the analogy of Leibniz's theory with his own system of point-centres of force, completely interrelated and functioning in harmony with one another. As his starting point for the derivation of this physical interpretation of the monadology, Boscovitch used another principle of Leibniz, the principle of continuity. The wider problems posed by the profound issues with which Leibniz was concerned in relation to his metaphysical theory of monads, such as the dualism of mind and matter, the origin of evil and the responsibility of man for his actions in a world of physical determinism, are today no nearer solution. What he had attempted was no less than a reconciliation of a mechanical determinist physics with a rationalist spiritual metaphysics that would allow freedom of the will for God and man.

References and bibliography

References

Aarsleff H 1975 Schulenburg's *Leibniz als Sprachforscher SL* **7** 123–34

Aiton E J 1962 The celestial mechanics of Leibniz in the light of Newtonian criticism *Annals of Science* **18** 31–41

—— 1964 The inverse problem of central forces *Ann. Sci.* **20** 81–99

—— 1972a Leibniz on motion in a resisting medium *Archive for history of exact sciences* **9** 257–74

—— 1972b *The vortex theory of planetary motions* (London and New York: Macdonald and Elsevier)

—— 1981 An unpublished letter of Leibniz to Sloane *Ann. Sci.* **38** 103–7

—— 1984 The mathematical basis of Leibniz's theory of planetary motion *SL Sonderheft* **13** 209–25

Aiton E J and Shimao E 1981 Gorai Kinzō's study of Leibniz and the *I ching* hexagrams *Ann. Sci.* **38** 71–92

Alexander H G 1956 *The Leibniz–Clarke correspondence* (Manchester: Manchester University Press)

Baruzi J 1905 *Revue de métaphysique et de morale* **13** 21–7

Becco A 1978 Leibniz et François-Mercure van Helmont: bagatelle pour des monades *SL Sonderheft* **7** 119–42

Birch T 1756–7 *The history of the Royal Society of London* 4 vols (reprinted 1968 New York: Johnson)

Blondel M 1972 *Le lien substantiel et la substance composée d'après Leibniz* (Paris: Nauwelaerts)

Bodemann E 1884 Briefwechsel zwischen Leibniz und Herzogin Elisabeth Charlotte von Orléans (1715–16) *ZHN* 1–66

—— 1888 Leibnizens Briefwechsel mit dem Herzog Anton Ulrich *ZHN* 73–244

—— 1889 *Die Leibniz-Handschriften* (reprinted 1966 Hildesheim: Olms)
—— 1895 *Der Briefwechsel des G. W. Leibniz* (reprinted 1966 Hildesheim: Olms)
—— 1895a *Briefe der Herzogin Elisabeth Charlotte von Orléans . . . und deren Gemahl* (Hanover and Leipzig)
Child J M 1920 *The early mathematical manuscripts of Leibniz* (London)
Cohen I B 1962 Leibniz on elliptical orbits *Journal of the history of medicine* **17** 72–82
Costabel P 1960 *Leibniz et la dynamique* (Paris: Hermann) (English translation by R E W Maddison 1973 London: Methuen)
Couturat L 1901 *La logique de Leibniz* (reprinted 1961 Hildesheim: Olms)
—— 1903 *Opuscules et fragments inédits de Leibniz* (reprinted 1966 Hildesheim: Olms)
Doebner R 1881 Leibnizens Briefwechsel mit dem Minister von Bernstorff *ZHN* 205–380
Eberhard J A 1795 *Gottfried Wilhelm Freyherr von Leibnitz* (reprinted in Eberhard J A and Eckhart J G 1982 *Leibniz-Biographien* Hildesheim: Olms)
Eckhart J G 1779 *Lebensbeschreibung des Freyherrn von Leibniz* (reprinted in Eberhard J A and Eckhart J G 1982 *Leibniz-Biographien* Hildesheim: Olms)
Fellmann E A 1973 *G. W. Leibniz. Marginalia in Newtoni Principia Mathematica* (Paris: Vrin)
Femiano S 1982 Über den Briefwechsel zwischen Michelangelo Fardella und Leibniz *SL* **14** 154–83
Forberger R 1964 J. D. Crafft. Notizen zu einer Biographie *Jahrbuch für Wirtschaftsgeschichte* Teil II/III (Berlin: Akademie Verlag) 63–79
Gerhardt C I 1888 *Archiv für Geschichte der Philosophie* **1** 375–81
Gerland E 1906 *Leibnizens nachgelassene Schriften physikalischen, mechanischen und technischen Inhalts* (Leipzig)
Gorai Kinzō 1929 *The influence of Confucianism on German political thought* (in Japanese) (Tokyo)
Grieser R 1969 Korrespondenten von G. W. Leibniz. Johannes Teyler, *SL* **1** 208–27
Guerrier W 1873 *Leibniz in seinem Beziehungen zu Russland und Peter dem Grossen* (St Petersburg and Leipzig)
Guhrauer G E 1839 *Kur-Mainz in der Epoche von 1672* (Hamburg)
—— 1846 *Gottfried Wilhelm Freiherr von Leibniz: Eine Biographie* 2 vols (reprinted with Appendix 1966 Hildesheim: Olms)
Haase R 1982 *Der Briefwechsel zwischen Leibniz und Conrad Henfling* (Frankfurt am Main: Klostermann)
Hankins O 1972 An epicedium by Leibniz on the death of a queen *SL* **4** 1–18
Hankins T L 1965 Eighteenth century attempts to resolve the *vis viva* controversy *Isis* **56** 281–97

Harnack A 1900 *Geschichte der königlich Preussischen Akademie der Wissenschaften zu Berlin* 3 vols (reprinted 1970 Hildesheim: Olms)
Heinekamp A 1979 Der Briefwechsel zwischen Leibniz und Christian Thomasius *SL* **11** 92–7
Hess H J 1978 Die unveröffentlichten Naturwissenschaftlichen und Technischen Arbeiten von G. W. Leibniz aus der Zeit seines Parisaufenthaltes *SL Supplementa* **17** 183–217
—— 1980 Bücher aus dem Besitz von Christiaan Huygens in der Niedersächsischen Landesbibliothek, Hanover *SL* **12** 1–51
Hestermeyer W 1969 *Paedagogia Mathematica* (Paderborn: Schöningh)
Hofmann J E 1969 Leibniz und Ozanams Problem *SL* **1** 103–26
—— 1973 Leibniz und Wallis *SL* **5** 245–81
—— 1974 *Leibniz in Paris 1672–1676* (London: Cambridge University Press)
Hohnstein O 1908 *Geschichte des Herzogtums Braunschweig* (reprinted 1979 Hanover-Döhren: Harra v Hirschleydt)
Horst U und Gottschalk J 1973 Über die Leibniz'schen Pläne zum Einsatz seiner Horizontalwindkunst im Oberharzer Bergbau and ihre missglückte Durchführung *SL Supplementa* **12** 35–59
L'Hospital G F A de 1696 *Analyse des infiniment petits* (new edition 1768 Avignon)
Iltis C 1970 D'Alembert and the *vis viva* controversy *Studies in history and philosophy of science* **1** 135–44
Jurgens M and Orzschig J 1984 Korrespondenten von G. W. Leibniz. Christophe Brosseau *SL* **16** 102–12
Kalinowski G 1977 La logique juridique de Leibniz *SL* **9** 168–89
Kangro H 1969 Joachim Jungius und Gottfried Wilhelm Leibniz *SL* **1** 175–207
Keill J 1708 *Philosophical Transactions* **26** (published 1710) 174–88
Knobloch E 1972 Die Entscheidende Abhandlung von Leibniz zur Theorie linearer Gleichungssysteme *SL* **4** 163–80
—— 1973 *Die mathematische Studien von G. W. Leibniz zur Kombinatorik* (*SL Supplementa* **11**. See also the Textband with the same title, *SL Supplementa* **16**, published in 1976.)
—— 1974 Studien von Leibniz zum Determinanten Kalkül *SL Supplementa* **13** 37–45
—— 1976 *Ein dialog zur Einführung in die Arithmetik und Algebra* (Stuttgart-Bad Cannstatt: frommann-holzboog)
—— 1980 *Der Beginn der Determinantentheorie* (Hildesheim: Gerstenberg)
—— 1982 Zur Vorgeschichte der Determinantentheorie *SL Supplementa* **22** 96–118
Korcik A 1967 La 'Defensio Trinitatis contra Wissowatium' de Leibniz *Organon* (Warsaw) **4** 181–6
Koyré A and Cohen I B 1961 The case of the missing Tanquam *Isis* **52** 555–66

Kroker E 1898 Leibnizens Vorfahren *Neues Archiv für sächsische Geschichte* **19** 315–38
Lackmann H 1969 Der Erbschaftsstreit um Leibniz' Privatbibliothek *SL* **1** 126–36
Lohne J A 1966 Fermat, Newton, Leibniz und das anaklastische Problem *Nordisk Matematisk Tidskrift* **14** 5–24
Loosen R and Vonessen F 1968 *G. W. Leibniz. Zwei Briefe über das binäre Zahlensystem und die Chinesische Philosophie* (Stuttgart: Belser)
Ludovici C G 1737 *Ausführlicher Entwurf eines vollständigen Historie der Leibnitzischen Philosophie* 2 vols (Leipzig)
Mccullough L 1978 Leibniz and traditional philosophy *SL* **10** 254–70
Mackensen L von 1974 Leibniz als Ahnherr der Kybernetik—ein bisher unbekannter Leibnizischer Vorschlag einer 'machina arithmeticae dyadicae' *SL Supplementa* **13** 255–68
Marchi P 1974 The controversy between Leibniz and Bernoulli on the nature of the logarithms of negative numbers *SL Supplementa* **13** 67–75
Meinecke F 1959 *Die Entstehung des Historismus* (Munich: Oldenbourg)
Miller S T J and Spielman J P 1962 Cristobal Rojas y Spinola *Transactions of the American Philosophical Society* **52** (part 5)
Mittelstrass J 1970 *Neuzeit und Aufklärung* (Berlin: De Gruyter)
Moll K 1978 *Der junge Leibniz I* (Stuttgart-Bad Cannstatt: frommann-holzboog)
—— 1982 *Der junge Leibniz II* (Stuttgart-Bad Cannstatt: frommann-holzboog)
Müller K 1966 Gottfried Wilhelm Leibniz, in Totok W and Haase C (eds) *Leibniz. Sein Leben—sein Wirken—seine Welt* (Hanover)
Mungello D E 1977 *Leibniz and Confucianism. The search for accord* (Honolulu: University Press of Hawaii)
Münzenmayer H P 1976 Leibniz' Inventum Memorabile. Die Konzeption einer Drehzahlregelung vom März 1686 *SL* **8** 113–19
—— 1979 Der Calculus Situs und die Grundlagen der Geometrie bei Leibniz *SL* **11** 274–300
Needham J 1959 *Science and civilisation in China* vol **3** (London: Cambridge University Press)
Quillet J 1979 Disputation métaphysique sur le principe d'individuation de G. W. Leibniz *Les études philosophiques* **1** 79–105
Ravier E 1937 *Bibliographie des oeuvres de Leibniz* (reprinted 1966 Hildesheim: Olms)
Ritter P 1916 Bericht eines Augenzeugen über Leibnizens Tod und Begräbnis *ZHN* **81** 247–52
Robinet A 1981 Suarez im Werk von Leibniz *SL* **13** 76–96
Rosemont H and Cook D J 1977 *G. W. Leibniz. Discourse on the natural theology of the Chinese* (Honolulu: University Press of Hawaii)
Ross G M 1974 Leibniz and the Nuremberg Alchemical Society *SL* **6** 222–48

Saint-Germain B de 1859 *Protogée ou de la formation des révolutions de globe par Leibniz* (Paris)
Schmiedecke A 1969 Leibniz' Beziehungen zu Zeitz *SL* **1** 137–44
Schnath G 1952 *Der Königsmarck-Briefwechsel. Korrespondenz der Princessin Sophie Dorothea von Hannover mit dem Grafen Philipp Christoph Königsmarck* (Hildesheim: August Lax)
—— 1978 *Geschichte Hannovers im Zeitalter der neunten Kur und der englischen Sukzession* vol **3** (Hildesheim: August Lax)
Schulenburg S von der 1973 *Leibniz als Sprachforscher* (Frankfurt am Main: Klostermann)
Scriba C J 1964 The inverse method of tangents *Archive for history of exact sciences* **2** 113–37
Spangenberg E 1826 *Historisch- topographisch- statistische Beschreibung der Stadt Celle* (reprinted 1979 Hanover-Döhren: Harro v Hirschleydt)
Spiess O 1955 *Der Briefwechsel von Johann Bernoulli* vol **1** (Basel: Birkhäuser)
Stiegler L 1968 Leibnizens Versuche mit dem Horizontalwindkunst auf dem Harz *Technikgeschichte* **35** 265–92
Uffenbach Z K 1753 *Merkwürdige Reisen durch Niedersachsen, Holland und England* vol **1** (Ulm and Memmingen)
Utermöhlen Gerda 1979 Leibniz' Antwort auf Christian Thomasius' Frage Quid sit substantia? *SL* **11** 82–91
Voisé W 1967 Leibniz's model of political thinking *Organon* **4** 187–205
Westfall R S 1980 *Never at rest. A biography of Isaac Newton* (Cambridge: Cambridge University Press)
Whiteside D T 1970 The mathematical principles underlying Newton's 'Principia Mathematica' *Journal of the history of astronomy* **1** 116–38
Widmaier Rita 1981 Die Rolle der chinesischen Schrift in Leibniz' Zeichentheorie *SL* **13** 278–98
Yates F A 1972 *The Rosicrucian Enlightenment* (London: Routledge and Kegan Paul)
Zacher H J 1973 *Die Hauptschriften zur Dyadik von G. W. Leibniz* (Frankfurt am Main: Klostermann)

Bibliography

The literature on Leibniz is so extensive that a complete presentation would occupy a large volume. The following selection, intended as a supplement to the sources listed above, contains classics and recent works of particular interest besides translations of some of Leibniz's major works and selections of his writings in English translation.

Further bibliographical information may be found in Ravier (1937), Müller (1967) and the annual bibliographies since 1969 in *Studia*

Leibnitiana. Information on the Leibniz manuscripts in the Niedersächsische Landesbibliothek may be found in Bodemann (1889, 1895) and Rivaud (1914). Special issues (Sonderheften) of *Studia Leibnitiana* and the series *Studia Leibnitiana Supplementa* include monographs and proceedings of International Symposia and Congresses promoted by the Leibniz-Society in Hanover.

Select bibliographies may also be found in Mittelstrass and Aiton (1973) and in Mittelstrass (1984).

Belaval Y 1960 *Leibniz critique de Descartes* (Paris: Gallimard)

Bernstein H R 1981 Passivity and inertia in Leibniz's dynamics *SL* **13** 97–113

Bos H J M 1974 Differentials, higher-order differentials and the derivative in the Leibnizian calculus *Archive for history of exact sciences* **14** 1–90

Buchdahl G 1969 *Metaphysics and the philosophy of science* (Oxford: Blackwell)

Cassirer E 1902 *Leibniz' System in seinen wissenschaftlichen Grundlagen* (reprinted 1962 Hildesheim: Olms)

Cohen I B 1982 Newton's copy of Leibniz's *Theodicy Isis* **73** 410–14

Fox M 1970 Leibniz's metaphysics of space and time *SL* **2** 29–55

Gale G 1974 Leibniz and some aspects of field dynamics *SL* **6** 28–48

Grua G 1948 G. W. Leibniz *Textes inédits* 2 vols (Paris)

Gueroult M 1967 *Leibniz. Dynamique et métaphysique* (Paris: Aubier-Montaigne)

Hall A R 1980 *Philosophers at war. The quarrel between Newton and Leibniz* (Cambridge: Cambridge University Press)

Heinekamp A 1969 *Das Problem des Guten bei Leibniz* (Bonn: Bouvier)

Hostler J 1975 *Leibniz's moral philosophy* (London: Duckworth)

Huggard E M and Farrer A 1951 *G. W. Leibniz. Theodicy* (London)

Ishiguro Hidé 1972 *Leibniz's philosophy of logic and language* (London: Duckworth)

Latta R 1898 *G. W. Leibniz. The monadology and other philosophical writings* (Oxford)

Loemker L E 1969 *G. W. Leibniz. Philosophical papers and letters* 2nd edn (Dordrecht: Reidel)

Lucas P G and Grint L 1961 *G. W. Leibniz. Discourse on metaphysics* 2nd edn (Manchester: Manchester University Press)

Mahnke D 1925 *Leibnizens Synthese von Universalmathematik und Individualmetaphysik* (reprinted 1964 Stuttgart-Bad Cannstatt)

Mason H T 1967 *The Leibniz–Arnauld correspondence* (Manchester: Manchester University Press)

Mittelstrass J (ed) 1984 *Enzyklopädie Philosophie und Wissenschaftstheorie* vol **2** (Mannheim: Bibliographisches Institut)

Mittelstrass J and Aiton E J 1973 *Dictionary of Scientific Biography* vol **8** pp 150–60, 166–8 (New York: Scribner)

Müller K 1967 *Leibniz-Bibliographie* 2nd edn 1984 ed A Heinekamp (Frankfurt am Main: Klostermann)

Mungello D E 1982 Die Quellen für das Chinabild Leibnizens *SL* **14** 233–43

O'Briant W H 1979 Russell on Leibniz *SL* **11** 159–222

Parkinson G H R 1966 *Leibniz. Logical Papers* (London: Oxford University Press)

Poser H 1969 *Zur Theorie der Modalbegriffe bei G. W. Leibniz SL Supplementa* **6**

Remnant P and Bennett J 1981 *G. W. Leibniz. New Essays on human understanding* (London: Cambridge University Press)

Riley P 1981 *G. W. Leibniz. The political writings* 2nd edn (London: Cambridge University Press)

Rivaud A 1914 *Catalogue critique des manuscrits de Leibniz, Fasc. II: 1672–1676* (Poitiers)

Russell B 1900 *A critical exposition of the philosophy of Leibniz* 2nd edn 1937 (London: Allen and Unwin)

Russell L J 1976 Leibniz's philosophy of science *SL* **8** 1–17

Schrecker P and Schrecker A 1965 *G. W. Leibniz. Monadology and other philosophical writings* (Indianapolis)

Stammel H 1982 *Der Krafftbegriff in Leibniz' Physik* (Inaugural-Dissertation, Mannheim)

Voisé W 1971 Meister und Schüler: Erhard Weigel und Gottfried Wilhelm Leibniz *SL* **3** 55–67

Wiener P P 1951 Leibniz. *Selections* (New York: Scribner)

Woolhouse R S (ed) 1981 *Leibniz: metaphysics and philosophy of science* (London: Oxford University Press)

Index